MODERN
DEVELOPMENTS IN
CATALYSIS

MODERN
DEVELOPMENTS IN
CATALYSIS

Editors

Graham Hutchings
Cardiff University, UK

Matthew Davidson
University of Bath, UK

Richard Catlow
University College London & Cardiff University, UK

Christopher Hardacre
University of Manchester, UK

Nicholas Turner
University of Manchester, UK

Paul Collier
Johnson Matthey Technology Centre, UK

World Scientific

NEW JERSEY · LONDON · SINGAPORE · BEIJING · SHANGHAI · HONG KONG · TAIPEI · CHENNAI · TOKYO

Published by

World Scientific Publishing Europe Ltd.

57 Shelton Street, Covent Garden, London WC2H 9HE

Head office: 5 Toh Tuck Link, Singapore 596224

USA office: 27 Warren Street, Suite 401-402, Hackensack, NJ 07601

Library of Congress Cataloging-in-Publication Data

Names: Hutchings, Graham (Graham John), 1951–
Title: Modern developments in catalysis / Graham Hutchings (Cardiff University, UK)
    [and three others].
Description: New Jersey : World Scientific, 2016.
Identifiers: LCCN 2016015329 | ISBN 9781786341211 (hc : alk. paper)
Subjects: LCSH: Catalysts. | Chemistry, Physical and theoretical.
Classification: LCC QD501 .M776654 2016 | DDC 541/.395--dc23
LC record available at https://lccn.loc.gov/2016015329

British Library Cataloguing-in-Publication Data
A catalogue record for this book is available from the British Library.

First published 2017 (hardcover)
Reprinted 2022 (in paperback edition)
ISBN 978-1-80061-199-3 (pbk)

Desk Editors: Chandrima Maitra/Mary Simpson

Typeset by Stallion Press
Email: enquiries@stallionpress.com

Foreword

Catalysis will play a major role in tackling the grand challenges of the 21st Century, such as climate change and the increase in demand and subsequent environmental issues which will come from the predicted rise in global population. According to the UN, the world population is projected to grow from its current level of 7.3–9.7 billion by 2050 (*Source*: World Population Prospects: The 2015 Revision, UN New York, 2015) which will put enormous pressure on both natural resources and the environment. At the same time, the world will become increasingly urbanised and 66% of the world's population is projected to live in cities by 2050 (World Urbanization Prospects: The 2014 Revision, United Nations, New York), which will lead to increased demands for construction and bulk chemicals for example. Over this period, the impact of climate change will become more disruptive with global temperatures predicted to rise between 1°C and 5°C (*Source*: IPCC Climate Change 2014: Synthesis Report Summary for Policy Makers). Catalysts make processes more efficient and effective and each of these challenges will require advances in catalytic technology in diverse sectors from energy, to water, food production, functional materials, bulk and intermediate materials and pharmaceuticals/fine chemicals. This will only come about through improved understanding and targeted research programmes with collaboration between academia and industry.

Catalysis already plays a leading role in many processes that contribute to human well-being including energy generation (e.g. pollution control and fuel cells), food production (e.g. ammonia synthesis for fertilisers), transportation (e.g. fuels and emissions control), construction (e.g. polymers and plastics, adhesives and coatings), healthcare and well-being (pharmaceuticals and functional materials) and water (waste water treatment). In the future, there will be an increasing demand for catalysts that make existing processes greener with less waste and use less energy and fewer raw materials to make the same mass of products. This will also mean new pathways to existing materials and products, for example, from renewable feedstocks, and also routes to entirely new classes of materials with as yet unimagined properties. In order to maximise the benefits from research and create the greatest contribution to all these areas and more, new programmes such as the UK Catalysis Hub are required to deepen understanding and develop new catalytic processes by using a design led approach.

The UK catalysis Hub utilises the world class capabilities at the STFC RAL campus, which include the Diamond Lightsource for synchrotron based techniques such as spectroscopy and scattering/diffraction; the ISIS Neutron Spallation Source for neutron based techniques such as INS, QENS and total scattering; and the Central Laser Facility for ultrafast spectroscopy and imaging. These offer world class capabilities for the study of catalysts and enable high quality studies of structure and reactivity under *in situ* and *operando* conditions. In addition to this the people at the RAL site are unparalleled in their capabilities and drive to make research projects succeed. *In situ* and *operando* methods are central to the efforts of the UK Catalysis Hub and this coupled with computational modelling is enabling advances to be made already.

From an industrial perspective, catalysis is at the heart of industrial processes and allied with process engineering is responsible for 90% of chemical processes (DECHEMA). New catalysed processes are being generated all the time from Lucite's alpha process (http://www.soci.org/chemistry-and-industry/cni-data/2009/20/a-winning-process) to work with Cardiff University and Johnson Matthey on VCM production using a non-toxic gold based catalyst which has replaced a toxic mercury-based one (reference: http://www.matthey.com/innovation/innovation_in_action/vcm-catalyst). The work going on at the UK Catalysis Hub is very relevant to industry and is supported by companies from sectors across industry, with the themes of catalyst design, energy, environment, chemical transformations and bio-feedstocks all having strong links to actual industrial processes.

This book highlights many powerful examples of how catalysis can impact society and also how catalysis science is making use of the most advanced capabilities and techniques to shed light on how catalytic processes work. Work like this is dependent on having high quality collaborative programmes in place, like the UK Catalysis Hub, which bring together academia and industry to focus on the most relevant and pressing problems. This work will continue and for the future, catalysis will see the development of more combined multiple techniques and approaches leading to increased knowledge of catalysts under working conditions. This will provide even more powerful tools to probe materials at the atomic level during chemical transformations and drive innovation for the next generation of catalysts.

About the Editors

C. Richard A. Catlow has long standing experience in the development and application of both experimental and computer modelling techniques in catalysis and molecular sciences. He holds approximately £2.5M of current EPSRC funding and has extensive experience in the field of HPC simulation techniques. He has been PI of the EPSRC funded Materials Chemistry HPC consortium for 15 years and has wide experience in managing large flexible consortium grants including a portfolio partnership grant (2005–2010), a High Performance Computing Consortium grant (2008–2013), and is currently the PI of the Centre for Catalytic Science (2011–2016).

Graham J. Hutchings is the Director of the Cardiff Catalysis Institute and is the inaugural Director of the UK Catalysis Hub. The UK Catalysis Hub will coordinate and strengthen research efforts in catalytic science, allowing the UK to remain a world-leader in the field and tackle major global challenges. There will be a strong emphasis on energy sustainability, environmental protection and innovative catalytic processes to support the UK chemical industry. One of Prof. Graham Hutchings' major scientific achievements is the pioneering work of using gold as an active catalyst, which still remains today as an important area of research.

Christopher Hardacre's research is focused on the understanding of heterogeneously catalysed reactions including water gas shift catalysis, the use of transients to determine gas phase mechanisms, liquid phase hydrogenation and oxidation of pharmaceuticals, low temperature fuel cells and clean energy production. Of particular interest is the development of techniques to probe reaction mechanisms at short time scales in the gas phase and the understanding of solvent effects in liquid phase reactions. Strong interactions exist between his group and the theory group of Prof. Peijun Hu (QUB) in order to develop DFT methods to predict new catalysts and validate the proposals made. He has also developed a strong research group in ionic liquids within the Queen's University Ionic Liquids Laboratory (QUILL) University-Industry research centre with interests in heterogeneously catalysed reactions, structural determination of ionic liquids, and species dissolved therein, analytical aspects, electrochemistry and prediction of physical properties of ionic liquids.

Matthew G. Davidson is Whorrod Prof. of Sustainable Chemical Technologies and director of the Centre for Sustainable Chemical Technologies at the University of Bath. His research interests are in the application of catalysis to the sustainable manufacture of fuels, materials and chemicals. Following a PhD and Research Fellowship at Cambridge, he held lectureships in Cambridge and Durham before being appointed to a Chair at the University of Bath. He is a Fellow of the Royal Society of Chemistry and a recipient of the Harrison Memorial Prize of the Royal Society of Chemistry and a Royal Society Industry Fellowship. He currently serves on the REF 2014 Chemistry Panel and holds over £13M of funding from research councils and industry.

Nicholas J. Turner obtained his DPhil in 1985 with Prof. Sir Jack Baldwin and from 1985–1987 was a Royal Society Junior Research Fellow, spending time at Harvard University with Prof. George Whitesides. He was appointed lecturer in 1987 at Exeter University and moved to Edinburgh in 1995, initially as a Reader and subsequently Prof. in 1998. In October 2004 he joined Manchester University as Prof. of Chemical Biology where his research group is located in the Manchester Institute of Biotechnology Biocentre (MIB: www.mib.ac.uk). He is Director of the Centre of Excellence in Biocatalysis (CoEBio3) (www.coebio3.org) and also a co-founder and Scientific Director of Ingenza (www.ingenza.com), a spin-out biocatalysis company based in Edinburgh and more recently Discovery Biocatalysts. He is a member of the Editorial Board of *ChemCatChem* and

Advanced Synthesis and Catalysis. His research interests are in the area of biocatalysis with particular emphasis on the discovery and development of novel enzyme catalysed reactions for applications in organic synthesis. His group are also interested in the application of directed evolution technologies for the development of biocatalysts with tailored functions

Paul Collier is a Senior Research Scientist at the Johnson Matthey Technology Centre, Sonning Common, UK. He is interested in all aspects of heterogeneous catalysis, especially gas phase catalysis. Dr Paul collier spends approximately one day a week at the Harwell Campus interacting with the UK Catalysis Hub and Diamond Light source. Paul completed his PhD at Liverpool University in 1996 before undertaking a postdoctoral research position at Cardiff University focusing on the direct synthesis of Hydrogen peroxide, as well as investigating other catalytic systems. Following this he went to work For Johnson Matthey.

About the Contributors

Angel Caravaca (born in Toledo, Spain, in 1985) studied Chemical Engineering at the University of Castilla-La Mancha (Ciudad Real, Spain). He holds a PhD by the same university in Chemical and Environmental Engineering, and Materials Science. Once he finished his PhD, he continued his career abroad. Firstly, he worked as a Post-doctoral researcher under the supervision of the internationally recognized Prof. Vayenas (University of Patras, Greece), who discovered one of the latest and most interesting phenomena in Catalysis: the phenomenon of Electrochemical Promotion of Catalysis (EPOC). After almost a year, he moved to the United Kingdom, where he worked as a *Core* Post-doctoral researcher for the EPSRC-funded "UK Catalysis Hub". His work was performed at Queen's University of Belfast (Northern Ireland, 18 months) and the Research Complex at Harwell (England, 6 months). Currently, Dr. Caravaca is working at the CEA (*Commissariat à l'énergie atomique et aux énergies* alternatives) in Marcoule (France). His research experience covers several fields in catalysis, electro-catalysis/solid oxide electrochemistry and photocatalysis, for their applications to a wide variety of environmental (de-NO$_x$, VOCs oxidation) and hydrogen production processes (methane/natural gas/alcohols steam reforming, alcohols/bio-alcohols electro-reforming and photoreforming, water electrolysis, etc.).

Francisco R. García-García is a Lecturer in Chemical Reaction Engineering at the School of Engineering at University of Edinburgh. His research seeks sustainable solutions to today's emission control and energy production challenges by mimicking nature strategies. The aim of his research group is to design, develop and fabricate multifunctional catalytic reactors inspired by how biological cell works, which allows the integration of multi-process in a single device. Dr Francisco R. García-García holds an MSc in Chemistry by the Autónoma University of Madrid (Madrid, Spain, 2004–2008) and a PhD in Chemical Engineering by the Institute of Catalysis and Petroleum-chemistry, CSIC (Madrid, Spain, 1997–2002). He gained his first post-doctoral experience working at the Department of Chemical Engineering of Imperial College London (2009–2013). In this period, I focused in the design and development of catalytic multifunctional reactors for hydrogen production. Afterwards, I worked as a Senior Scientist at Johnson Matthey in the Emissions Control Department (2013). Despite having a very rewarding experience working in industry, he soon realized that he preferred to be involved in more fundamental science and he moved back to the academia. Hence, he joined the UK Catalysis Hub as a research fellow working at the Chemical Engineering Department at Cambridge University (2014–2015), and at the Chemical Engineering Department at Newcastle University (2015–2016). During this time his research focused in chemical looping reforming for syngas and hydrogen production. Dr. Francisco R. García-García is recognized for his knowledge in the area of gas phase heterogeneous catalysis, new materials development, membrane technology and chemical looping in the interphase between chemistry and chemical engineering.

Nachal Subramanian is a Postdoctoral Researcher in the UK Catalysis Hub at Harwell, working under the theme 'Catalysis by Design' led by Prof. Richard Catlow. Her research interests include heterogeneous catalysis, fuel reforming, syngas chemistry, hydrogen production, materials synthesis and characterization, reaction kinetics and *in-situ* spectroscopy. Dr. Subramanian holds a PhD in Chemical Engineering from Louisiana State University, USA (May 2011), followed by postdoctoral experience at Georgia Institute of Technology, USA and Praxair India Ltd. (Development Specialist, Applications R&D). Dr. Subramanian has authored several research papers/book chapters and is an active peer reviewer and freelance technical editor.

Mike Bowker leads the Heterogeneous Catalysis and Surface Science group in Cardiff, consisting of 9 academic members of staff and 70 researchers. He founded the Wolfson Nanoscience Laboratory at Cardiff in 2006 and is Deputy Director of the recently established Cardiff Catalysis Institute. His research has focused on surface structure/reactivity and catalysis, ranging from theoretical studies of the effect of sintering on product yields, to selective oxidation catalysis on oxide nanomaterials, to studies of adsorption on well-defined surfaces. He has used STM for 20 years to study various aspects of surface structure and reactivity, pioneering the use of high temperature, atomic resolution STM in this field. The group continues to focus on aspects of surface science and catalysis, now extending to include nanofabrication, nanoengineering and bio-surface interactions.

Emma K. Gibson is a UCL researcher based in the Research Complex at Harwell using *in situ* and *operando* methods for catalyst characterisation. Previous projects have included using Raman spectroscopy to investigate impregnation and drying processes in extrudate preparation and *operando* FTIR analysis of steam reforming catalysts. Emma has recently been working to improve *in situ* cells for combined XAFS-DRIFTS measurements and has taken over management of the UK Catalysis Hub's Block Allocation Group (BAG) of beamtime on the B18 beamline at the Diamond Light Source.

Ian P. Silverwood is an instrument scientist at the ISIS neutron source working mainly with the quasielastic spectrometer IRIS. He has previously used infrared, Raman and inelastic neutron scattering to investigate catalytic reactions and has experience with measurement under non-traditional conditions, such as microwave-heating, supercritical CO_2 and spectroelectrochemistry.

Peter P. Wells is a lecturer at the University of Southampton, a joint appointment with Diamond Light Source, and was previously Associate Director of the UK Catalysis Hub. His research focuses on the development of *operando* spectroscopic methods and controlled catalyst preparation, both essential tools in understanding structure-function relationships. These studies make extensive use of synchrotron radiation facilities, both in the UK and elsewhere in Europe, and strong collaboration with the UK Catalysis Hub.

Stewart F. Parker joined the ISIS Facility in 1993 having spent the previous eight years with BP Research. His interests are the application of vibrational spectroscopy to problems in chemistry, including polymers, hydrogen storage materials and catalysis. In 2014, he was promoted to be ISIS Catalysis Scientist. He has pursued both pure and applied research in the catalysis field and collaborated with a range of investigators from industry and academia and has close ties to the UK Catalysis Hub. He has published over 300 papers on neutron scattering and co-authored the authoritative reference book on vibrational spectroscopy with neutrons.

Amanda G. Jarvis graduated from the University of St. Andrews (UK) in 2007 with a Masters in Chemistry. During her studies, she spent a year working in the petrochemicals industry in Calgary, Alberta (SynOil Fluids Inc./Core Laboratories Canada Ltd.), and undertook her final year project in the laboratory of Prof. David Cole-Hamilton. Following her undergraduate studies, she went on to receive a PhD from the University of York (UK) under the supervision of Prof. Ian Fairlamb, on 'Multidentate phosphine-alkene ligands and their late-transition metal complexes'. Amanda joined the group of Dr. Philippe Dauban (ICSN, Gif-sur-Yvette) as a post-doctoral research fellow in 2011 and worked on the development of Rh(II)-catalysed nitrene reactions. In 2013, she moved to Professor Paul Kamer's group (University of St Andrews, UK) to work on the development of artificial metalloenzymes for catalysis. In 2015, she received a Marie Curie Individual Fellowship to continue working in Prof. Paul Kamer's group on Artificial Metalloenzymes for the Oxidation of Alkanes (ArtOxiZymes).

Paul C. J. Kamer did his PhD in organic chemistry at the University of Utrecht. As a postdoctoral fellow of the Dutch Cancer Society (KWF) he spent 1 year at the California Institute of Technology and 1 year at the University of Leiden, where he worked on the development of phosphorothioate analogues of nucleotides. He subsequently moved to the University of Amsterdam to work in the field of homogeneous with Prof. Piet van Leeuwen, where he was appointed full professor of homogeneous catalysis in January 2005. In 2005, he also received a Marie Curie Excellence Grant to move his activities to the University of St. Andrews where he is currently Professor of Inorganic Chemistry. His current research interests are (asymmetric) homogeneous catalysis, organometallic chemistry, combinatorial synthesis, and artificial metalloenzymes.

Ben Davis got his BA (1993) and DPhil (1996) from the University of Oxford. He then spent two years as a postdoctoral fellow in the laboratory of Prof. Bryan Jones at the University of Toronto, exploring protein chemistry and biocatalysis. In 1998, he returned to the UK to take up a lectureship at the University of Durham. In the autumn of 2001, he moved to the Dyson Perrins Laboratory, University of Oxford and received a fellowship at Pembroke College, Oxford. He was promoted to Full Professor in 2005.

His group's research centres on the chemical understanding and exploitation of biomolecular function (Synthetic Biology, Chemical Biology and Chemical Medicine) with an emphasis on carbohydrates and proteins. In particular, the group's interests encompass synthesis and methodology; target biomolecule synthesis; inhibitor/probe/substrate design; biocatalysis; enzyme and biomolecule mechanism; biosynthetic pathway determination; protein engineering; drug delivery; molecular biology; structural biology; cell biology; glycobiology; molecular imaging and in vivo biology.

Ben Davis was co-founder of Glycoform, a biotechnology company that from 2002 to 2011 investigated the therapeutic potential of synthetic glycoproteins and of Oxford Contrast a company investigating the use of molecular imaging for brain disease. In 2003, he was named among the top young innovators in the world by Technology Review, the Massachusetts Institute of Technology (MIT)'s magazine of innovation in the TR35 awards and was a finalist in the BBSRC Innovator of the Year competition in 2010. He was elected to the Royal Society in 2015.

Charlie Fehl hails from southeast Michigan, where began his research career in the lab of Bruce Palfey studying flavoenzyme structure and function. He graduated with a BS in biochemistry in 2009 from the University of Michigan. He completed his PhD in medicinal chemistry at the University of Kansas in 2014 under the joint mentorship of Professors Jeffrey Aubé and Emily E. Scott, where he developed expertise in the organic synthesis, drug design, biochemistry, and structural biology of the cancer target CYP17A1. He then moved to the University of Oxford, where he is currently conducting postdoctoral research in the Benjamin G. Davis group. Charlie's research focuses on the biocatalytic abilities of glycosyltransferases and cytochromes P450 for the manipulation of heterocyclic substrates. He is also identifying organic reactions compatible with proteins, toward chemo-enzymatic routes for improved efficiency and access to chemical space in biocatalysis.

Maria Palm-Espling obtained her Master's degree in chemistry at Umeå University (Sweden), focusing on organic and medicinal chemistry. After working half a year as synthetic chemist for the Swedish Defense Research Agency she went on to PhD studies in the group of Prof. Pernilla Wittung-Stafshede at Umeå University. The thesis work centred around biophysical studies of metal binding to proteins, and was successfully defended in October 2013. In 2014, she joined the group of Prof. Benjamin G. Davis as a postdoctoral researcher at the University of Oxford (UK), where she worked on artificial metalloenzymes. Now Maria work as a senior clinical chemist at Falun regional hospital (Sweden).

Alexios Grigoropoulos obtained his PhD degree at the University of Athens, Greece, in the synthesis and spectroscopic investigation of transition metal complexes as biomimetic analogues of various enzymes. In 2007, he joined Dr Szilagyi's group as a post-doc (Montana State University, USA), where he studied the covalent character of metal-sulfur bonds by combing X-Ray Absorption spectroscopy and computational chemistry methods. In 2011, he moved back to the University of Athens and joined Dr Paraskevopoulou's group, where he carried out the synthesis of homogeneous catalysts in olefin hydroformylation and metathesis polymerization. From 2013, he has been working as a core post-doc for the UK Catalysis Hub in the group of Professor Rosseinsky FRS in the immobilization of transition metal homogeneous catalysts inside the pores of Metal-Organic Frameworks.

Prabhjot K. Saini did her PhD at Imperial College London following a Masters in Chemistry. She then proceeded to do Postdoctoral research with Robert Davis and Chris Braddock in Sustainable Polymer Chemistry/ Inorganic Chemistry before working with Prof. Charlotte Williams in New Catalysts and Processes for Oxygenated Polymers with the UK Catalysis Hub.

Charlotte K. Williams is Professor of Polymer Chemistry and Catalysis at Imperial College London. Her research interests include the development and application of polymerization catalysts to make oxygenated polymers, including polyesters and polycarbonates. She discovered a series of dinuclear homogeneous catalysts for carbon dioxide/epoxide copolymerization and has applied the same catalysts in a switchable catalysis allowing the preparation of block sequence controlled copolymers. She is the author of around 100 research papers and her research has been recognised by, amongst others, the RSC Corday Morgan Prize 2016, the BEPS Outstanding Young Scientist Award 2011, RSC Energy Environment and Sustainability Early Career Award 2009 and the RSC Meldola Medal 2005.

Catherine Davies completed her undergraduate and then PhD in heterogeneous catalysis at Cardiff University. She now works as a UK Catalysis Hub core PDRA at Cardiff in the area of automotive exhaust clean-up, in particular catalytic soot oxidation.

Peter Miedziak was awarded his PhD in 2009 from Cardiff University. He now works at Cardiff as a UK Catalysis Hub core PDRA in the field of photocatalysis.

Simon Kondrat completed his undergraduate degree at Warwick University followed by his PhD at Cardiff University. He now works as a UK Catalysis Hub core PDRA at Cardiff specialising in catalyst materials preparation and characterisation.

Vinod Kumar Puthiyapura have completed post graduate degree in Applied Chemistry (2007–2009) from Cochin University of Science and

Technology (CUSAT), India and his PhD (2014) from Newcastle University, UK on the topic: "Development of anode catalysts for Proton Exchange Membrane Water Electrolyser/Fuel cell". Upon completion of PhD he moved to Belfast, Northern Ireland and worked at Queens University of Belfast as Post doc research fellow for 2 years (2014–2016). He then moved to The University of Manchester and has been working as Research Associate since then within the 'Energy Theme' of UK-Catalysis Hub project. His research mainly focuses on the development of Pt based multi-metallic electro catalysts for direct alcohol fuel cell (DAFC) and characterization of the catalyst using various *in-situ* and *ex-situ* techniques (e.g.: FTIR, CV, RDE, MEA). Dr. Puthiyapura is also interested in the investigation of photocatalyst for small organic molecule oxidation reaction and development of catalyst support material for the fuel cell anode.

Hasliza Bahruji received BSc (2002) and MSc (2005) from Universiti Teknologi Malaysia and PhD (2011) from Cardiff University. She was a lecturer in Universiti Pendidikan Sultan Idris, Malaysia (2005–2008) before appointed as research associate in School of Chemistry, Cardiff University in 2012. She is currently a researcher for Cardiff Catalysis Institute and Catalysis Hub UK. Her research interest in heterogeneous catalysis for carbon dioxide utilisation and photocatalytic water splitting for hydrogen production.

Shaoliang Guan received his PhD in Chemistry from Cardiff University in 2015 under the supervision of Prof. Gary Attard. His PhD thesis was about determining the catalyst active sites in hydrogenation reactions by cyclic voltammetry and *in-situ* Surface enhanced Raman spectroscopy. He has been working as a PDRA with Prof. Philip Davies and Dr. David Willock in Cardiff University, and Prof. David Lennon in the University of Glasgow since July 2014. His current research topic is towards closing the chlorine cycle, funded by UK Catalysis Hub. He is trying to investigate the oxy-chlorination of CO by theoretical modelling, using UHV surface technique systems, and monitoring the real catalytic reaction using industrial supplied catalyst.

Jayesh S. Bhatt obtained MSci degree in Astrophysics from Queen Mary & Westfield College, London, and PhD in Physics from University College

London. His PhD work involved developing a stochastic approach to heterogeneous catalytic reaction kinetics under ultra-low concentrations, for which he received the C.N. Davies Award from the Aerosol Society. His subsequent research has focussed on heterogeneous nucleation, agglomeration kinetics and diffusion in porous media.

Arunabhiram Chutia obtained his PhD in theoretical and computational chemistry from Graduate School of Engineering at Tohoku University. Currently, he is a research associate at University College London with Prof. C. Richard A. Catlow FRS and full-time based at the UK Catalysis Hub in Rutherford Appleton Laboratory. He employs electronic structure methods and actively collaborates with experimentalist to understand catalytic processes at atomic scale with a motivation to design novel catalysts.

Marc-Olivier Coppens is Ramsay Memorial Chair and Head of Department of Chemical Engineering at UCL, since 2012, after academic posts at Rensselaer (the USA) and TU Delft (the Netherlands). He is also Director of the EPSRC "Frontier Engineering" Centre for Nature Inspired Engineering, and a member of the UK Catalysis Hub. Coppens is internationally recognised for pioneering work on nature-inspired chemical engineering: learning from fundamental mechanisms behind desirable traits in nature to develop creative and effective solutions to engineering challenges. He is a Fellow of IChemE and the first International Director of the AIChE's Catalysis and Reaction Engineering Division. Since 2016, he is Editor of *Chemical & Engineering Processing: Process Intensification.*

Carla Tagliaferri is currently a research associate at the Chemical Engineering Department at UCL and works as process engineer at Advanced Plasma Power. Her research interests include environmental impacts of developing energy systems and catalytic processes.

Dan Brett is Co-Director of the Electrochemical Innovation Lab at UCL. He specialises in the materials and engineering of electrochemical power conversion and storage systems, with a particular interest in developing novel diagnostics techniques for understanding their operation.

Paola Lettieri is Academic Director for UCL EAST and Vice-Dean (Strategic Projects) in the Faculty of Engineering Sciences. She specialises in the Life Cycle Assessment (LCA) of energy systems, chemical processes and waste management.

Sara Evangelisti did her undergraduate in Mechanical Engineering from the University of Pisa before moving to the University of Rome to complete an MSc in Design for Sustainable Development. She then went on to complete her PhD in Environmental and Energy Technologies at Sapienza University of Rome (Italy) in 2013. She moved to University College London to take a Research Assistant position with Prof. Paola Letteri working on a cradle-to-grave life cycle assessment of PEM Fuel cell systems as part of the UK Catalysis Hub.

Contents

Introduction

Catalysis is an important process that underpins everyday life for many of us. Without catalysis most of the manufactured goods and many food-stuffs would not be possible. Overall catalysis contributes up to 40% of GDP worldwide. Hence, it is important that we gain an improved under-standing of how catalysts function so that improved catalysts can be pre-pared. Indeed, there is a continuous need to identify new catalysts for ever more challenging reactions. It is against this background that UK scien-tists interested in catalysis have come together to form the UK Catalysis Hub. This is a consortium of *ca.* 40 Universities funded by the Engineering and Physical Sciences Research Council in the UK. The Hub has a central location and is based in the Research Complex at Harwell. This location is key since it permits access to the world class facilities available at Harwell; namely, the Diamond Light Source, the ISIS Neutron Source and the Central Laser Facility. The Hub was established in early 2013 and was officially opened in April 2013. The Hub provides a focal point for cataly-sis research in the UK. The central aim is to establish project teams to tackle difficult problems that are not feasible to be tackled by one researcher alone. This has proved to be very successful and we now have our 30 collaborative projects in operation. The Hub currently focuses on five themes of research:

- Catalysis design
- Environmental catalysis
- Catalysis and energy

- Chemical transformations
- Biocatalysis and biotransformations

At this stage of the Hub development we thought it was appropriate for the research associates to write a series of detailed reviews setting out current thinking in a number of key areas. The topics range from complex reactions to the intricacies of catalyst preparation for supported nanoparticles. The topics span the areas of current interest to the UK Catalysis Hub and we hope will be of interest to any general reader in the field of catalysis.

Chapter 1

Sustainable Hydrogen and/or Syngas Production: New Approaches to Reforming

Nachal Subramanian[*,†], **Angel Caravaca**[†,‡],
Francisco R. García-García[§,¶]
and Mike Bowker[†,#]

** Department of Chemistry, University College London,
London WC1E 6BT*

*† UK Catalysis Hub, Research Complex at Harwell, Rutherford
Appleton Laboratory, Harwell, Oxford Oxon OX11 0FA*

*‡ School of Chemistry and Chemical Engineering,
Queen's University Belfast, Belfast BT9 5AG*

*§ Department of Chemical Engineering and Biotechnology,
University of Cambridge, Cambridge CB2 3RA*

*¶ School of Chemical Engineering and Advanced Materials,
Newcastle University, Merz Court,
Newcastle upon Tyne NE1 7RU*

*# Cardiff Catalysis Institute, School of Chemistry,
Cardiff University, Cardiff CF10 3AT*

The use of non-renewable resources together with the high-energy consumption and low selectivity makes current reforming processes unsustainable solutions for syngas and/or hydrogen production. New reforming

1

technology is needed in order to control and manage CO_2 emissions, circumvent the current high-energy consumption and enhance the selectivity of the reforming process. The UK Catalysis Hub has proposed three different approaches to reforming which are envisaged to overcome the above-mentioned issues. The aim of this chapter is to present a critical review of recent approaches in reforming processes for hydrogen and/or syngas production, with particular focus on catalytic reforming of alcohols, chemical looping reforming (CLR) technology, and novel photocatalytic reforming routes. Likewise, the last section of the chapter summaries the challenges and current achievements of the UK Catalysis Hub projects in this area.

1 Introduction

Hydrogen production appears to be one of the most promising technologies for meeting future global energy needs as it is an environmentally clean and efficient fuel, compared to the conventional petroleum-based fuels.[1-8] The current industrial processes for H_2 production are all based on non-renewable fossil fuels via syngas route. Syngas is produced industrially by steam reforming (SRM) and dry reforming (DRM) of methane (Eqs. (1.1) and (1.2)) with H_2/CO ratios equal to 3:1 and 1:1, respectively.[9,10]

$$CH_4 + H_2O \rightarrow CO + 3H_2 \quad \Delta H^\circ = +206 \text{ kJ/mol}, \quad (1.1)$$

$$CH_4 + CO_2 \rightarrow 2CO + 2H_2 \quad \Delta H^\circ = +247 \text{ kJ/mol}. \quad (1.2)$$

High H_2/CO ratios such as that from SRM process are ideal for hydrogen production. Today, 96% of the global hydrogen production is based on SRM of fossil fuels followed by the water gas shift (WGS) reaction according to the following reaction.[11]

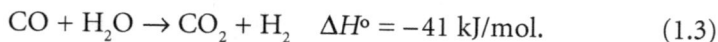

$$CO + H_2O \rightarrow CO_2 + H_2 \quad \Delta H^\circ = -41 \text{ kJ/mol}. \quad (1.3)$$

During the WGS reaction, the carbon monoxide produced in the SRM is reoxidised with steam to produce more hydrogen and carbon dioxide.[12] The actual hydrogen economy produces about 250 million tons of carbon dioxide emissions per year. Thus, the conventional processes based on fossil fuels for hydrogen and/or syngas production result in large CO_x emissions and in turn severe environmental concerns. Hence, the production of clean and high purity hydrogen (and/or syngas) at low cost is a key requirement for the development of a suitable hydrogen economy.

Biomass, on the other hand, is proving to be a promising renewable source for the sustainable production of hydrogen and/or syngas due to environmental concerns and the decline in fossil fuel reserves.[8,13–29] However, at present, the processes used for the conversion of biomass, such as enzymatic decomposition, pyrolysis and gasification suffer from low hydrogen production rates and complex processing steps.[1,4,26,30–32] Hence, there is a need for the development of an efficient reforming process, and catalytic reforming of biomass-derived alcohols and polyols (such as ethanol, glycerol, sorbitol, ethylene glycol, etc.) appears to be one promising approach.

Reforming of bio-alcohols is attracting more research attention due to its potential to control and manage CO_2 emissions. The recent interest in biomass being an alternative and renewable fuel has increased the annual production of different bio-alcohols such as ethanol, glycerol and butanol. It is estimated that 4 billion gallons of glycerol will be produced by 2016.[11] In addition, the carbon dioxide produced during the reaction is consumed for biomass growth which nearly closes the carbon loop. Also, the non-toxicity of some of the bio-alcohols, along with their ability to remain as liquids at room temperature, makes them suitable for use as hydrogen storage and a convenient source of energy.[12]

Independent of the nature of the source used for hydrogen production, reforming reactions tend to be highly endothermic and as a result, energy supply to the process stream is always a key consideration. In order to overcome this issue, novel means of energy input to the reforming reaction such as photocatalytic reforming has been proposed. Compared with traditional reforming processes, photocatalytic reforming is an attractive up-and-coming technology due to its potential of operating close to ambient conditions by using sunlight.[33]

Partial oxidation of methane (POM), as represented by Eq. (1.4), produces hydrogen and carbon monoxide with a H_2/CO ratio equal to 2:1, as needed for a Fischer–Tropsch reaction, and it is ideal for most downstream processes. However, POM reaction has not been commercialised owing to the risk of explosion of premixed methane and oxygen and the high cost of supplying pure oxygen.

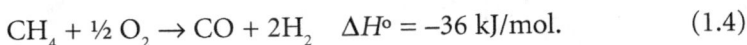

$$CH_4 + \tfrac{1}{2} O_2 \rightarrow CO + 2H_2 \quad \Delta H^\circ = -36 \text{ kJ/mol.} \quad (1.4)$$

Alternative processes which also produce syngas with a H_2/CO ratio equal to 2:1 and are recently gaining interest include the chemical looping steam reforming of methane (CL-SRM) and chemical looping dry reforming of methane (CL-DRM).[34] In the chemical looping process, the lattice oxygen of a solid metal oxide (oxygen carrier), instead of gaseous oxygen, is used to partially oxidise the methane to hydrogen and carbon monoxide. The regeneration of the reduced metal oxides can be performed in a subsequent step by oxidation with steam or carbon dioxide, CL-SRM and CL-DRM respectively.[34] The CL-SRM process has the advantage over conventional reforming processes of producing separate streams of syngas and hydrogen. On the other hand, the CL-DRM process is very interesting from an environmental point of view, since two greenhouse effect molecules (methane and carbon dioxide) are converted into more valuable products (hydrogen and carbon monoxide).

The use of non-renewable resources together with the high-energy consumption and low selectivity makes current reforming processes unsustainable solutions for syngas and hydrogen production. New reforming technology is needed in order to control and manage CO_2 emissions, circumvent the current high-energy consumption and enhance the selectivity of the reforming process. The transition from current reforming technologies to more sustainable ones is set to become a reality in upcoming decades. However, in the meantime, efforts must be focused on making this transition as sustainable as possible. This chapter provides an overview of some of the important challenges of current reforming industry and summarises the state-of-the-art of three new approaches for reforming: catalytic reforming of biomass-derived alcohols and polyols, chemical looping reforming (CLR) and photocatalytic reforming.

2 Catalytic Reforming of Bio-alcohols as New Sources for H_2 Production

The focus on mitigating global climate change and replacing petroleum-based energy sources is growing rapidly which has boosted the research interest towards alternative and renewable energy strategies. As discussed before, biomass derived alcohols are one important alternative energy strategy for the development of a sustainable hydrogen economy.[13–29]

Compared to all the processes used for the conversion of biomass to hydrogen, catalytic reforming processes seem to offer higher hydrogen production rates and simpler processing steps.[30–32]

The reforming processes used to produce hydrogen and syngas from oxygenated compounds can be classified into: steam reforming (SR), aqueous phase reforming (APR), partial oxidation (PO), auto thermal reforming (ATR), dry reforming (DR) and supercritical water gasification (SCWG).[35]

SR is the most common industrial reforming process to produce hydrogen[36] since it leads to complete conversions and high hydrogen yields. However the major disadvantage of this process is its high endo-thermicity, due to which a large amount of external heat is required.[37] PO occurs in the presence of air, it is highly exothermic and hence no additional heat is required.[36] However this process results in very low hydrogen yields and purity. ATR is a combination of PO and SR pro-cesses.[36] Air and water are co-fed into the reactor which enables an auto-thermal reformer to react at the thermal neutral point.[37] This process can lead to higher hydrogen yields and selectivities compared to PO, and the oxygen present can inhibit coke formation on the catalyst surface. But the conversions are much lower than traditional SR.[36] The DR method is beneficial to the environment as CO_2 can be converted into inert carbon and removed from the biosphere cycle,[38,39] however lower hydrogen yields have been reported. SCWG requires very high temperatures and pressures (~900°C and 240 atm). This increases the operating costs and undesirable side reactions, despite the ability to produce high yields of hydrogen. APR is a newer technique that operates at much lower tem-peratures than other methods, and the reaction occurs in liquid phase as opposed to SR.[40] This has significant advantages in terms of operating costs and side reactions.

Of all the processes, SR and APR are the most studied processes due to their simplicity and high yields of hydrogen. Dumesic *et al.* developed a new single-reactor APR process to efficiently produce hydrogen from model biomass glucose and biomass-derived polyols (e.g. ethylene glycol, glycerol, sorbitol) at mild conditions around 227°C.[41–46] It is carried out at low temperatures which results in reduced operating costs and mini-mises undesirable side reactions. Low temperatures also facilitate the

WGS reaction thereby leading to low levels of CO in the product stream, thereby resulting in highly purified hydrogen. APR is particularly useful for reforming biomass resources with high water contents as the need for water vaporisation is reduced compared to conventional SR.[1,4]

Several biomass-derived oxygenated compounds have been used in APR process, such as methanol, sorbitol, glycerol, ethylene glycol and ethanol.[2,4] Glycerol is of particular interest because it is obtained as a main by-product (10 wt.%) of biodiesel production by transesterification of vegetable oils or animal fats.[1,29,35,36,47–50] With increasing biodiesel production, the crude glycerol is also produced in substantial amounts and one of the promising ways to utilise this crude glycerol is to produce hydrogen by reforming process.[1,29,35]

SR (or APR) of a polyol (e.g. glycerol) occurs according to the following stoichiometric reaction[1–4,6,24,29,35,47,51]:

$$C_3H_8O_3 + 3H_2O = 7H_2 + 3CO_2 \quad \Delta H^O_{298} = +123 \text{ kJ/mol.} \quad (1.5)$$

Equation (1.5) is further constituted of the glycerol decomposition reaction (Eq. (1.6)) and the WGS reaction (Eq. (1.7)).

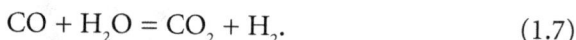

$$C_3H_8O_3 = 4H_2 + 3CO, \quad (1.6)$$

$$CO + H_2O = CO_2 + H_2. \quad (1.7)$$

The further reaction of CO and/or CO_2 with H_2 would result in methanation or Fischer–Tropsch reactions, and the other side reactions might include methane DR and/or decomposition, carbon monoxide disproportionation (Boudouard reaction), and carbon gasification.

SR is thermodynamically favoured at low pressures and high temperatures in order to increase the hydrogen yield. The most important parameters in the SR process are temperature, steam-to-carbon (S/C) ratio and catalyst-to-feed ratio. The feed (glycerol + water mixture) is vaporised in a preheater before entering the reactor. The major drawbacks of this method are high energy consumption, higher CO production (since WGS reaction is not favourable at high temperatures), and also rapid catalyst deactivation compared to the APR process in which the energy to vaporise the feed can be eliminated.

The APR process is carried out in the liquid phase at higher pressures and significantly lower temperatures (~227°C), where the WGS reaction is favourable, making it possible to generate hydrogen with low amounts of CO in a single reactor.[4] Also use of higher pressures (15–50 bar) facilitates the effective purification of H_2-rich effluent by adsorption or membrane technology. This process demands less energy inputs as the need to vaporise the reactants (both water and the oxygenated hydrocarbon) is eliminated.[1,4] Low temperatures also minimise undesirable side reactions and carbon deposition/sintering. Thus, APR process is a promising option for conversion of biomass into H_2.

2.1 Mechanism and catalysts for reforming

Luo *et al.*[40] describe the different reaction pathways that lead to various liquid and gaseous by-products during APR of glycerol. The selectivity to hydrogen remains challenging mainly due to the subsequent/parallel reactions in both liquid and gas phase. Methanation, methane DR, carbon monoxide disproportionation, and/or carbon gasification could lead to gaseous by-products and in parallel, glycerol could undergo dehydration and/or hydrogenation/dehydrogenation to produce liquid by-products (such as methanol, ethanol, acetone, acetic acid, propylene glycol, diglycerol, etc.). Hence, catalysts must be chosen keeping these issues in mind, if hydrogen production is the primary objective.

The catalytic pathway for the production of H_2 by APR of glycerol or similar polyols involves the cleavage of C–C bonds as well as C–H and/or O–H bonds to form adsorbed species on the catalyst surface.[6,44,47] Cleavage of these bonds occurs readily over Group VIII metals, such as Pd and Rh. Undesired by-products arising from parallel and series pathways are due the cleavage of C–O bonds and hydrogenation of adsorbed CO or CO_2. Thus, a catalyst active for APR should be active for C–C bond cleavage and promote WGS reaction for the removal of adsorbed CO species, but must not favour C–O bond cleavage and hydrogenation of CO or CO_2.[1,2,44]

Group VIII metals, particularly Pt, Pd and Ni, have been shown to be effective catalysts due to their ability to break C–C bonds and also to promote the WGS reaction.[2,5,51] Among all the metals, it is reported that Pt is the best monometallic catalyst in terms of activity and selectivity for APR. Cortright *et al.* studied the APR of sugars and alcohols for H_2 production using a Pt-based catalyst at 227°C and found the selectivity to H_2 increasing in the order of reactants — glucose < sorbitol < glycerol < ethylene glycol < methanol. Low operating temperatures resulted in high H_2 selectivity, low alkane selectivity, but low biomass conversions.[14] Davda *et al.* reported the H_2 selectivity to decrease in the order of Pd > Pt > Ni > Ru > Rh catalysts supported on silica for the APR of ethylene glycol.[41]

Several bimetallic catalysts, including PtNi, PtCo, PtFe and PdFe show higher activities than monometallic catalysts. While Ni catalysts are active for APR, they favour methanation and exhibit severe deactivation.[2] Addition of promoters such as Sn has been reported to improve the Ni catalyst stability towards carbon formation and deactivation. Also, the stability of Ni-based catalysts can be improved by using a Raney-Ni catalyst.[16] Perovskites and $NiAl_2O_4$ spinels have also been reported to assist in better stability of Ni catalysts.[47,52] Ceria, as a promoter or support, inhibits secondary dehydration reactions that lead to unsaturated hydrocarbon formation which act as coke precursors.[29] To increase hydrogen yield, a moderate Ni reduction degree, high Ni dispersion, and small Ni particle size are shown to be beneficial.[53]

The nature of the support such as the acidity and basicity has a remarkable influence on the activity as well as the gas distribution of the products.[4-6] Some commonly studied supports are SiO_2, Al_2O_3, MgO, TiO_2, CeO_2, ZrO_2, CeZrO, zeolites, perovskites and hydrotalcites.[1-3,5,29,47] Silica-supported Rh, Ru and Ni catalysts showed low H_2 selectivity, but high alkane selectivity.[44] In addition, Ni/SiO_2 showed significant deactivation at higher temperatures of 225°C. On the other hand, silica-supported Pt and Pd catalysts exhibited higher H_2 selectivities and lower rates of alkane production. Other supports such as TiO_2, carbon, or Al_2O_3 have been found to improve the activity and selectivity of monometallic Pt-based catalysts. Ceria supports appear to increase the stability of Ni catalysts by promoting carbon removal from the metallic surface due to its high oxygen storage capacity.[1,29,54,55] Shao *et al.* reported that the addition of ZrO_2 into CeO_2 further improved the redox properties and oxygen storage capacity, thereby promoting WGS and SR reactions.[29]

Addition of promoters also greatly influences the activity and selectivity of monometallic catalysts. Some promoters reported in literature are Sn, Ca, La, K, Sr, Co, Fe, Mg, Pt, Ce, Zr and Cu.[4,24] The activity of a Pt/Al_2O_3 catalyst improved further by adding Ni, Co or Fe as promoters.[44] The APR process of glucose, sorbitol, glycerol and ethylene glycol using Ni–Cu and Ni–Sn bimetallic catalysts have been reported to show considerable hydrogen selectivities and yields equal to or even higher than noble metal catalysts. Cu and Sn are not involved in C–C bond cleavage or dehydrogenation steps, but they promote WGS reaction thereby increasing H_2 production.[51] Huber *et al.* also studied the APR of biomass derived oxygenates using tin-promoted Raney–Ni catalyst and found that Raney–NiSn performed similar to Pt-based catalysts. Addition of Sn significantly decreased the rate of methane formation, without inhibiting the rate of H_2 production.[16,44]

Many reviews are available in the literature which describe in detail the progress made in the catalytic reforming of glycerol (or similar polyols and alcohols) to hydrogen, thermodynamical considerations, catalysts used, etc.[4,35,44,49,56] Table 1 gives an overview of several catalysts used for the APR process for different biomass feedstocks.[51]

Table 1: Overview of APR catalysts for different biomass feedstocks.

Feedstock	Catalysts	
	Metal	Support
Glucose	Pt, Ni, Ru, Rh–Pd, Ir	Al_2O_3, TiO_2, ZrO_2, SiO_2–Al_2O_3
Fructose	Pt	Al_2O_3
Sorbitol	Pt, Raney Ni–Sn, Ni–Pd	Al_2O_3, TiO_2, CeO_2
Xylitol	Pt, Pt–Re	Al_2O_3, TiO_2
Glycerol	Pt, Ni–Sn, Ni–Cu, Ni–B	Al_2O_3, SiO_2, ASA
Ethylene glycol	Pt, Ni, Ru, Rh–Pd, Pt–Co, Ni–Sn	Al_2O_3, SiO_2, CNK–3, Al Hydrotalcite
Ethanol	Pt, Ni, Ru	Al_2O_3, SiO_2, TiO_2
Methanol	Pt	Al_2O_3
Acetic acid	Pt	Al_2O_3
Biomass hydrolysate	Pt, Raney Ni	Al_2O_3

Data Source: Ref. [51].

2.2 Variations in catalytic steam reforming

2.2.1 Sorption-enhanced reforming process (SERP)

Catalytic reforming with *in situ* CO_2 removal, using a sorbent, increases H_2 production and decreases CH_4 and CO production.[15,35,57-68] The co-generated CO_2 can be selectively removed from the product gas mixture by using a sorbent. When a CO_2 sorbent is added along with the catalyst, the CO_2 is converted to a solid carbonate and thus directly produces high purity H_2. It is known from thermodynamic analysis that H_2 production from a polyol, like glycerol, is limited by WGS reaction. But this constraint can be overcome by concurrent WGS and carbonation (of sorbent) for CO_2 removal. In an ideal scenario, 100% CO_2 removal leads to the generation of 7 mols H_2 and 0 mols CH_4 and CO from glycerol according to Eq. (1.5).

The major limitation of SERP is that with time, the sorbent saturates and eventually reduces the CO_2 removal efficiency. Hence there must be a way to regenerate the sorbent and to extend the operation time. This can be achieved by simultaneous reaction and regeneration of sorbent using two slow moving bed reactors (reformer and regenerator). This process is termed as continuous sorption-enhanced reforming process (CSERP).[69] The integration of continuous SR, WGS, CO_2 capture and H_2 separation in one single reactor would lead to high purity H_2 production and reduced operating costs.

Several natural and synthetic CO_2 sorbents such as dolomite, lime, CaO, hydrotalcites, Li_2ZrO_3, $LiSiO_4$ and Na_2ZrO_3 have been studied.[57-68] CaO is reported to show the highest CO_2 sorption capacity at high temperatures.[70] Hydrotalcites are also used as sorbents, but they have poor CO_2 sorption capacity and low reaction rates compared to CaO. Some synthetic oxide sorbents such as Li_2ZrO_3, $LiSiO_4$ and Na_2ZrO_3 have good capacities at high temperatures, but they are very expensive.[59,60,66] Hence, natural sorbents are most commonly used.

2.2.2 Sorption-enhanced chemical looping reforming process (SECLR)

A process called CLR is used for the generation of H_2 and CO through cyclic reduction and oxidation of a solid oxygen carrier (like NiO or CeO_2).[53] In this process, a part of the fuel in the reformer may get completely oxidised to CO_2 and H_2O by the oxygen provided via the solid oxygen carrier, but most fuel should be partially oxidised to H_2 and CO. The use of an oxygen carrier overcomes the issue of coking and allows the reforming reaction to be performed at a lower temperature by the close coupling of the endothermic reforming reaction with the exothermic oxidation.[35,53,71] However, the key issue in the system performance is the oxygen carrier material and its stability. The selection of the material is based on sufficient oxygen transport capacity, high reactivity for reduction and oxidation reactions, SR and WGS reactions, attrition resistance and low cost.[35]

Chemical-looping processes can be combined with sorption-enhanced reforming processes into one single process (SECLR).[53] Three fluidised bed reactors are used: (i) a reforming reactor operated at low temperatures where the hydrocarbon feed is partially oxidised by the oxygen (from the oxygen carrier) and steam, and CO_2 is captured by a sorbent, (ii) a calcination reactor where the CO_2 sorbent is regenerated, (iii) an air reactor operated at high temperatures where the oxygen carrier is re-oxidised. Unlike SERP, the SECLR process is self-sufficient with heat. The chemical looping process will be explained later in detail (Sec. 3) as a new approach to reforming.

2.3 Summary

This section has reviewed the current literature in catalytic reforming of biomass-derived alcohols/polyols to hydrogen. Compared to all the processes used for the conversion of biomass to hydrogen, the catalytic reforming process seems to offer higher hydrogen production rates and simpler processing steps. A review of the most energy efficient APR process, the mechanism involved, the catalysts utilised, and recent extensions of the processes is provided. The APR process proves to be a promising approach for hydrogen production. It occurs at lower temperatures and in the liquid phase, thereby minimising undesirable side reactions and eliminating the need for water vaporisation as compared to conventional SR. This results in reduced operating costs and low levels of CO in the product stream, thereby resulting in highly purified hydrogen. Recent variations to the process include sorption-enhanced reforming and CLR. The former process uses a sorbent to assist in selectively removing the co-generated CO_2 from the product gas mixture, resulting in increased H_2 production and decreased CH_4 and CO production. Chemical looping involves the use of a solid oxygen carrier which overcomes the issue of coking and allows the reforming reaction to be performed at a lower temperature by the close coupling of the endothermic reforming reaction with the exothermic oxidation.

3 CLR

CLR of methane process uses the lattice oxygen of a solid metal oxide to produce syngas with a H_2/CO ratio equal to 2:1. The regeneration of the reduced metal oxides can be performed in a subsequent step by oxidation with steam or carbon dioxide, CL-SRM and CL-DRM processes, respectively. The CL-SRM processes allow the integration of two different process, syngas and hydrogen production, in a single device whereas in the CL-DRM process, two greenhouses molecules (methane and carbon dioxide) are converted into more valuable products (syngas).

A critical problem with in chemical looping process is that the oxygen carriers' materials undergo significant chemical, thermal and physical stresses as the oxides cycle between the fully-oxidised and fully-reduced condition. As a result, their longevity tends to be limited. In this respect, the feasibility of different transition metals (TMs) oxides as oxygen carriers for chemical looping process have been investigated. In order to avoid these issues, TM oxides are usually combined with an inert porous support which provides higher surface area for reaction, mechanical strength and attrition resistance. An ideal oxygen carrier material for both CL-DRM and CL-SRM processes should exhibit: (i) high reactivity during both reduction and oxidation steps, (ii) high and stable capacity to donate oxygen over thousands of successive redox cycles, (iii) resistance to attrition and (iv) low cost of production at large scales.

NiO-based oxygen carriers have generally been shown to be the most reactive material for the chemical looping processes.[34,72–75] This can be explained due to the ability of NiO to activate the strong C–H bond (i.e. 439 kJ/mol) of methane molecule, together with its high oxygen store capacity (i.e. 21.4 wt.%) and high melting point (i.e. 1,955°C). In order to prevent its deactivation over a large number of cycles, NiO is usually combined with different inert substrate such as YSZ,[76] Al_2O_3,[77] SiO_2, TiO_2 and ZrO_2.[78,79] However, the main drawback of using Ni-based oxygen carriers in large scale chemical looping processes is NiO's high price and toxicity.

The use of CuO as an oxygen carrier in chemical looping processes is the growing interest in the scientific community due to its high reactivity and oxygen storage capacity (i.e. 20.1 wt.%). However, due to the unacceptably low melting point (i.e. 1083°C), CuO particles easily agglomerated at the reaction conditions and may not be suitable as an oxygen carrier.[75,80,81] In order to increase its thermal stability, CuO is mixed with different refractory oxides such as Al_2O_3, ZrO_2 and MgO.[82–85] A further problem at high temperatures ($T > 870$°C) is that CuO can decompose to Cu_2O, which lower its oxygen storage capacity from 20.1 wt.% to 11.2 wt.%.[75,78,86] Nevertheless, the oxygen gas released during CuO decomposition can be use in a subsequent step in a process that is known as a chemical looping with oxygen uncoupling (CLOU).

The low oxygen storage capacity of Mn_3O_4 and Co_3O_4, 10.1 wt.% and 6.6 wt.%, respectively, has limited their use as a oxygen carriers materials for chemical looping process.[73,75,81,87] However, Mn_3O_4 and Co_3O_4 have been successfully employed as promoters of other TM oxides such as Fe_2O_3 and NiO_2. It is well known that a small amount of Mn_3O_4 and Co_3O_4 has a positive effect on the performance of other TM oxides in terms of both reaction kinetics and cyclic stability.[87–90]

Recently, $CaSO_4$-based materials have been proposed as interesting alternatives to transition metal oxides oxygen carriers for chemical looping processes.[91,92] As a novel oxygen carrier, $CaSO_4$ has several advantages over transitional metal oxides oxygen carriers such as its low price and low environmental impact. Furthermore, $CaSO_4$ oxygen storage capacity (i.e. 47.1 wt.%) is significantly higher than that in any other transition metal oxides. Although, preliminary results have shown that $CaSO_4$-based materials may be interesting candidates as oxygen carriers in chemical looping processes, they have the drawback of low reactivity, deactivation at high temperatures and sulphur release from side reactions.

Fe_2O_3-based oxygen carriers are very attractive due to their low cost (i.e. market price between NiO and Fe_2O_3 differs by a factor of 300) and high oxygen storage capacity (30.1 wt.%). Beside its very competitive price, it has been reported that iron ores and wastes from the steel industry can be successfully used as a oxygen carriers in chemical looping combustion processes.[72,93–95] The main drawback of Fe_2O_3 is that sintering and coking reactions can occur when there is a substantial proportion of Fe present. Nevertheless, by careful control of the support for the Fe_2O_3, the activity of the carrier can be adjusted so as to substantially favour reforming of the methane and discourage coking reactions. In addition, it is well known that mixing Fe_2O_3 with different refractory oxides such as CeO_2, Al_2O_3, SiO_2, TiO_2, ZrO_2 and $MgAl_2O_4$ improves its resistance to sintering at high temperatures.

Among non-NiO-based metal oxides, Fe_2O_3/CeO_2-based materials are the most attractive for CL-DRM and CL-SRM processes, owing to the elevated oxygen storage capacity and unique redox properties of CeO_2 and the positive effect that Fe^{3+} ions have on the oxygen carrying capacity and mobility of CeO_2.[96,97] It has been reported that the introduction of trivalent ions such as Fe^{3+} into the CeO_2 lattice lower the energy barrier for oxygen migration.[98,99] Although, Fe_2O_3/CeO_2 based mixed oxides are less active and show lower selectivity towards syngas than Ni-based metal oxides during both CL-DRM and CL-SRM processes, they have the advantage of being non-toxic and more abundant in nature.[34,100]

A schematic representation of ideal CL-SRM and CL-DRM process using a Fe_2O_3-based oxygen carrier is shown in Figures 1(A) and 1(B), respectively. Both processes can be divided in two main steps: reduction and oxidation of the Fe_2O_3-based oxygen carrier. The reduction step involves multiple sub-reactions in which methane reduces the Fe_2O_3 oxide to Fe producing syngas (Eqs. (1.8)–(1.10)).

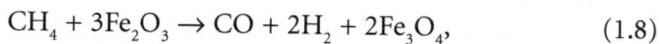

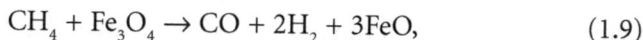

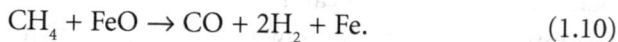

$$CH_4 + 3Fe_2O_3 \rightarrow CO + 2H_2 + 2Fe_3O_4, \tag{1.8}$$

$$CH_4 + Fe_3O_4 \rightarrow CO + 2H_2 + 3FeO, \tag{1.9}$$

$$CH_4 + FeO \rightarrow CO + 2H_2 + Fe. \tag{1.10}$$

The regeneration of the oxygen carrier is performed in a subsequent step by oxidation with carbon dioxide, (Eqs. (1.11) and (1.12)), or steam, (Eqs. (1.13) and (1.14)), known as CL-DRM and CL-SRM processes, respectively.

$$CO_2 + Fe \rightarrow CO + FeO, \tag{1.11}$$

CL-SRM

(A)

CL-DRM

(B)

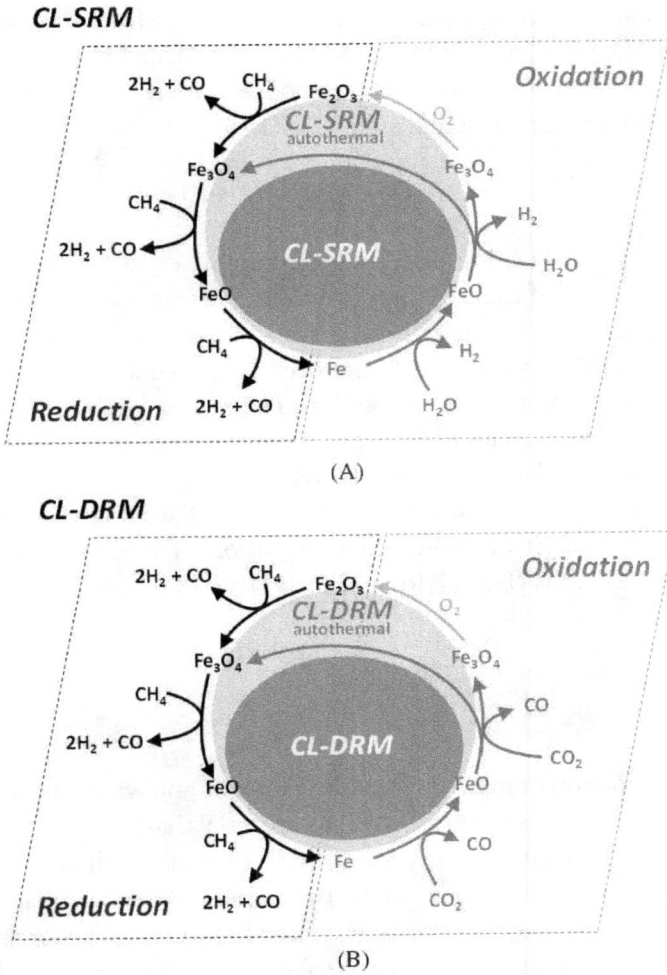

Figure 1: Schematic representation of the CL-SRM and CL-DRM processes using iron (III) oxide as an oxygen carrier, A and B, respectively.

$$CO_2 + 3FeO \rightarrow CO + Fe_3O_4, \tag{1.12}$$

$$H_2O + Fe \rightarrow H_2 + FeO, \tag{1.13}$$

$$H_2O + 3FeO \rightarrow H_2 + Fe_3O_4. \tag{1.14}$$

It should be noted that whereas Fe and FeO can be oxidised to magnetite (Fe_3O_4) either with carbon dioxide or steam, the oxidation of magnetite to haematite (Fe_2O_3) in this way is not thermodynamically feasible. However, an air oxidation step can be used instead (Eq. (1.15)).

$$2Fe_3O_4 + \tfrac{1}{2}O_2 \rightarrow 3Fe_2O_3. \tag{1.15}$$

Note that CL-DRM and CL-SRM are highly endothermic processes, and, consequently, the energy supplied to the process is an important consideration. Since oxidation with air is a highly exothermic reaction, autothermal CL-DRM and CL-SRM have been proposed as a novel means of energy input to the reforming reaction. The key research challenge is to demonstrate that such an approach is practical as the overall process has only been described at a conceptual level.

The main challenge of both CL-DRM and CL-SRM processes is to control the selectivity, avoiding total oxidation of methane and carbon deposition reactions (Eqs. (1.16) and (1.17)) respectively.

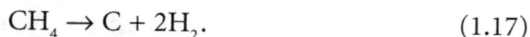

$$CH_4 + 2O_2 \rightarrow CO_2 + 2H_2O, \tag{1.16}$$

$$CH_4 \rightarrow C + 2H_2. \tag{1.17}$$

Total oxidation of methane to carbon dioxide and water due to oxygen surface groups present on the Fe_2O_3/CeO_2 materials has been reported.[101–103] Moreover, carbon deposition via decomposition of methane (DM) can occur over the Fe_2O_3/CeO_2 materials when there is a substantial proportion of Fe present on its surface.[103] According to García-García *et al.*, higher syngas selectivity and lower coke formation can potentially be achieved by controlling the oxidation state of the Fe_2O_3/CeO_2-based mixed oxide during the reduction step of the CL-DRM process. Hence, limiting the extent of the reduction step during the CL-DRM process can eliminate the carbon deposition while significantly increasing syngas selectivity by avoiding low Fe oxidation stages.

3.1 Summary

This section reports the recent advances in CLR of methane processes for syngas and hydrogen production. The reactivity of the four most studied oxygen carriers materials in the activation of the methane molecule C–H bond follows the order of NiO > CuO > Fe_2O_3 > Mn_2O_3. However, the high price and toxicity of NiO and the low melting point of CuO avoid their use as oxygen carriers in CLR processes. Although, Fe_2O_3 is less active and show lower selectivity towards syngas than NiO during both CL-DRM and CL-SRM processes, it has the advantage of being non-toxic, of low price and having high oxygen storage capacity. Double Fe_2O_3-based metal oxides provide promising materials for CL-DRM and CL-SRM processes showing an excellent overall performance with good reactivity and selectivity. CL-SRM and CL-DRM process present several advantages over conventional reforming process such production of syngas with a H_2/CO ratio equal to 2:1 and separate production of syngas and hydrogen or carbon monoxide streams, respectively. In addition, autothermal CL-DRM and CL-SRM using air during the oxidation step has been proposed as a novel means of energy input to the process. The main challenge of both CL-DRM and CL-SRM processes is to control the selectivity avoiding total oxidation of methane and carbon deposition reactions. Thus, the product selectivity during the CL-DRM can be linked to a specific material oxidation state which is varying monotonically during the reaction time. Hence, low iron oxidation states can be avoided by limiting the duration of the reduction step of the CL-DRM process, in an attempt to overcome carbon deposition. Likewise, high iron oxidation states can be avoided by using the appropriate buffer gas mixture during the oxidation step, which allow to avoid total oxidation of methane. In this respect, it seems that an oxidation state operating window could potentially be used to improve future chemical looping processes.

4 Photocatalytic Reforming

Among the current energy sources, solar energy is one of the most promising since it is free, easily accessible and inexhaustible. Lately, researchers have attempted to convert solar energy to a more useful energy form, such as chemical energy.[104,105] In this sense, a novel approach to the reforming technologies is the "photocatalytic reforming of alcohols", commonly known as *photoreforming of alcohols*. The general aim of this approach is to use solar energy for the activation of the catalysts involved in the reforming of biomass-derived alcohols, such as methanol, ethanol, glycerol and sugars.

As a matter of fact, the photoreforming process is the result of an improvement in the photocatalytic water splitting for direct H_2 and O_2 production. The origin of the photocatalytic water splitting is the work published by Fujishima and Honda in the early 1970s.[106] This publication deals with the "Electrochemical Photolysis of Water" in an electrolytic cell. Upon irradiation of a TiO_2 electrode (anode, for water oxidation and O_2 evolution), while maintaining a Pt counter electrode in the dark (cathode, for hydrogen evolution), stoichiometric amounts of hydrogen and oxygen were generated. Even though this work does not report pure photocatalysis, it has been cited in most of the publications addressing the photocatalytic water splitting and the photoreforming of organic molecules, due to the fact that it stimulated significant work in the field.

Photoreforming processes take place over semiconductor materials as catalysts. As in every other photocatalytic process based on solid semiconductor materials, the first step in the photocatalytic water splitting is the activation of a semiconductor by photoirradiation with light of energy equal or higher than the band gap of the semiconductor material. It leads to the migration of electrons from the valence band (VB) to the conduction band on the surface. It also gives rise to the formation of charge transfer holes (h^+) in the VB.[107] In the end, the presence of electrons leads to reduction reactions in the conduction band (such as the H_2 evolution reaction) and oxidation reactions in the VB (as the water oxidation and further O_2 evolution).

As we will explain later in detail, one of the main drawbacks of the water splitting process is the recombination of the O_2 and H_2 generated by the photoirradiation of the semiconductor material. Indeed, if we produce a catalyst with a lowered activation barrier to the water dissociation process, we also produce a material with an even lower activation barrier for the back-reaction, the recombination.[33] In other words, if a catalyst is good for the water splitting process, it will be even better for the recombination of the O_2 and H_2 produced. On the other hand, in the absence of suitable charge scavengers, the recombination of e^- and h^+ generated by the photoirradiation of the catalyst is usually facile. Both, the recombination of the reaction products, and the recombination of the photogenerated charge carriers, decrease dramatically the efficiency of the water splitting process, decreasing therefore the overall hydrogen production. In order to overcome these issues, one alternative is to add sacrificial agents to consume the O– derived species produced, avoiding the further O_2 evolution and its recombination with H_2. Moreover, these sacrificial agents act as hole scavengers, partially avoiding the recombination of e^- and h^+.

Among the wide range of sacrificial agents that can be used to produce H_2 by means of the photoreforming process, we will focus this chapter in the photoreforming of alcohols, as biomass derived products. For instance, bio-ethanol is produced by fermentation of biomass materials.[108] Sugar cane, potatoes, corns and other starch-rich materials can be effectively converted to ethanol by fermentation. Moreover, bio-methanol can also be produced from biomass. However, the production of methanol is a cost-intensive chemical process. It involves the PO of the biomass source to produce syngas, followed by the high pressure methanol synthesis reaction. That is why, in current conditions, only waste biomass such as old wood or bio-waste is used to produce methanol.[109] On the other hand, as previously mentioned, currently about 10% of glycerol is produced during the conversion of vegetable oils or animal fats into biodiesel fuel through the catalytic transesterification process.[110]

Hence, it seems that the photoreforming of biomass-derived products offers the possibility to produce hydrogen, one of the most important energy vectors nowadays, by the valorisation of biomass-derived alcohols and the use of renewable solar energy. Moreover, photoreforming processes usually take place at room temperature, which decrease the energy requirements in comparison with conventional SR processes.

Even though the production of methanol from biomass is a really cost-intensive process, the photoreforming of methanol is the most commonly used reaction for the photoreforming of organic molecules. It is probably due to the fact that methanol is the simplest alcohol and works reasonably well in this reaction. Regardless the hydrogen production from photoreforming of methanol could be used for practical proposes, the amount of high quality studies published on this matter has allowed to deeply understand the basics of the photoreforming process *per se.* Moreover, it has led to the development of novel materials that can be successfully used for the photoreforming of more complex alcohols (such as ethanol and glycerol). This is the reason why this section will mainly review the recent approaches towards the photoreforming of methanol, as a representative of the alcohols photoreforming technology.

In general, an efficient photocatalyst needs to have the following characteristics[111]:

(a) To be able to absorb light in the UV–Vis region of the solar spectrum (which is about 50% of the solar energy), and to use this energy to generate charge transporters (electron–hole pairs).
(b) To be able to separate the e⁻–h⁺ pairs, allowing the charge carriers to participate in the half reactions. As above mentioned, it is well known that photogenerated e⁻ and h⁺ can recombine, reducing in this case the hydrogen generation. It has been reported that, for any photocatalyst to achieve the commercial stage, it has to display an efficiency of overall energy capture of about 15% in the visible region of the electromagnetic spectrum.
(c) To be characterised by an electronic structure which makes the half-reactions of interest thermodynamically feasible. Figure 2 shows the importance of the position of the energy bands in the semiconductor photocatalyst (e.g. TiO_2). The position of the conduction band (CB, in which the e⁻ are present) potential in the photocatalyst should be lower than that of the H^+/H_2 couple. In addition, the potential of the

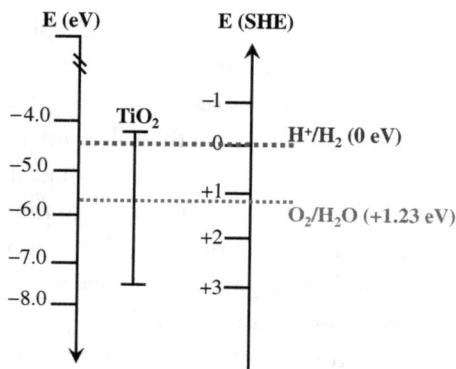

Figure 2: Positions of electronic bands of TiO_2 and energy levels of some redox couples. The energy scale is reported referring either to the vacuum level or to the Standard Hydrogen Electrode (SHE).

Data source: Ref. [111].

VB in which the h[+] are present) has to be higher than that of the H_2O/O_2 couple (for the water splitting reaction) and the alcohols oxidation (for the photoreforming process).

(d) To be characterised by surface active sites that makes possible the occurrence of these reactions.

In this sense, since the discovery of hydrogen production via water splitting by Fujishima and Honda over titanium dioxide (TiO_2),[106] significant research has been focusing on photoreforming with TiO_2-based materials, as TiO_2 exhibits a superior photocatalytic activity and is photochemically stable in aqueous solution,[112] together with its abundant availability, non-toxicity, low-cost and environmental friendly nature.[113,114] This is why, compared to any other photocatalyst, TiO_2 dominates in the overall scientific literature (including patents), exceeding 80% of the overall amount.[115] TiO_2 exists commonly in three phases: rutile, anatase and brookite. Rutile is thermodynamically stable while anatase and brookite are metastable. However, TiO_2 has a large band gap energy (3.0 eV for rutile and 3.2 eV for anatase), which only allows the absorption of light in the UV range. In other words, it does not absorb visible light. Hence, the photoconversion efficiency of TiO_2 is much lower than the acceptable solar-to-hydrogen efficiency (10%) for benchmark applications.[116] Moreover, even under the presence of a sacrificial agent such as methanol, the photogenerated electrons and holes in TiO_2 may experience a rapid recombination, limiting the further photocatalytic efficiency of the system towards the photoreforming process.[117] In this sense, it seems that small improvements might be achieved by chemical, morphological and textural modifications of this material, which clearly has the potential to significantly improve its performance in view of a higher hydrogen production. Thus, different strategies have been adopted to either enhance the efficiency of TiO_2 under UV irradiation to overcome the quick recombination of photogenerated electrons and holes (modifying the structure of the TiO_2 material, supporting metal/metal oxide nanoparticles on the semiconductor surface), or to broaden the absorption capability of TiO_2 to the visible range (doping with metal/non-metal ions, coupling the TiO_2 with low band gap semiconductors), as described in the following sections:

4.1 Modification of TiO$_2$ properties

As previously mentioned, TiO$_2$ exists in three phases: anatase, rutile and brookite. Even though it has been reported that brookite nanocrystals could have higher photocatalytic activities as compared to rutile and anatase, it is rarely studied due to the difficulties in synthesising this material. Then, rutile and anatase have been often used in the study of hydrogen from photocatalytic water splitting. In spite of the lots of work focusing on rutile, the anatase form is the most active polymorph for catalytic and photocatalytic applications,[118] mainly due to its higher surface area. It could be attributed to the fact that rutile is usually formed by high temperature calcination of the anatase phase.[33]

However, the photocatalytic activity of anatase is strongly limited by the fast electron–hole recombination. One way to extend the lifetime of the photogenerated charge carriers is by coupling different TiO$_2$ phases, allowing the charge transfer between them, resulting in a more effective charge separation. A well-known example is P25, which consists of a mixed phase of anatase and rutile in a ratio around 80/20%. In the anatase–rutile heterojuntions, upon UV irradiation, photogenerated electrons from the anatase phase will be transferred to the rutile phase due to its lower conduction band energy.[119] It results in a more efficient separation of e$^-$ and h$^+$, and this may be the reason why P25 exhibit a higher photocatalytic activity than both, anatase and rutile. Other theories have been given to the better performance of P25, but till date there is not a conclusive explanation about it. In spite of this, most of the studies in the field of photoreforming have been carried out using P25 as a photocatalyst.

Another way to improve the activity of TiO$_2$ photocatalysts is by playing on its morphology. For instance, the morphology of nanoparticles has a direct influence on the photoactivity by affecting important parameters as: the specific surface area, the hydrophilic character and, more importantly, the availability of charge carrier by modifying the internal electric field close to the surface.[120] Working at the (nano) material level for the development of more efficient TiO$_2$ photocatalysts is of importance to improve the efficiency of photocatlytic reactions and for targeting viable processes. Among them, there has been a significant effort to elaborate 1D and 2D TiO$_2$-based nanostructures.

In this sense, we can find in literature several works about the application of 1D TiO$_2$ nanotubes,[120–122] 1D TiO$_2$ nanorods[123] or 2D TiO$_2$ nanosheets[124] to the photoreforming of methanol in batch liquid phase reactors. The aim of these new structures with enhanced charge conductivity is to transfer the photogenerated carriers through a shorter and faster route due to the 1D or 2D structure. For instance in TiO$_2$ nanotubes, the charge carriers easily move along the longitudinal direction of the tubular nanostructure, which is favourable to improve the separation ability of photoinduced electron/hole pairs. The results obtained show that this unique morphology plays a very important role in prolonging the lifetime of the charge carrier, which in the end leads to a higher hydrogen production.

4.2 Supporting metallic/metal oxide nanoparticles on the titania surface to promote electron and hole transfer reactions at the TiO$_2$/substrate interface

One of the most extended ways to prolong the lifetime of the photogenerated charge carriers (e$^-$ and h$^+$) is by doping the TiO$_2$ semiconductor with a metal. The metal acts as an electron sink, reducing the charge carriers recombination, and thereby prolonging the e$^-$–h$^+$ pair lifetime. It can be attributed to the creation of a Schottky barrier due to the interaction between the metal and the TiO$_2$ semiconductor. It allows the electrons from the conduction band in TiO$_2$ to be injected into the Fermi level of the metals, which in turn improves the reaction chemistry.[125]

With regard to the methanol photoreforming, different systems consisting of noble and non-noble metals have been used to enhance the final hydrogen production. Most of these studies have been carried out in a batch reactor, in the liquid phase and at room temperature. The role of noble metals such as Pt, Pd and Au has been demonstrated to be very similar. Precious metals generally have more stable metallic states, resisting oxidation. It has been proposed that this property is what makes them useful as active co-catalysts for TiO$_2$ in methanol photoreforming.

For instance, supporting Pt nanoparticles on TiO$_2$ has been demonstrated to be one of the most efficient methods to enhance the hydrogen production in this reaction under UV irradiation.[114,126–130] The e$^-$–h$^+$ pairs generated in the TiO_2 particles upon irradiation migrate to their surfaces (Eq. (1.18)). According to literature, based on the products detected, it is generally accepted that methanol is photo-oxidised to CO_2 on the photogenerated h^+, via the formation of stable reaction intermediates, such as formaldehyde and formic acid, by means of reactions as described by Eqs. (1.19)–(1.22)[129]:

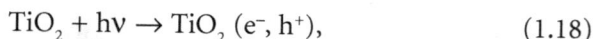

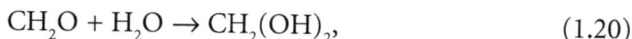

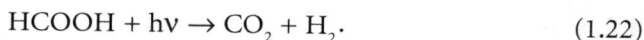

$$TiO_2 + hv \rightarrow TiO_2 \ (e^-, h^+), \tag{1.18}$$

$$CH_3OH + hv \rightarrow CH_2O + H_2, \tag{1.19}$$

$$CH_2O + H_2O \rightarrow CH_2(OH)_2, \tag{1.20}$$

$$H_2(OH)_2 + hv \rightarrow HCOOH + H_2, \tag{1.21}$$

$$HCOOH + hv \rightarrow CO_2 + H_2. \tag{1.22}$$

On the other hand, the photogenerated e^- will be injected to the Pt nanoparticles, followed by the reduction of a H^+ from water and/or from methanol to produce molecular H_2 (Eq. (1.23)).

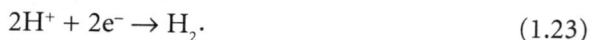

$$2H^+ + 2e^- \rightarrow H_2. \tag{1.23}$$

Moreover, important parameters have been studied such as the metal loading and the calcination temperature.[114,127] Both of them affect the morphological properties of the catalyst. Usually with the increase of Pt loading, the hydrogen production rate gradually increases. It could be attributed to the role of Pt as electron traps, reducing the further recombination of photogenerated electrons and holes. As the Pt loading increases, the amount of e^- traps increases too, enhancing the overall photocatalytic activity towards the hydrogen production. However, optimum Pt loadings around 1% have been found in literature. Above this loading, the TiO_2 surface may be covered by excessive Pt particles, decreasing the access of the UV light to the TiO_2 surface.[131] In addition, for high Pt loadings the Pt particles may act as recombination centres for e^- and h^+, resulting in the decrease of the photocatalytic activity.[132] On the other hand, the influence of the calcination temperature could be related with the Pt particle size and with the optimum anatase/rutile ratio, as observed by Wei *et al.*[114]

Together with Pt, Au has been thoroughly studied for the photoreforming of methanol.[121,133–138] Most of this studies have been carried out upon irradiation of UV light to allow the excitation of the TiO_2[121,133,136,138] and under similar conditions than the studies above mentioned for Pt. Under these conditions, the promotional effect of Au could be attributed to the formation of a Schottky barrier at the metal/TiO_2 interface. It decreases the recombination of the photogenerated charge carriers, increasing therefore the overall H_2 evolution. However, it has been reported elsewhere that Au/TiO_2 catalysts also present high visible light photocatalytic activity in the photoreforming of methanol.[134,137] It is due to the broad absorption originating from the surface plasmon band of AuNPs expanding from 400 nm to 700 nm. In this case, Au may play simultaneously several roles including light absorption and photosensitization of TiO_2. The Au surface plasmon band absorption maximum appears at 560 nm. The reaction mechanism for alcohols photoreforming over Au/TiO_2 photocatalyst[137] could be understood as follows: the light absorption in Au nanoparticles leads to its excitation, followed by the e^- injection to the TiO_2 conduction band, where the H_2O reduction takes place, and hence the H_2 evolution. Then, the oxidation of methanol takes place in the photogenerated holes on the Au surface.

On the other hand, the addition of oxidised metals to the TiO_2 support has also been studied. This is the case of Pd,[139] Ni[125,140,141] and Cu.[111,125] The higher photocatalytic effect found for these materials compared to the plain TiO_2 could be attributed, again, to a lower recombination of the photogenerated electrons and holes. As previously explained, at the metal/TiO_2 interfaces, UV irradiation induces Fermi level equilibration via charge distribution creating a Schottky barrier between the TiO_2 and the metal interface. That is why the electrons migrate to the metal surface. However, in metal oxide/TiO_2 interfaces, this is not the case. The case of copper modified TiO_2 catalysts is one of the most studied in this matter for the photoreforming of methanol.[111,122,125,142,143] It has been reported that, under UV irradiation conditions, CuO_x/TiO_2 catalysts show an enhanced photocatalytic activity in comparison with TiO_2. It could be attributed to the migration of the e^- from the conduction band of TiO_2 to the metal oxides, followed by its further reduction, lowering this way the recombination of electrons and holes in the semiconductor surface. Moreover, it has been found[142] that the presence of Cu^{2+} decrease the photocatalytic activity in comparison with the Cu^+. In other words, the Cu^+/Cu component is responsible for photocatalytic H_2 generation, whereas the Cu^{2+} may hamper activity as it could compete with the protons for the photogenerated electrons.

On the other hand, some studies carried out by Prof. Bowker's research group proposed a different theory regarding the role of the metal in the photoreforming process.[33,144-148] The methanol photoreforming over Pd/TiO_2 catalysts could be summarised in the following reaction scheme (Eqs. (1.24)–(1.27)):

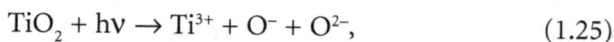

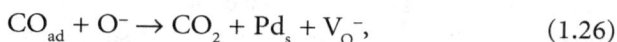

$$CH_3OH + Pd_s \rightarrow CO_{ad} + 2H_2, \tag{1.24}$$

$$TiO_2 + h\nu \rightarrow Ti^{3+} + O^- + O^{2-}, \tag{1.25}$$

$$CO_{ad} + O^- \rightarrow CO_2 + Pd_s + V_O^-, \tag{1.26}$$

$$V_O^- + Ti^{3+} + O^{2-} + H_2O \rightarrow TiO_2 + H_2. \tag{1.27}$$

According to this, the main dehydrogenation reactions take place on the metal and, in the absence of UV light, the reaction would stop at this point, with the metal sites blocked by reaction intermediates, such as CO (Eq. (1.24)). It is well known that CO strongly chemisorbs on precious metals at low temperatures. On the other hand, the UV irradiation of the catalyst leads to the production of highly activated oxygen species on the TiO_2 by band-gap excitation (Eq. (1.25)). These species attacks the CO and oxidise it to CO_2 (Eq. (1.26)), enabling another methanol molecule to adsorb and produce more H_2, restarting the cycle. They propose that this mechanism could apply for metals such as Pd, Pt, Ru, Rh, etc. However, they have demonstrated that the mechanism in the case of Au/TiO_2 is different than that for Pd/TiO_2. CO is bound much more weakly to the Au surface due to the low energy of d orbitals in Au. Thus, the blocking surface intermediate from methanol is probably different than that with Pd. The candidate species could be methoxy or formate, which further oxidise to CO_2 with the photogenerated oxygen species.

Then, it seems that the former theory considers that the enhancement in the hydrogen production over the metal/TiO_2 catalyst is due to the electronic modification that the metal causes in the whole system. However, the latter theory attributes this enhancement to the catalytic role of the metal itself. Up to date it is not clear which one of them takes place in a higher extent, but probably the overall increase in the hydrogen production is due to the combination of both, the electronic and the catalytic effect of the metal supported over the TiO_2 semiconductor.

4.3 Extending the absorption wavelength of TiO_2 to the visible range

The relatively large band gap of TiO_2 results in the absorption of the UV part of the solar spectrum. However, the solar energy only contains about 4–5% UV light. Therefore, the utility of TiO_2 might be dramatically improved by shifting the onset absorbance towards the visible region.

One of the alternatives in literature is to modify the electronic structure of TiO_2, while maintaining its excellent photocatalytic properties. In this sense, ever since Sato first discovered visible light response of anion doped TiO_2,[149] various metal ion doped TiO_2 have been reported by many researchers, but there is only limited success due to photo corrosion.[150] Moreover, the photogenerated electrons and holes are easily recombined in a very short time due to the smaller band gap of the ion-modified TiO_2, which could decrease the activity of the photocatalyst.[151] In recent years, various anion doped TiO_2 materials such as $N-TiO_2$,[150–154] $S-TiO_2$,[155–157] $P-TiO_2$,[158] $B-TiO_2$,[159] have been prepared by different methods. However, the substitutional doping of nitrogen have been found to be the most effective, decreasing the band gap by mixing of nitrogen 2p states with oxygen 2p states on the top of the VB.

For instance, the photoreforming of methanol has also been studied for $N-TiO_2$ catalysts in liquid phase and batch conditions.[150–154] As expected, an important red shifting of the absorbance peak is observed for these catalysts in comparison with the TiO_2 catalysts. This could be attributed by the fact that when N2p and O2p state of TiO_2 mixed, they may form a localised state just above the balance band of TiO_2. It narrows the band gap, making possible the H_2 production in the visible range of the spectra (l > 400 nm). Since the band gap is smaller due to the doping of nitrogen, the presence of metals such as Pt[150,151,153] have been proposed to decrease the e^- and h^+ recombination, as explained in the previous section.

On the other hand, one of the most promising methods to extend the light-absorbing property of TiO_2 is to couple TiO_2 with narrow band-gap semiconductors. This way, the second semiconductor acts as a sensitizer of TiO_2. Hence, under visible light irradiation, only the sensitizer is excited, and the electrons generated to their conduction band are injected into the inactivated TiO_2 conduction band (since TiO_2 only activates under UV irradiation). On the other hand, if the VB of the sensitizer is more cathodic than the VB of TiO_2, the holes generated in the narrow band-gap semiconductor cannot migrate to the TiO_2. It leads to an efficient charge carrier separation. The charge transfer between the two semiconductors happens in the heterojunctions between the two of them.[160]

Previously, we highlighted the high photocatalytic activity of TiO_2 doped with metal oxides as a co-catalyst under UV irradiation conditions. The enhancement observed in the catalytic activity was attributed to the more efficient separation of e^- and h^+ due to the role of the metal oxide as electron trap. However, it has also been reported in literature that low band gap metal oxides perform as p-type semiconductors. This is the case, for instance, of copper oxides. Cu_2O is one of the few p-type semiconductors which are inexpensive, non-toxic and readily available. It has been reported in literature that Cu_2O (with a band-gap of 2.1 eV) combined with TiO_2 could be activated under irradiation in the visible range for the photoreforming process. Then, by visible light absorption in the copper oxide, the photogenerated electrons migrate from the conduction band of Cu_2O to the conduction band of TiO_2, while the holes stay at the conduction band of Cu_2O. It favours the separation of the photogenerated e^- and h^+. Promising results have been found with this system for the photoreforming of methanol, ethanol and glycerol as hole scavengers,[161–163] obtaining high and stable hydrogen rates under the irradiation of visible light.

4.4 Summary

New approaches for the photoreforming of methanol (as a representative of the biomass-derived alcohols) have been reviewed in this section. We can summarise that the photoreforming of alcohols seems to be a suitable process with high potential for the production of hydrogen, with low energy requirements (it is usually carried out at room temperature), by the use of renewable solar energy. Among the different materials available for this process, TiO_2 is the most widely used photocatalyst for the photoreforming of alcohols. It is due to its superior photocatalytic activity and stability, together with its abundant availability, non-toxicity, low-cost and environmentally friendly nature.

Most research in the photoreforming field deals with the improvement of the TiO_2 catalysts in view of its further practical application. In this sense, the main areas of study are the following:

- The enhancement of the half-reactions by reducing the recombination of photogenerated electrons and holes. It involves modifying the structure of the TiO_2 material and supporting metal/metal oxide nanoparticles on the semiconductor surface.
- The extending of the absorption capability of TiO_2 into the visible range by doping with metal/non-metal ions, or by coupling the TiO_2 with low band-gap semiconductors.

5 Reforming in the UK Catalysis Hub at Harwell

This chapter provides an overview of some of the important challenges of current reforming technology and summarises the state-of-the-art of three new approaches for reforming: catalytic reforming of biomass-derived alcohols and polyols, CLR and photocatalytic reforming.

The UK Catalysis Hub at Harwell is strongly involved in the development of these three approaches to a sustainable reforming technology. Based at the Research Complex at Harwell (RCaH), the UK Catalysis Hub is uniquely placed to benefit from the state-of-the-art facilities on site (Diamond Light Source, Central Laser Facility and ISIS).

The catalytic reforming of biomass-derived alcohols and polyols to hydrogen, with particular focus on glycerol, is led by Prof. Richard Catlow and colleagues under the theme "Catalysis by design". The project involves the synthesis, characterization and catalytic testing of a range of materials for APR of glycerol (or similar polyol) to hydrogen. The overall theme is to develop novel and highly active/selective catalysts that will drive the reaction towards low-cost hydrogen production, while minimising undesirable by-products such as methane and carbon monoxide.

In doing so, the research focuses on conducting mechanistic investigations with the use of advanced spectroscopic techniques available on site in order to develop structure–activity relationships. The group is also involved in the development of *in situ* methodologies for combined catalysis and XAFS studies to understand the catalyst materials under realistic reaction conditions. Molecular level understanding of the catalysts is essential for the design of new and improved materials. Currently, the materials of particular interest for this project include alumina supported platinum group metals and surface modified supported metal nanoparticles.

The second and third approaches to reforming, chemical looping technology and photocatalysis, are carried out within the program "Catalysis for energy", led by Prof. Chris Hardacre.

One of the key research challenges for the UK Catalysis Hub is to demonstrate the viability of the CLR process as it has only been demonstrated at a conceptual level. Two different CLR approaches such as CL-SRM and CL-DRM are currently investigated in order to produce of syngas with a H_2/CO ratio equal to 2:1 and separate production of syngas and hydrogen or carbon monoxide streams, respectively. Among different oxygen carriers reported in the literature, CeO_2-based oxygen carriers seem to have a great potential for CLR processes, owing the redox properties and oxygen storage capacity of the CeO_2. Four different $Ce_{1-x}Met_xO_2$-based oxygen carriers series (Met = Fe, Zr, Tb and Pr) with a metal content range from 20 wt.% to 100 wt.% have been studied during the CL-DRM process. In order to have an overall picture of the physico-chemical properties of these materials, a multi-technique characterization approach was carried out. Preliminary results show that Fe_2O_3/CeO_2-based materials are the most active during the CL-DRM process. Moreover, by controlling the Fe_2O_3/CeO_2 ratio, the activity of the oxygen carrier material can be adjusted so as to substantially favour oxidation of methane to syngas and discourage both the total oxidation of methane and carbon deposition via DM reactions. Furthermore, controlling the oxidation state of the oxygen carrier was used to influence product selectivity. Such use of an oxidation state operating window could potentially be used to improve future chemical looping processes. In the near future, autothermal CL-DRM and CL-SRM using air during the oxidation step will be investigated as a novel energy input to the process.

With regards to the photocatalysis technology, the Catalysis Hub is working in different approaches. To this date, most of the studies carried out in this field have been performed in liquid phase and in batch reactors. However, in view of its further practical application, this technology could be improved by the development of continuous flow configurations. Moreover, working at higher temperatures in the gas phase under the concept of photocatalytic reforming would be desirable to enhance the kinetics of the general process towards the hydrogen production. Hence, one of the areas of study in this field is the development of novel configurations to study the photoreforming of biomass-derived alcohols in gas phase continuous flow reactors, and at higher temperatures.

On the other hand, as previously mentioned, most studies in photoreforming use biomass derived alcohols (such as methanol, ethanol or glycerol) and some sugars as sacrificial donors. However, even though lingo-cellulosic biomass is the most abundant type of biomass on the earth, the photoreforming of lingo-cellulosic materials has been rarely studied in the literature. Thus, the project also focuses on a new area of study with respect to photoreforming of cellulose, one of the main polymers present in lingo-cellulosic biomass.

References

1. R. L. Manfro, A. F. da Costa, N. F. P. Ribeiro and M. M. V. M. Souza, *Fuel Process. Technol.*, 2011, **92**(3), 330–335.
2. P. V. Tuza, R. L. Manfro, N. F. P. Ribeiro and M. M. V. M. Souza, *Renew. Energ.*, 2013, **50**, 408–414.
3. G. Wen, Y. Xu, H. Ma, Z. Xu and Z. Tian, *Int. J. Hydrogen Energ.*, 2008, **33**(22), 6657–6666.
4. R. R. Davda, J. W. Shabaker, G. W. Huber, R. D. Cortright and J. A. Dumesic, *Appl. Catal. B-Environ.*, 2005, **56**(1–2), 171–186.
5. A. O. Menezes, M. T. Rodrigues, A. Zimmaro, L. E. P. Borges and M. A. Fraga, *Renew. Energ.*, 2011, **36**(2), 595–599.
6. J. A. Calles, A. Carrero, A. J. Vizcaino and L. Garcia-Moreno, *Catal. Today*, 2014, **227**, 198–206.
7. R. L. Manfro and M. M. V. M. Souza, *Cataly. Lett.*, 2014, **144**(5), 867–877.
8. G. Wen, Y. Xu, Y. Wei, R. Pei, K. Li, Z. Xu and Z. Tian, *Chinese J. Catal.*, 2009, **30**(8), 830–835.
9. D. Pakhare and J. Spivey, *Chemi. Soc. Rev.*, 2014, **43**(22), 7813–7837.

10. G. Jones, J. G. Jakobsen, S. S. Shim, J. Kleis, M. P. Andersson, J. Rossmeisl, F. Abild-Pedersen, T. Bligaard, S. Helveg, B. Hinnemann, J. R. Rostrup-Nielsen, I. Chorkendorff, J. Sehested and J. K. Nørskov, *J. Catal.*, 2008, **259**(1), 147–160.
11. M. Momirlan and T. N. Veziroglu, *Int. J. Hydrogen Energ.*, 2005, **30**(7), 795–802.
12. D. S. Newsome, *Catal. Rev.*, 1980, **21**(2), 275–318.
13. A. C. Caetano de Souza and J. L. Silveira, *Renew. Sust. Energ. Rev.*, 2011, **15**(4), 1835–1850.
14. R. D. Cortright, R. R. Davda and J. A. Dumesic, *Nature*, 2002, **418**(6901), 964–967.
15. J. Fermoso, L. He and D. Chen, *J. Chem. Technol. Biotechnol.*, 2012, **87**(10), 1367–1374.
16. G. W. Huber, J. W. Shabaker and J. A. Dumesic, *Sci.*, 2003, **300**(5628), 2075–2077.
17. S. M. Kim and S. I. Woo, *Chemsuschem*, 2012, **5**(8), 1513–1522.
18. D. I. Kondarides, V. M. Daskalaki, A. Patsoura and X. E. Verykios, *Cataly. Lett.*, 2008, **122**(1–2), 26–32.
19. D. B. Levin and R. Chahine, *Int. J. Hydrogen Energ.*, 2010, **35**(10), 4962–4969.
20. M. d. O. Melo and L. A. Silva, *J. Brazi. Chem. Soc.*, 2011, **22**(8), 1399–1406.
21. B. Meryemoglu, A. Hesenov, S. Irmak, O. M. Atanur and O. Erbatur, *Int. J. Hydrogen Energ.*, 2010, **35**(22), 12580–12587.
22. B. Meryemoglu, B. Kaya, S. Irmak, A. Hesenov and O. Erbatur, *Fuel.*, 2012, **97**, 241–244.
23. G. Sadanandam, K. Ramya, D. B. Kishore, V. Durgakumari, M. Subrahmanyam and K. V. R. Chary, *Rsc Adv.*, 2014, **4**(61), 32429–32437.
24. K. Seung-hoon, J. Jae-sun, Y. Eun-hyeok, L. Kwan-Young and M. D. Ju, *Catal. Today*, 2014, **228**, 145–151.
25. K. Shimura and H. Yoshida, *Energ. Environ. Sci.*, 2011, **4**(7), 2467–2481.
26. V. K. Skoulou, P. Manara and A. A. Zabaniotou, *J. Anal. Appl. Pyrolysis*, 2012, **97**, 198–204.
27. C. Wu, Z. Wang, P. T. Williams and J. Huang, *Sci. Rep.*, 2013, **3**.
28. J. Xuan, M. K. H. Leung, D. Y. C. Leung and M. Ni, *Renew. Sust. Energ. Rev.*, 2009, **13**(6–7), 1301–1313.
29. S. Shao, A.-W. Shi, C.-L. Liu, R.-Z. Yang and W.-S. Dong, *Fuel Process. Technol.*, 2014, **125**, 1–7.
30. S. Guo, L. Guo, J. Yin and H. Jin, *J. Supercrit. Fluid.*, 2013, **78**, 95–102.

31. T. Valliyappan, D. Ferdous, N. N. Bakhshi and A. K. Dalai, *Top. Catal.*, 2008, **49**(1–2), 59–67.
32. F. A. P. Voll, C. C. R. S. Rossi, C. Silva, R. Guirardello, R. O. M. A. Souza, V. F. Cabral and L. Cardozo-Filho, *Int. J. Hydrogen Energ.*, 2009, **34**(24), 9737–9744.
33. M. Bowker, *Green Chem.*, 2011, **13**(9), 2235–2246.
34. J. Adanez, A. Abad, F. Garcia-Labiano, P. Gayan and L. F. de Diego, *Prog. Energ. Combust. Sci.*, 2012, **38**(2), 215–282.
35. B. Dou, Y. Song, C. Wang, H. Chen and Y. Xu, *Renew. Sust. Energ. Rev.*, 2014, **30**, 950–960.
36. S. Adhikari, S. D. Fernando and A. Haryanto, *Energ. Convers. Manage.*, 2009, **50**(10), 2600–2604.
37. T. Pairojpiriyakul, W. Kiatkittipong, W. Wiyaratn, A. Soottitantawat, A. Arpornwichanop, N. Laosiripojana, E. Croiset and S. Assabumrungrat, *Int. J. Hydrogen Energ.*, 2010, **35**(19), 10257–10270.
38. K. W. Siew, H. C. Lee, J. Gimbun and C. K. Cheng, *Int. J. Hydrogen Energ.*, 2014, **39**(13), 6927–6936.
39. K. W. Siew, H. C. Lee, J. Gimbuna and C. K. Cheng, *J. Energ. Chem.*, 2014, **23**(1), 15–21.
40. N. Luo, X. Fu, F. Cao, T. Xiao and P. P. Edwards, *Fuel*, 2008, **87**(17–18), 3483–3489.
41. R. R. Davda, J. W. Shabaker, G. W. Huber, R. D. Cortright and J. A. Dumesic, *Appl. Catal. B*, 2003, **43**(1), 13–26.
42. P. J. Dietrich, T. Wu, A. Sumer, J. A. Dumesic, J. Jellinek, W. N. Delgass, F. H. Ribeiro and J. T. Miller, *Top. Catal.*, 2013, **56**(18–20), 1814–1828.
43. G. W. Huber, J. N. Chheda, C. J. Barrett and J. A. Dumesic, *Sci.*, 2005, **308** (5727), 1446–1450.
44. G. W. Huber and J. A. Dumesic, *Catal. Today*, 2006, **111**(1–2), 119–132.
45. E. L. Kunkes, R. R. Soares, D. A. Simonetti and J. A. Dumesic, *Appl. Catal. B*, 2009, **90**(3–4), 693–698.
46. J. W. Shabaker, G. W. Huber and J. A. Dumesic, *J. Catal.*, 2004, **222**(1), 180–191.
47. C. A. Franchini, W. Aranzaez, A. M. Duarte de Farias, G. Pecchi and M. A. Fraga, *Appl. Catal. B*, 2014, **147**, 193–202.
48. M. Gupta and N. Kumar, *Renew. Sust. Energ. Rev.*, 2012, **16**(7), 4551–4556.
49. M. Stelmachowski, *Ecol. Chem. Eng. S*, 2011, **18**(1), 9–30.
50. P. D. Vaidya and A. E. Rodrigues, *Chem. Eng. Technol.*, 2009, **32**(10), 1463–1469.
51. Y. Wei, H. Lei, Y. Liu, L. Wang, L. Zhu, X. Zhang, G. Yadavalli, B. Ahring and S. Chen, *J. Sust. Bioenerg. Sys.*, 2014, **4**, 113–127.

52. E. A. Sanchez and R. A. Comelli, *Int. J. Hydrogen Energ.*, 2014, **39**(16), 8650–8655.

53. B. Dou, Y. Song, C. Wang, H. Chen, M. Yang and Y. Xu, *Appl. Energ.*, 2014, **130**, 342–349.

54. S. Adhikari, S. D. Fernando, S. D. F. To, R. M. Bricka, P. H. Steele and A. Haryanto, *Energ. Fuel.*, 2008, **22**(2), 1220–1226.

55. B. Zhang, X. Tang, Y. Li, Y. Xu and W. Shen, *Int. J. Hydrogen Energ.*, 2007, **32**(13), 2367–2373.

56. X. Fan, R. Burton and Y. Zhou, *Open Fuel. Energ. Sci. J.*, 2010, **3**, 17–22.

57. H. Chen, T. Zhang, B. Dou, V. Dupont, P. Williams, M. Ghadiri and Y. Ding, *Int. J. Hydrogen Energ.*, 2009, **34**(17), 7208–7222.

58. A. L. da Silva and I. L. Mueller, *Int. J. Hydrogen Energ.*, 2011, **36**(3), 2057–2075.

59. B. Dou, V. Dupont, G. Rickett, N. Blakeman, P. T. Williams, H. Chen, Y. Ding and M. Ghadiri, *Bioresour. Technol.*, 2009, **100**(14), 3540–3547.

60. B. Dou, G. L. Rickett, V. Dupont, P. T. Williams, H. Chen, Y. Ding and M. Ghadiri, *Bioresour. Technol.*, 2010, **101**(7), 2436–2442.

61. V. Dupont, M. V. Twigg, A. N. Rollinson and J. M. Jones, *Int. J. Hydrogen Energ.*, 2013, **38**(25), 10260–10269.

62. J. Fermoso, L. He and D. Chen, *Int. J. Hydrogen Energ.*, 2012, **37**(19), 14047–14054.

63. L. He, J. M. S. Parra, E. A. Blekkan and D. Chen, *Energ. Environ. Sci.*, 2010, **3**(8), 1046–1056.

64. I. Iliuta, H. R. Radfarnia and M. C. Iliuta, *Aiche J.*, 2013, **59**(6), 2105–2118.

65. Y. Li, W. Wang, B. Chen and Y. Cao, *Int. J. Hydrogen Energ.*, 2010, **35**(15), 7768–7777.

66. W. Wang and Y. Cao, *Int. J. Hydrogen Energ.*, 2011, **36**(4), 2887–2895.

67. X. Wang, N. Wang and L. Wang, *Int. J. Hydrogen Energ.*, 2011, **36**(1), 466–472.

68. Y.-J. Wu, P. Li, J.-G. Yu, A. F. Cunha and A. E. Rodrigues, *Chemi. Eng. Sci.*, 2014, **118**, 83–93.

69. B. Dou, C. Wang, H. Chen, Y. Song and B. Xie, *Int. J. Hydrogen Energ.*, 2013, **38**(27), 11902–11909.

70. X. Wang, M. Li, S. Li, H. Wang, S. Wang and X. Ma, *Fuel Process. Technol.*, 2010, **91**(12), 1812–1818.

71. W. Wang *J. Energ. Inst.*, 2014, **87**(2), 152–162.

72. M. Rydén, E. Cleverstam, A. Lyngfelt and T. Mattisson, *Int. J. Greenhouse Gas Control*, 2009, **3**(6), 693–703.

73. M. Johansson, T. Mattisson and A. Lyngfelt, *Energ. Fuel.*, 2006, **20**(6), 2399–2407.

74. A. R. Oller, M. Costa and G. Oberdörster, *Toxicol. Appl. Pharmacol.*, 1997, **143**(1), 152–166.

75. T. Mattisson, A. Järdnäs and A. Lyngfelt, *Energ. Fuel.*, 2003, **17**(3), 643–651.

76. M. Ishida and H. Jin, *Ind. Eng. Chem. Res.*, 1996, **35**(7), 2469–2472.

77. H. Jin, T. Okamoto and M. Ishida, *Ind. Eng. Chem. Res.*, 1999, **38**(1), 126–132.

78. B. M. Corbella, L. F. de Diego, F. García-Labiano, J. Adánez and J. M. Palacios, *Environ. Sci. Technol.*, 2005, **39**(15), 5796–5803.

79. S. R. Son and S. D. Kim, *Ind. Eng. Chem. Res.*, 2006, **45**(8), 2689–2696.

80. L. F. de Diego, P. Gayán, F. García-Labiano, J. Celaya, A. Abad and J. Adánez, *Energ. Fuels*, 2005, **19**(5), 1850–1856.

81. P. Cho, T. Mattisson and A. Lyngfelt, *Fuel*, 2004, **83**(9), 1215–1225.

82. S. Y. Chuang, J. S. Dennis, A. N. Hayhurst and S. A. Scott, *Combust. Flame*, 2008, **154**(1–2), 109–121.

83. Q. Song, W. Liu, C. D. Bohn, R. N. Harper, E. Sivaniah, S. A. Scott and J. S. Dennis, *Energ. Environ. Sci.*, 2013, **6**(1), 288–298.

84. J. S. Dennis, C. R. Müller and S. A. Scott, *Fuel*, 2010, **89**(9), 2353–2364.

85. J. S. Dennis and S. A. Scott, *Fuel*, 2010, **89**(7), 1623–1640.

86. B. M. Corbella, L. F. de Diego, F. García-Labiano, J. Adánez and J. M. Palacios, *Ind. Eng. Chem. Res.*, 2006, **45**(1), 157–165.

87. H. Jin, T. Okamoto and M. Ishida, *Energ. Fuel.*, 1998, **12**(6), 1272–1277.

88. G. Azimi, H. Leion, T. Mattisson and A. Lyngfelt, *Energ. Procedia*, 2011, **4**(0), 370–377.

89. M. M. Hossain and H. I. de Lasa, *AIChE J.*, 2007, **53**(7), 1817–1829.

90. A. Shulman, E. Cleverstam, T. Mattisson and A. Lyngfelt, *Energ. Fuel.*, 2009, **23**(10), 5269–5275.

91. Q. Song, R. Xiao, Z. Deng, W. Zheng, L. Shen and J. Xiao, *Energ. Fuel.*, 2008, **22**(6), 3661–3672.

92. H. Tian, Q. Guo and J. Chang, *Energ. Fuel.*, 2008, **22**(6), 3915–3921.

93. N. Berguerand and A. Lyngfelt, *Energ. Procedia.*, 2009, **1**(1), 407–414.

94. H. Leion, Y. Larring, E. Bakken, R. Bredesen, T. Mattisson and A. Lyngfelt, *Energ. Fuel.*, 2009, **23**(10), 5276–5283.

95. R. Xiao, Q. Song, M. Song, Z. Lu, S. Zhang and L. Shen, *Combust. Flame*, 2010, **157**(6), 1140–1153.

96. K. Otsuka, Y. Wang, E. Sunada and I. Yamanaka, *J. Catal.*, 1998, **175**(2), 152–160.

97. F. J. Pérez-Alonso, M. López Granados, M. Ojeda, P. Terreros, S. Rojas, T. Herranz, J. L. G. Fierro, M. Gracia and J. R. Gancedo, *Chem. Mater.*, 2005, **17**(9), 2329–2339.

98. G. Balducci, J. Kašpar, P. Fornasiero, M. Graziani and M. S. Islam, *J. Phys. Chem. B*, 1998, **102**(3), 557–561.

99. E. Mamontov, T. Egami, R. Brezny, M. Koranne and S. Tyagi, *J. Phys. Chem. B*, 2000, **104**(47), 11110–11116.

100. A. Shafiefarhood, N. Galinsky, Y. Huang, Y. Chen and F. Li, *Chem. Cat. Chem.*, 2014, **6**(3), 790–799.

101. K. Li, H. Wang, Y. Wei and M. Liu, *J. Rare Earths*, 2008, **26**(5), 705–710.

102. K. Li, H. Wang, Y. Wei and M. Liu, *J. Rare Earths*, 2008, **26**(2), 245–249.

103. K. Watanabe, T. Miyao, K. Higashiyama, H. Yamashita and M. Watanabe, *Catal. Commun.*, 2011, **12**(11), 976–979.

104. R. W. Coughlin and M. Farooque, *Nature*, 1979, **279**(5711), 301–303.

105. D. Wang, S. Czernik, D. Montané, M. Mann and E. Chornet, *Ind. Eng. Chem. Res.*, 1997, **36**(5), 1507–1518.

106. A. Fujishima and K. Honda, *Nature*, 1972, **238**(5358), 37–38.

107. M. R. Hoffmann, S. T. Martin, W. Choi and D. W. Bahnemann, *Chemi. Rev.*, 1995, **95**(1), 69–96.

108. M. Ni, D. Y. C. Leung and M. K. H. Leung, *Int. J. Hydrogen Energ.*, 2007, **32**(15), 3238–3247.

109. A. Demirbas, *Pro. Energ. Combust. Sci.*, 2007, **33**(1), 1–18.

110. T. Hirai, N. O. Ikenaga, T. Miyake and T. Suzuki, *Energ. Fuel.*, 2005, **19**(4), 1761–1762.

111. L. Clarizia, D. Spasiano, I. Di Somma, R. Marotta, R. Andreozzi and D. D. Dionysiou, *Int. J. Hydrogen Energ.* 2014.

112. O. Khaselev and J. A. Turner, *Sci.*, 1998, **280**(5362), 425–427.

113. H. Kim and W. Choi, *Appl. Catal. B: Environ.*, 2007, **69**(3–4), 127–132.

114. P. Wei, J. Liu and Z. Li, *Ceram. Int.* 2013, **39**(5), 5387–5391.

115. S. Protti, A. Albini and N. Serpone, *Phys., Chem. Chemi. Phys.*, 2014, **16**(37), 19790–19827.

116. H. Lu, B. Zhao, R. Pan, J. Yao, J. Qiu, L. Luo and Y. Liu, *RSC Adv.*, 2014, **4**(3), 1128–1132.

117. T. Bak, J. Nowotny, M. Rekas and C. C. Sorrell, *Int. J. Hydrogen Energ.*, 2002, **27**(10), 991–1022.

118. C. Xu, W. Yang, Q. Guo, D. Dai, M. Chen and X. Yang, *J. Am. Chemi. Soci.*, 2014, **136**(2), 602–605.

119. R. I. Bickley, T. Gonzalez-Carreno, J. S. Lees, L. Palmisano and R. J. D. Tilley, *J. Solid State Chem.*, 1991, **92**(1), 178–190.

120. D. D'Elia, C. Beauger, J. F. Hochepied, A. Rigacci, M. H. Berger, N. Keller, V. Keller-Spitzer, Y. Suzuki, J. C. Valmalette, M. Benabdesselam and P. Achard, *Int. J. Hydrogen Energ.*, 2011, **36**(22), 14360–14373.

121. A. F. Feil, P. Migowski, F. R. Scheffer, M. D. Pierozan, R. R. Corsetti, M. Rodrigues, R. P. Pezzi, G. Machado, L. Amaral, S. R. Teixeira, D. E. Weibel and J. Dupont, *J. Braz. Chemi. Soci.*, 2010, **21**(7), 1359–1365.

122. H. Dang, X. Dong, Y. Dong, Y. Zhang and S. Hampshire, *Int. J. Hydrogen Energ.*, 2013, **38**(5), 2126–2135.

123. M. Lafjah, A. Mayoufi, E. Schaal, F. Djafri, A. Bengueddach, N. Keller and V. Keller, *Catal. Today*, 2014, **235**, 193–200.

124. Y. Zou, S. Z. Kang, X. Li, L. Qin and J. Mu, *Int. J. Hydrogen Energ.*, 2014, **39**(28), 15403–15410.

125. H. Bahruji, M. Bowker, P. R. Davies, J. Kennedy and D. J. Morgan, *Int. J. Hydrogen Energ.*, 2015, **40**(3), 1465–1471.

126. M. P. Languer, F. R. Scheffer, A. F. Feil, D. L. Baptista, P. Migowski, G. J. Machado, D. P. De Moraes, J. Dupont, S. R. Teixeira and D. E. Weibel, *Int. J. Hydrogen Energ.*, 2013, **38**(34), 14440–14450.

127. E. P. Melián, C. R. López, A. O. Méndez, O. G. Díaz, M. N. Suárez, J. M. Doña Rodríguez, J. A. Navío and D. Fernández Hevia, *Int. J. Hydrogen Energ.*, 2013, **38**(27), 11737–11748.

128. R. S. Khnayzer, L. B. Thompson, M. Zamkov, S. Ardo, G. J. Meyer, C. J. Murphy and F. N. Castellano, *J. Phys. Chem. C*, 2012, **116**(1), 1429–1438.

129. T. A. Kandiel, R. Dillert, L. Robben and D. W. Bahnemann, *Catal. Today*, 2011, **161**(1), 196–201.

130. T. A. Kandiel, R. Dillert and D. W. Bahnemann, *PhotoChemi. Photobiol. Sci.*, 2009, **8**(5), 683–690.

131. Y. Ikuma and H. Bessho, *Int. J. Hydrogen Energ.*, 2007, **32**(14), 2689–2692.

132. X. Fu, J. Long, X. Wang, D. Y. C. Leung, Z. Ding, L. Wu, Z. Zhang, Z. Li and X. Fu, *Int. J. Hydrogen Energ.*, 2008, **33**(22), 6484–6491.

133. J. A. Ortega Méndez, C. R. López, E. Pulido Melián, O. González Díaz, J. M. Doña Rodríguez, D. Fernández Hevia and M. Macías, *Appl. Catal. B: Environ.*, 2014, **147**, 439–452.

134. J. Cai, X. Wu, S. Li, F. Zheng, L. Zhu and Z. Lai, *ACS Appl. Mater. Interfaces*, 2015, **7**(6), 3764–3772.

135. T. Sreethawong and S. Yoshikawa, *Catal. Commun.*, 2005, **6**(10), 661–668.

136. S. Oros-Ruiz, R. Zanella, R. López, A. Hernández-Gordillo and R. Gómez, *J. Hazard Mater.*, 2013, **263**, 2–10.

137. P. Montes-Navajas, M. Serra and H. Garcia, *Catal. Sci. Technol.*, 2013, **3**(9), 2252–2258.

138. O. Rosseler, M. V. Shankar, M. K. L. Du, L. Schmidlin, N. Keller and V. Keller, *J. Catal.*, 2010, **269**(1), 179–190.

139. B. S. Kwak, J. Chae, J. Kim and M. Kang, *Bull. Kor. Chem. Soc.*, 2009, **30**(5), 1047–1053.

140. D. Jing, Y. Zhang and L. Guo, *Chem. Phys. Lett.*, 2005, **415**(1–3), 74–78.

141. E. P. Melián, M. N. Suárez, T. Jardiel, J. M. D. Rodríguez, A. C. Caballero, J. Araña, D. G. Calatayud and O. G. Díaz, *Appl. Catal. B: Environ.*, 2014, **152–153**(1), 192–201.

142. M. Jung, J. Scott, Y. H. Ng, Y. Jiang and R. Amal, *Int. J. Hydrogen Energ.*, 2014, **39**(24), 12499–12506.

143. S. Xu and D. D. Sun, *Int. J. Hydrogen Energ.*, 2009, **34**(15), 6096–6104.

144. L. S. Al-Mazroai, M. Bowker, P. Davies, A. Dickinson, J. Greaves, D. James and L. Millard, *Catal. Today*, 2007, **122**(1–2), 46–50.

145. H. Bahruji, M. Bowker, P. R. Davies and F. Pedrono, *Appl. Catal. B: Environ.*, 2011, **107**(1–2), 205–209.

146. M. Bowker, *Catal. Lett.*, 2012, **142**(8), 923–929.

147. M. Bowker, L. Millard, J. Greaves, D. James and J. Soares, *Gold Bull.*, 2004, **37**(3–4), 170–173.

148. J. Greaves, L. Al-Mazroai, A. Nuhu, P. Davies and M. Bowker, *Gold Bull.*, 2006, **39**(4), 216–219.

149. S. Sato, *Chem. Phys. Lett.*, 1986, **123**(1–2), 126–128.

150. K. M. Parida, S. Pany and B. Naik, *Int. J. Hydrogen Energ.*, 2013, **38**(9), 3545–3553.

151. W. C. Lin, W. D. Yang, I. L. Huang, T. S. Wu and Z. J. Chung, *Energ. Fuel.*, 2009, **23**(4), 2192–2196.

152. A. Kachina, E. Puzenat, S. Ould-Chikh, C. Geantet, P. Delichere and P. Afanasiev, *Chem. Mater.*, 2012, **24**(4), 636–642.

153. T. Sreethawong, S. Laehsalee and S. Chavadej, *Catal. Commun.*, 2009, **10**(5), 538–543.

154. O. D. Jayakumar, R. Sasikala, C. A. Betty, A. K. Tyagi, S. R. Bharadwaj, U. K. Gautam, P. Srinivasu and A. Vinu, *J. Nanosci. Nanotechnol.*, 2009, **9**(8), 4663–4667.

155. G. Yang, Z. Yan and T. Xiao, *Appl. Surf. Sci.*, 2012, **258**(8), 4016–4022.

156. W. Zhang, S. Wang, J. Li, F. Ma and X. Yang, *Asian J. Chem.*, 2015, **27**(3), 1111–1116.

157. W. Zhang, S. Wang, J. Li and X. Yang, *Catal. Commun.*, 2015, **59**, 189–194.

158. L. Lin, W. Lin, J. L. Xie, Y. X. Zhu, B. Y. Zhao and Y. C. Xie, *Appl. Catal. B: Environ.*, 2007, **75**(1–2), 52–58.

159. F. Wang, Y. Jiang, A. Gautam, Y. Li and R. Amal, *ACS Catal.*, 2014, **4**(5), 1451–1457.
160. S. Biswas, M. F. Hossain, T. Takahashi, Y. Kubota and A. Fujishima, *Phys. Stat. Solidi (A) Appl. Mater. Sci.*, 2008, **205**(8), 2028–2032.
161. Y. Li, B. Wang, S. Liu, X. Duan and Z. Hu, *Appl. Surf. Sci.*, 2015, **324**(0), 736–744.
162. K. Lalitha, G. Sadanandam, V. D. Kumari, M. Subrahmanyam, B. Sreedhar and N. Y. Hebalkar, *J. Phys. Chem. C*, 2010, **114**(50), 22181–22189.
163. S. Zhang, B. Peng, S. Yang, Y. Fang and F. Peng, *Int. J. Hydrogen Energ.*, 2013, **38**(32), 13866–13871.

Chapter 2

In Situ and *Operando* Measurement of Catalysts at Synchrotron X-ray and Neutron Sources

Emma K. Gibson* and Ian P. Silverwood†
Editors: Peter P. Wells* and Stewart F. Parker†

* UCL, Department of Chemistry, 20 Gordon St. London, UK
† ISIS Facility, STFC Rutherford Appleton Laboratory,
Chilton, Didcot, UK

Large scale research facilities such as the Diamond Light Source and ISIS neutron source exist to offer access to techniques that are impractical in other settings. The time available is restricted and it is therefore imperative that experiments are well planned and executed. This work aims to help by discussing the techniques and equipment available for *in situ* and *operando* measurements with X-ray and neutron probes to encourage greater use of these facilities. The benefits of each technique for *in situ* measurement will be briefly discussed and the constraints they demand on reactor design will be mentioned in detail. Examples illustrating existing cell design and their use to elucidate catalytic reactions will be explored with reference to the literature. A description of progress in linking X-ray and neutron probes with secondary techniques will be included and trends and future directions will be identified.

1 Introduction

Before providing a detailed discussion of the techniques available at synchrotron X-ray and neutron sources used to perform *in situ* and *operando* measurements, it is useful to first define these terms. An *in situ* experiment usually refers to the characterisation of a catalyst under working conditions. This is in contrast to *ex situ* characterisations which are performed on the catalyst before or after reaction. An *operando* experiment in addition to being *in situ*, also simultaneously measures the catalytic activity and selectivity, usually by analysis of the outlet of the reactor by GC or mass spectrometry.[1] In this way, the *operando* experiment is performed at conditions where the catalyst is truly "at work".

2 Neutrons

2.1 Introduction

Sir James Chadwick discovered the neutron in 1932, and received the Nobel Prize three years later. Fifty nine years after this, the value of neutron scattering as an analytical tool was recognised with a second Nobel Prize awarded jointly to B. N. Brockhouse and C. G. Shull for their development of neutron scattering for spectroscopy and diffraction, respectively. Key properties of the neutron are a rest mass similar to the proton, no electrical charge and quantum spin of 1/2. A neutron in equilibrium with its surroundings at 293 K has a de Broglie wavelength of approximately 1.8 Å and an energy of 204 cm^{-1}. These values are well suited for the measurement of materials, and neutrons are in many cases an ideal probe, although they interact relatively weakly with matter. Adoption of neutron techniques is unfortunately inhibited by the challenges in generating sufficient flux for the sensitivity of available detectors. Furthermore, there are no energy sensitive detection methods, which leads to loss of flux in experiments that demand a monoenergetic beam. Practical neutron sources for scientific investigations are therefore restricted to large facilities, either based around a nuclear reactor or particle accelerator. Even with such relatively high flux sources, experimental measurements often take a number of days.

2.2 Neutron sources

Neutrons are generated at reactor sources by fission of ^{235}U. A large amount of energy is also released, and cooling is a key consideration. Most research reactors are therefore sited at the bottom of a pool of heavy water (D_2O). The water acts both to cool the reactor and as a moderator to slow the neutrons to useful energies through multiple inelastic collisions. Neutrons reach approximate thermal equilibrium in the pool, with a Maxwellian distribution around 300 K. To obtain different neutron energies, smaller specialised moderators may be installed in the pool to thermalise neutrons with different energy distributions. A typical low energy moderator (cold source) would be a volume of liquid deuterium at 25 K, and a high energy (hot source) could be a block of graphite heated to around 2400 K by the reactor.

Spallation sources, as shown in Figure 1, use particle accelerators to generate a high energy proton beam that bombards a heavy metal target. This excites the metal nucleus, which then decays by evaporating nucleons, primarily neutrons. These may either leave the target or trigger further reactions within it. For each proton impacting the target, approximately 15 neutrons are produced. Spallation sources generally produce an intense, pulsed beam of neutrons. The exception to this is the SINQ source at the Paul Scherrer Institute in Switzerland, which operates in a continuous mode, and generates a neutron beam similar in character to a reactor source. Spallation sources generate much less waste heat than a reactor source, which relaxes the need for cooling. The high energy neutrons produced are under-moderated in small moderators which means that they do not achieve thermal equilibrium. These neutrons are known as epithermal or non-Maxwellian neutrons and spallation sources therefore provide a greater proportion of high energy neutrons. The neutron pulse is considerably more intense than that obtained from a reactor, although the time-averaged flux is lower. However, as all neutron energies are generated in a narrow time window, time of flight measurements can be easily exploited.

2.3 Neutron interactions with matter

When a neutron encounters an atom, it may be absorbed, scattered or simply not interact. Since neutrons are uncharged, they do not interact

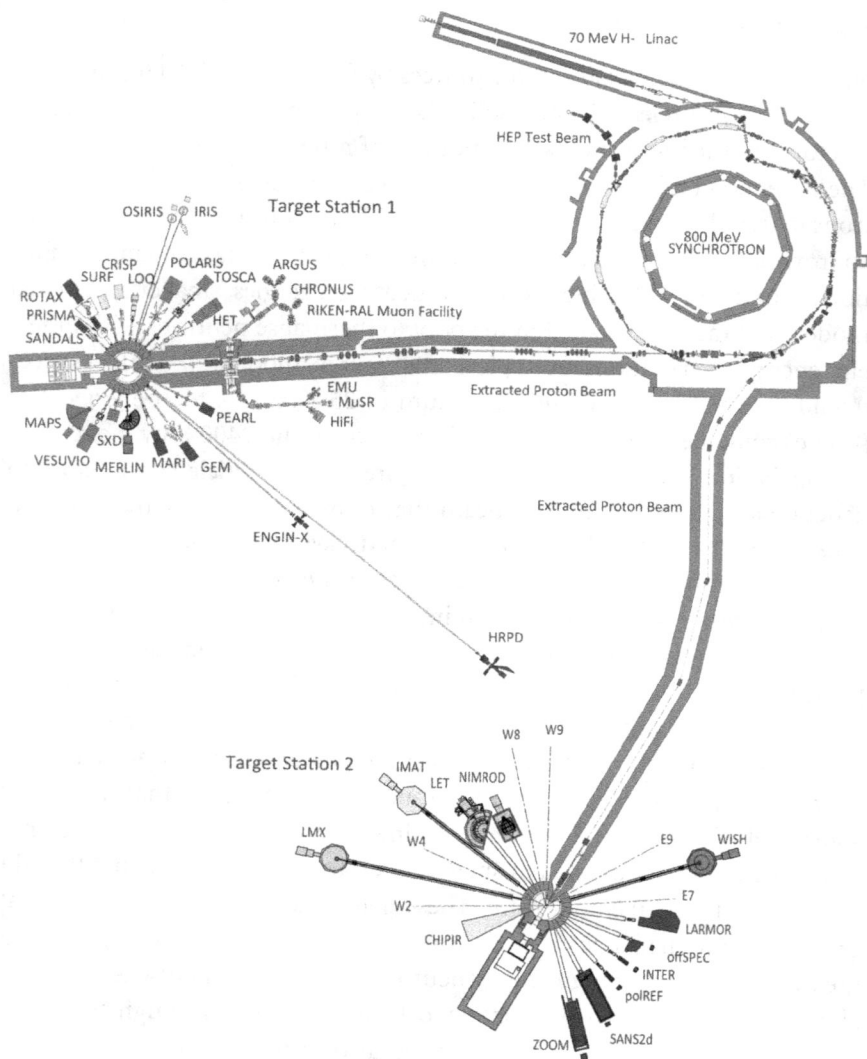

70 MeV H- Linac

HEP Test Beam

800 MeV
SYNCHROTRON

OSIRIS   IRIS   Target Station 1

CRISP   POLARIS   ARGUS
SURF   LOQ   TOSCA   CHRONUS
ROTAX   RIKEN-RAL Muon Facility
PRISMA   HET
SANDALS

EMU
MuSR
MAPS   PEARL   HiFi

SXD
VESUVIO   MERLIN   MARI   GEM

Extracted Proton Beam

ENGIN-X

Extracted Proton Beam

HRPD

W8   W9

Target Station 2   IMAT
LET   NIMROD

LMX   W4

W2

CHIPIR

E9   WISH

E7

LARMOR

offSPEC

INTER

polREF

ZOOM   SANS2d

Figure 1:   Layout of ISIS spallation source showing particle accelerator, both target stations and instruments. The synchrotron generates a 50 Hz pulsed proton beam with 1 in 5 pulses extracted to target station 2. (Credit: STFC)

strongly with the electron cloud that occupies the majority of the atomic volume. Their penetrating nature thus has advantages in designing sample holders that are compatible with extreme conditions. For many systems, it is desirable to measure inside a metal container without the need for windows and this is trivially easy for neutron techniques. Weak interaction with the sample may also be a disadvantage and large sample volumes are often necessary. Neutron measurement is therefore inherently a bulk technique, and this can be viewed as an advantage or not, depending on the desired information.

The likelihood of an interaction between sample and neutron depends on the atom involved. The strength of scattering can be represented in a similar way to the Beer–Lambert law in optical spectroscopy, as $J_f = J_i \exp(-d_sC\sigma)$ where J_i and J_f are the initial and final neutron flux, d_s is the sample thickness, C is concentration of the scattering atom and σ is the strength of scattering. The scattering strength of an atom has units of area, and is known as the cross section. As neutrons interact weakly, cross sections are small and are conventionally presented in barn (b), where $1\,b = 10^{-28}\,m^2$. There is no adequate theory that allows calculation of cross sections, and these have to be obtained experimentally. The likelihood of an atom absorbing a neutron is determined by its absorption cross section, which varies with neutron energy and will not be considered here, besides acknowledging it is of importance in instrumental design (particularly shielding) and can lead to induced radioactivity in samples.

Scattering from a sample can be either coherent or incoherent. Coherent scattering can be thought of as scattering from multiple nuclei in a coordinated fashion. The scattered neutron waves interfere constructively with each other to provide a strong angular dependence. Incoherent scattering occurs independently from individual nuclei and no interference results. Scattering can also be associated with a change in (kinetic) energy. Coherent elastic scattering forms the basis of neutron diffraction; incoherent elastic scattering occurs in all directions equally and generally appears as background. Coherent inelastic scattering gives information about collective external modes such as phonons and magnons, and incoherent inelastic scattering probes the local environment, providing information about internal vibrational modes and diffusive motion. It should be noted that different isotopes of the same element display different scattering cross sections, and this may be manipulated to great benefit. For example, while (^{1}H) has an incoherent cross section of over 80 b, deuterium (^{2}H) has an incoherent cross section of 2.05 b. This variance of scattering strength is of key importance in both diffraction and spectroscopy with neutrons and will be further discussed in later sections.

When a neutron is scattered, its momentum, direction and/or energy will change. To fully quantify the event, the energy exchange and scattering angle must be known. Readers are directed elsewhere for a rigorous treatment of neutron scattering but reference to the relevant quantities is made here to improve understanding of the literature. Q refers to the scattering vector, (momentum change) and the change in kinetic energy is referred to as either ω, or ΔE. Scattering intensity (S) must be recorded as a function of these two properties, and is rigorously represented as $S(Q, \omega)$ or $S(Q, E)$. For diffraction measurements, $S(Q)$ is often reported, with $\omega = 0$ inferred.

Despite the cost and inconvenience of using central facilities, neutron scattering is highly complementary to other techniques, and can be used to reinforce or validate other measurements. For difficult samples, neutrons may be the only capable probe, and their unique properties make neutron scattering a powerful device in the researcher's toolkit.

2.3.1 Inelastic scattering

Inelastic Neutron Scattering (INS) investigates the transfer of energy between a neutron and sample, and provides similar data to infrared and Raman spectra regarding vibrational energy levels, both internal and external modes. All vibrational modes are INS active in theory, and could be observed given sufficient neutron flux. However, hydrogen has a uniquely high incoherent scattering cross section, and a useful assumption in INS is that in a hydrogenous material, only vibrational modes involving hydrogen motions are visible. It should be remembered that this refers to ^1H, and deuteration can be profitably exploited to "switch off" selected vibrational modes. Infrared and Raman spectroscopy interact with the electron density in materials and are generally more sensitive to heavier elements. Neutrons interact solely with atomic nuclei and are sensitive to isotopic substitution. Where hydrogenous molecules are interacting with materials that have intense interfering optical absorbance or fluorescence, INS may be the only way to obtain vibrational information. Vibrational modes which are inactive in IR and Raman can be seen with INS, and overtones and combination bands may appear intense compared to conventional measurement.

Although INS is generally compared to infrared and Raman spectroscopies, the measurable energy transfer range using neutrons lies across a broader range. Techniques at the extremes are generally referred to using other names, but are still concerned with inelastic scattering events. At low energies, Quasi-Elastic Neutron Scattering (QENS) can be used to investigate translation and rotation in an energy range similar to far-IR and microwave spectroscopy (0–400 cm^{-1}). Neutron Compton Scattering (NCS) also known as Deep Inelastic Neutron Scattering (DINS) extends the range above INS to energies comparable with UV–Vis and soft X-rays.

Obtaining synthetic infrared and Raman spectra from computational models is made difficult due to the complex coupling between the electromagnetic wave and electron configuration of the material. This complication is not present with INS, and mode intensities are directly related to the displacement of the atoms and their scattering cross sections. Consequently data obtained by INS is currently the best test of computational models available.

2.3.2 Quasi-elastic scattering

QENS concerns small energy changes that correspond to low energy motions. In spectroscopic measurements, the strongest band will appear around zero energy transfer. For a perfectly elastic collision in an ideal instrument, no energy transfer would occur, and the spectrometer would measure an infinitely narrow line at this point. For a real system, the band is broadened by the resolution of the spectrometer and other experimental imperfections. Where there is energy transfer to and from within a range close to zero, this band will appear broadened further, and its shape will be affected. For analysis of a system where there is motion of a hydrogen-containing species, the experimental data can be fitted to a delta function convoluted with a number of Lorentzian functions. The Lorentzian peaks show a broadening that varies with momentum transfer (Q) for each motion. With free diffusion in a liquid for example, the peak width is proportional to Q^2. Various models exist to fit other circumstances that will not be expanded upon here. In a catalytic context, this has been generally applied to measuring diffusion in zeolites and similar materials.

2.3.3 Elastic scattering

Crystallography using neutrons is largely analogous to that done with X-rays. However, the interaction between the probe and the sample is different in two major aspects. Firstly, X-ray photons scatter from electron density, where neutrons predominantly scatter from nuclei. This is because neutrons have no charge and therefore do not interact electrostatically. A major consequence of this is that X-ray scattering increases with atomic number, whereas neutron scattering does not. The coherent neutron scattering cross section varies across elements in a complex manner, and there can be dramatic differences between isotopes of the same element. Neutron diffraction can therefore be very useful in differentiating structures where two elements close in atomic number are present in a sample, or in identifying the positions of light atoms where the X-ray scattering pattern is dominated by heavy atoms. Neutrons also have spin and interact with magnetic moments, including those caused by the electron cloud around an atom. Neutrons are therefore a powerful tool in the characterisation of magnetic materials, although this is of limited utility in the field of catalysis. Although diffraction is widely appreciated for the analysis of structure in materials with long-range order, it is also possible to obtain information from disordered materials. Through manipulation of the collected diffraction pattern, it is possible to obtain a Pair Distribution Function (PDF) which describes the probability of encountering another scattering centre as a function of distance from an original scatterer. In a chemical system, this can be related to bond lengths and angles, although complex systems can rapidly become intractable without an *a priori* model.

2.4 Practical considerations

2.4.1 Sample Geometry, Size and Presentation

Three major geometries are used in scattering measurements; their selection is dependent on the nature of the sample and instrument. As measurements are limited by the neutron flux, a sample that has a profile that matches that of the beam will encounter the greatest number of neutrons. To ensure that the measurement made corresponds to the desired interaction between neutron and sample, the possibility of multiple scattering must be reduced. If large quantity of a strong scatterer is placed in the beam, then the probability of multiple scattering events increases, which is detrimental to the measurement. The commonly accepted rule is therefore that the ideal sample will scatter 10% of the incident neutrons, with 90% transmission.

Flat plate (slab) cells hold the sample between two parallel windows and have the benefit of simple manufacture. These are the preferred geometry where neutrons are primarily detected at angles close to that of beam transmission. It allows the beam to be filled and the sample depth can be varied to the requirements of the sample easily. It is possible to decrease multiple scattering, which leads to noise in the data, to its optimum value in the direction of interest. If it is required to collect data angles close to perpendicular to the beam, then the edges of a flat-plate "shadow" the detectors as the sample is much thicker in this direction and additional scattering will occur. For these measurements, cylindrical or annular (hollow cylinder) geometries are preferred. A cylindrical sample geometry eliminates the edge effects caused by a slab, but a sample's thickness is then defined by its volume. Where this is an issue, cylindrical inserts of different diameter can create annuli of controlled sample thickness to efficiently use the full neutron beam flux. The amount of sample required for a measurement depends on its cross section. For inelastic scattering measurements, this can be approximated to the amount of hydrogen, and a reasonable quality spectrum can be routinely obtained with 1% of a mole of hydrogen, or 6×10^{21} H atoms on the TOSCA spectrometer at ISIS, although 10 times this amount is preferable. More sensitive instruments are available. For diffraction, non-hydrogenous powder samples would typically be around 1 g. A few grams are needed for QENS where the mobile species does not contain hydrogen, and less than a gram when it does.

2.4.2 Sample environment

To avoid loss of neutrons through undesired scattering with air, neutrons are transmitted to the sample position in evacuated tubes and beam guides. Control of the environment may be as simple as sealing the sample inside a container to prevent evaporation, but many complex

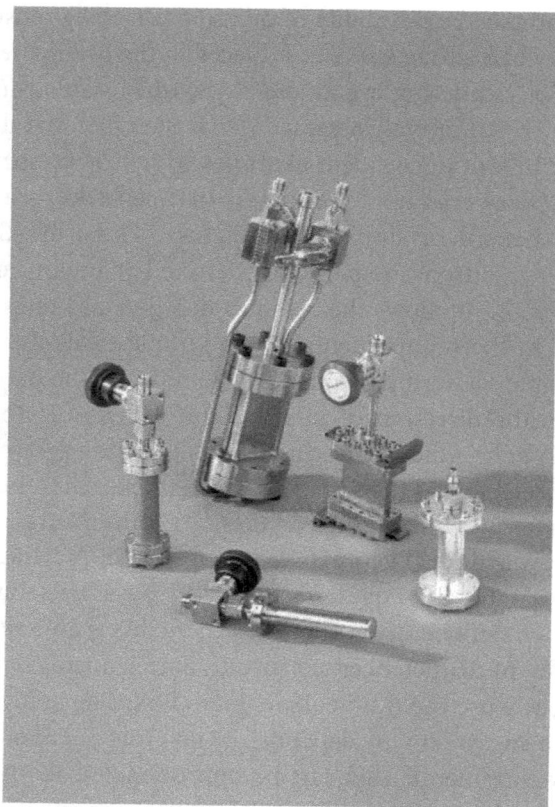

Figure 2:   Selection of gas handling cells suitable for catalysis in flat plate and circular/annular geometry. To provide a sense of scale, the gas tubing on all cells is ¼″ O.D. (Credit: STFC)

environments are available to effect measurable variation in systems under study. Control of temperature from mK to kK and pressure from mbar to kbar is routine at neutron sources. Sample cans can be manufactured from many desirable engineering materials, see Figure 2. Simple cells are generally aluminium or vanadium due to their low cross sections. Vanadium is an almost completely incoherent scatterer and as such is vastly preferred for diffraction studies. As scattering lengths can be both negative and positive, it is possible to generate "null scattering alloys" which have a mathematically zero average scattering length. The most common is a 2.08 Ti/1 Zr mix, (pronounced tizer) which is mechanically robust, although incompatible with hydrogen. Quartz and stainless steel are also commonly used, although some grades of the latter may also suffer from hydrogen embrittlement. In this case, high performance alloys such as Inconel may be an acceptable alternative.

Sealing of sample containers must be compatible with the conditions of the experiment, both in terms of pressure and temperature. Seals may be broadly classified into three groups: gasket seals, coned metal seals and unsupported area seals. Gasket seals use an easily deformed material compressed between sealing faces that fills irregularities in the mating surfaces. Indium wire, and polymer o-rings are commonly used examples. CF flanges, originally developed for UHV applications, have also been used. These feature a machined knife edge to cut into a copper or aluminium gasket, and are the seal of choice for applications involving cycling between cryogenic and elevated temperatures.

Coned metal seals compress well-machined mating faces together using threaded fittings. The compressive forces deform the two cones into compliance with each other. For extreme pressures, they are limited to relatively narrow bores, but this sealing method is readily available up to 2″ diameter for moderate pressures in the form of commercially available Swagelok®-type fittings. These seals are compatible with a broad temperature range, although may gradually become unreliable after repeated mating and uncoupling.

Unsupported area seals use a close-fitting piston inside a highly polished bore, see Figure 3. Under high pressure, this deforms to seal in the tube. This may take the form of a hard primary seal sandwiched between two softer seals that may be mechanically compressed to form an initial seal and often known as a Bridgman seal. This initial soft seal allows pressurisation to the point at which the harder primary seal deforms to match the cylinder bore and forms the sealing plug. These seals are theoretically limited only by the strength of the materials used.

Arrangements suitable for diffraction measurements on materials under static pressure are well established, and can reach extreme pressures beyond those used in industrial catalytic reactions.[2] Fluid cells have been described that are suitable for PDF measurement in supercritical CO_2,[3] and *in situ* reaction measurement of hydrothermal synthesis has been described under conditions suitable for zeolite synthesis.[4] Apparatus

Figure 3: Selection of high pressure cells with unsupported area seals. (Credit: STFC)

suitable for handling of aggressive materials has also been described.[5] Gas flow reactors are also reported in the literature, with a quartz high temperature cell reported by Haynes *et al.* that can reach 1300 K.[6] The use of quartz limits the application of pressure, but flow through cells have been reported that allow measurement at both elevated temperature and pressure. A cell described by Turner *et al.* in 1999 incorporated custom valves to withstand temperatures up to 1273 K.[7] A more recent paper by Kandemir *et al.*[8] describes methanol synthesis over $Cu/ZnO/Al_2O_3$ measured at the Echidna beamline[9] at ANSTO in Australia, with a reactor designed for *in situ* use where the valves are outside the hot zone of the apparatus.

Detailed reports of systems for use with inelastic scattering are less common. Simple flow through cells described by Nicol,[10] are commonly used and take the form of a tube with valves at either end. This design was developed to provide a flat plate geometry and described by Silverwood *et al.*[11] Successful operation of this apparatus investigating catalysts for methane reforming,[12,13] Fischer–Tropsch[14] and beyond[15,16] led to further refinements[17] and is now routinely used for catalysis experiments at ISIS in the UK.[18]

2.5 *In situ* and *operando* diffraction for structural characterisation

Single crystal neutron diffraction demands a large crystal $\sim 1\,mm^3$ to obtain reasonable data. Its application to *in situ* catalysis is necessarily limited to situations where macroscopic crystals may be obtained and would not readily be applied to many catalytic situations. Although both zeolites[19] and Metal Organic Frameworks (MOFs)[20] have been characterised with neutrons as single crystals, there does not appear to have been any attempts to study these materials *in situ*. The structure of homogeneous catalysts have also been solved through single crystal neutron diffraction *ex situ*,[21] but the greatest concentration of relevant publications is in the bio-transformations field. For enzymes that involve hydrogen transfer, neutron diffraction is the most powerful tool to locate active hydrogen atoms.[22] Drugs that affect the catalytic reactions of enzymes are well studied[23] as are more general enzyme reactions.[24] These studies investigate the nature of the binding site through analysis of co-crystals, which are grown with a substrate interacting with the protein to provide details of the fit between "lock" and "key".

Where single crystals cannot be obtained or are not relevant for the desired study, powder diffraction is preferred. A recent review by Hansen and Kohlmann[25] describes Powder Neutron Diffraction (PND) for *in situ* and *operando* methods. Although this review has broader scope than just the field of catalysis, it demonstrates the capabilities of the technique and may act as inspiration for catalytic measurements. As *in situ* PND is such a well-established method, a thorough review of the literature is beyond the scope of this article and so only a few examples will be discussed.

Adsorption of small molecules in zeolites,[26,27] zeotypes[28,29] and MOFs[30,31] have all been described and the measurement exploits the atomic contrast provided by neutrons, particularly in the case of hydrogen.[32,33] X-ray scattering would be dominated by the heavier metal atoms making structural assignment difficult. In many cases, combining crystallographic data obtained with both X-rays and neutrons is beneficial.[20,33] The greater relative strength of neutron scattering from oxygen in metal oxides makes PND a good probe of these materials compared to XRD, and may be used to gain additional inisght.[34] As such oxides are common supports in heterogeneous catalysis and are often catalytic in their own right, much work has been done. The oxygen storage properties of ceria are important in automotive three-way catalysts, and tuning oxygen storage capacity (OSC) through the use of mixed oxides is of great interest. Neutron diffraction has been used to quantify reduction of Ce^{4+} to Ce^{3+} and identify the defects in the reduced material.[35,36] Materials with mobile oxygen that may find application in solid oxide fuel cells have also been characterised.[37] The transformations involved in the formation of an active catalyst have not been overlooked. Calcination,[38] reduction[39] and even de-alloying for the formation of Raney-type catalysts[40] have been studied by neutron diffraction. For *operando* catalyst measurement with industrially relevant conditions, the Haber–Bosch[41] process and methanol synthesis[42] have both been successfully investigated with PND by Kandemir *et al.* It was suggested that effective methanol synthesis can be correlated with crystal defects caused by a strong interaction between catalyst components, and that iron nitrides are not involved in ammonia production.

Although diffraction is most commonly understood in terms of long-range crystallographic order, short-range order will also lead to a diffraction pattern and can be used to gain information from disordered systems. Where the incident radiation has a wavelength that is similar to interatomic distances, it is possible to extract information about bond lengths within a molecule, but also about how larger ensembles are arranged (i.e. interatomic distances at longer range in amorphous solids such as glasses, or molecular packing in liquids). This is often expressed as a Pair Distribution Function (PDF), which describes the distribution of distances between pairs of particles. This has been used to investigate local structure in a number of catalytically relevant materials,[43–45] which could also theoretically be carried out using X-ray PDF. Choice of the scattering probe depends on the relative scattering strengths of the atoms in the sample and source intensity, and can be refined simultaneously.[46,47] There is a report of using this technique to study acetylene bound at the active site of a zeolite catalyst by Turner *et al.*[48] and more recently by Parker *et al.* to obtain the structure of adsorbed hydrogen on Raney nickel.[49] This is an active area of research that benefits from structural models to provide simulated data for comparison. Computational models are also helpful for the use of PDF fitting in the liquid phase. Investigation of hydration shells of metal atoms[50,51] and structure of ionic liquids have been reported,[52,53] and this is an exciting area in liquid phase catalysis.

Small Angle Neutron Scattering (SANS) investigates relatively large structures (~1–100 nm) by selecting neutrons that are only scattered through a small angle. This is another technique that is strongly complementary to X-ray methods, together with Small Angle X-ray Scattering (SAXS) they are collectively known as Small Angle Scattering (SAS). This is a coherent neutron technique, and good contrast is obtainable with light elements compared to X-rays. ^1H has a coherent scattering length of -3.374 b and ^2H has a value of 6.67 b.[54] By varying the ratio of these isotopes (e.g. mixing H_2O and D_2O) and through deuteration of samples, it is possible to change the contrast between different aspects of many samples, to facilitate understanding of the relevant morphology. SANS is often used in characterisation of soft materials, and enzymatic systems are commonly investigated.[55] Micelles and colloids are other common systems to be investigated with the technique which may occasionally be used in catalysis.[56] SANS can also be used to characterise the pore structure and particle size of solid catalysts.[57–60] A paper by Diaz *et al.* using the ingress of a contrast matched solvent into the pore network of coked cracking catalysts is an elegant example of what can be achieved with this technique.[61] Particle growth is also of interest in catalysis, and this may be followed with SANS. The formation of metal oxides,[62] the catalysed sol–gel transition in silica gels,[63] and growth of polymers via living radical polymerisation[64,65] have all been studied.

The distinction between different elastic scattering methods is arbitrary, as the techniques differ only in selecting which neutrons are collected through instrument selection, and the data treatment used. It is (theoretically at least) possible to collect all of the scattered neutrons and obtain all of the information at once. This concept is known as total scattering and allows collection of data at all length scales simultaneously. Whilst this is a powerful tool, there is a great increase in difficulty in data processing and effective interpretation. A powerful example of what can be

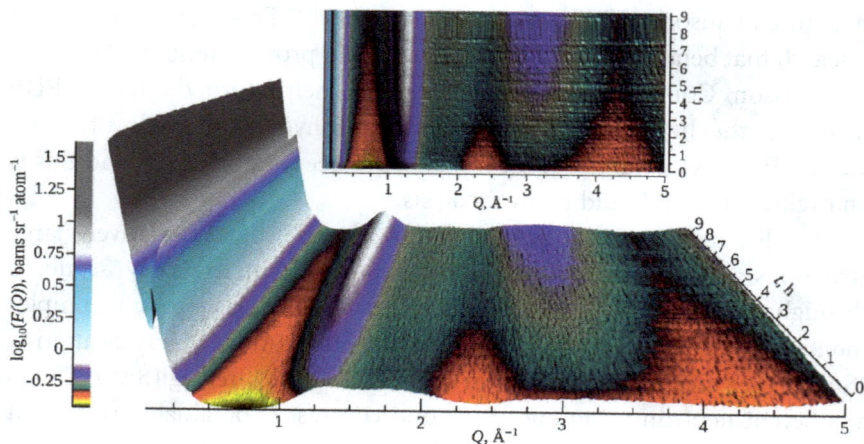

Figure 4: Operando total scattering measurement of benzene hydrogenation over Pt/SiO$_2$ (reproduced from Ref. [66] under CC BY 3.0 [http://creativecommons.org/licenses/by/3.0/]).

achieved is reported by Youngs *et al.* and describes an *operando* total scattering measurement of benzene hydrogenation with Pt/SiO$_2$ see Figure 4.[66] The observation of both pore size effects and interatomic distances were recorded simultaneously and demonstrate the influence of diffusion on the kinetics of reaction.

2.6 Quasielastic neutron scattering for motion and diffusion

Of all the motions that can be measured with QENS, arguably the most important from a catalytic point of view is diffusion within porous materials.[67,68] For these measurements, the motion is termed "self diffusion" as it corresponds to materials at dynamic equilibrium and is carried out without a chemical potential gradient. A review by Jobic and Theodorou illustrates the complementarity between QENS and molecular dynamics,[69] and increased computational power is constantly allowing more accurate models to be used.[70] As the observable data generated by the QENS measurement is limited, comparison with theory may be key to obtaining the optimal result. Nevertheless, it is often possible to obtain qualitative data about motions rapidly, and quantitative data with the expenditure of moderate effort. It is relatively easy for example, to discern jump diffusion, where a sorbate periodically jumps between cages where it mostly resides, and free diffusion, where the barrier for moving between cages is lower.

Hydrogen motion is also important in fuel cells, and has been widely studied using QENS.[71,72] There has also been recent work by Larese to investigate hydrogen mobility in a spillover context using QENS.[73–75] Finally, the motions within proteins can be investigated with this technique, and can provide mechanistic information for biocatalysis.[76]

2.7 Vibrational analysis through INS

INS is a very well-established tool in the characterisation of catalysts and the field benefits from having an excellent reference available in the book by Mitchell *et al.*[77] Although the technique is not surface sensitive, in a system with a hydrogen-containing species adsorbed on a metal, the signal will almost exclusively be from the adsorbate due to the strength of scattering from hydrogen. This makes spectroscopy of chemisorbed organic species feasible, although in a physisorbed system care is needed to ensure that desired sorbate is present and interacting with the catalyst. Dosing too heavily will lead to a spectrum dominated by the bulk organic, and too lightly will lead to a weak signal. Even with these restrictions, it is possible to avoid the pressure and materials gaps that may be present in other techniques. With the abundance of information available from vibrational spectroscopy, the breadth of research is unsurprising. Hydrogen is the simplest adsorbate for neutron studies,[78] but model catalytic intermediates formed from molecular adsorption are common.[79,11] With solid acid catalysts, the acid sites can be investigated beyond the range possible by IR and Raman,[80] both directly and through adsorption of probe molecules.[81] The major drawback of INS is that it is mostly carried out at cryogenic temperatures in order to minimise the Debye–Waller factor. This does not affect the vibrational transition energies that are involved, but limits the possibilities for *operando* measurement. However, Parker recently demonstrated the ability of a sensitive direct geometry instrument to measure a reaction with INS in real time.[82] For this measurement to be successful, a number of fortuitous coincidences were required. However, it must be stressed that this measurement was not possible 10 years earlier and in the future, it is likely that more reactions will become accessible to this technique as the instrumentation improves.

Carbon and carbides are another fertile area of study for neutrons, as their high absorbance makes optical spectroscopy difficult. In a catalytic context, these often take the form of active materials in their own right,[83] catalyst supports,[84–86] and also coke.[87] Workers in the Lennon group have investigated coking during methane reforming,[12,13,88,89] using INS to quantify and speciate the hydrogen present.[90] The same approach has been applied very successfully to Fischer–Tropsch catalysts in an attempt to clarify the role of the carbide phases that form on the active materials.[14,91]

There is relatively little work on INS for liquid phase catalysis[92] and bio-transformations.[93] Infrared and Raman characterisation is relatively easy for these systems, which makes the experimental challenges of neutron scattering less attractive. However, as the technique becomes more routine, and where theoretical modelling is important, increased activity may be expected.

2.8 Other techniques

Attenuation of the neutron beam can be used as an imaging technique, in a similar manner to medical X-rays, and is referred to as neutron radiography or neutron imaging. It has been used to visualise liquid film thickness in annular flow systems,[94] and many fuel cell assemblies.[95–97] This may find application in visualising flows in mixed phase reactors under extreme conditions due to the high contrast between hydrogen and other elements. This method has also been used to quantify the degree of coking within reactors in a non-invasive manner.[98]

Neutron reflectometry is a technique that can investigate the nature of thin films, and has been used in analysis of hydrogen adsorption for storage applications.[99,100] The strength of the neutron interaction with light elements and its short wavelength allows absorption (not adsorption) measurements to be made on these model catalyst systems. Although unlikely to become a regular tool for catalyst characterisation, it has the opportunity to find a niche in analysis of interfacial systems.

3 X-rays

3.1 Introduction

Techniques made available by using synchrotron radiation, such as X-ray diffraction (XRD), X-ray Absorption Fine Structure (XAFS) and even infrared spectroscopy have long been used to investigate catalysts or model systems. A full review of this large body of work is beyond the scope of this chapter, instead the focus here will be on *in situ* and *operando* measurements of catalytic systems performed using synchrotron radiation and in some cases combined with other techniques.

In the case of XAFS, synchrotron radiation provides a source of high intensity, high energy X-rays which can be tuned to the energy required to excite a core electron from the element of interest. As one example, the XAFS beamlines at the Diamond Light Source can provide X-rays in the energy range of 2–35 keV, which for catalysis is ideal, allowing measurement of all of the first-row transition metals in addition to others such as Ag, Au and Pd, and also lower atomic number elements such as S. The increased flux of photons provided by synchrotron sources, (especially third and fourth generation sources), the improved focus, collimation and the use of high performance detectors, allow XRD measurements to be performed not just on powders but on real catalyst pellets, in 3D with μm resolution.[101] Examples of these experiments, named μ-XRD-CT, under *operando* conditions will be discussed in the increasing sections. For infrared measurements, synchrotron sources are 100–1000 times brighter than a conventional infrared glow-bar source providing much improved spatial resolution.[102]

The following sections will explore current *operando* style measurements available using XAFS, XRD and IR available at synchrotrons, discuss the different reactor capabilities and their limitations and aims to outline future possibilities.

3.2 *In situ* liquid phase catalysis followed by XAFS

XAFS comprises X-ray Absorption Near Edge Spectroscopy (XANES) and Extended X-ray Absorption Fine Structure (EXAFS) and provides information on the oxidation state and local structure of the element studied.

The X-rays are tuned to scan across a range of energies near the absorption edge (which is element specific) of the element of interest, in this way the sample will absorb photons of sufficient energy to excite an electron from the core to an empty state or to the continuum. By measuring the intensity of the photons before and after the sample (in a transmission measurement), the change in absorption is observed, as in Eq. (2.1)

$$I = I_0 e^{-\mu t}. \tag{2.1}$$

Equation (2.1) shows Beers' Law, where I is the intensity of X-rays transmitted through the sample, I_0 the intensity of the incident X-rays, μ the absorption coefficient and t the sample thickness.[103]

XANES can be defined as the part of the spectrum from the edge to roughly 50 eV after the edge and gives information on the electronic properties of the samples, the oxidation state and coordination geometry of the element. One interpretation of the XANES region is to use it as a fingerprint against reference compounds to determine the valence state and coordination of the element. The EXAFS part of the spectrum is due to scattering of the outgoing electron from the electron clouds of nearby atoms. The outgoing electron can be imagined as a wave which scatters off neighbouring atoms producing a backscattered wave.[104] These two waves interfere generating an interference pattern that contains information about the type and number of neighbouring atoms and the distance to them. For a full explanation of EXAFS, please refer to Refs. [103] and [104] and the tutorials found in Ref. [103]. Since EXAFS does not depend on long-range order in the sample, amorphous materials, liquids and small metal nanoparticles can be studied.[105] One drawback from the point of view of catalysis, is that XAFS is an averaging technique, looking at all the atoms of the element of interest in the sample, so that the surface is not studied in isolation. However, this can in some instances be surface sensitive when the element of interest is only in the surface layer, as will be described in one of the following examples.[106]

When studying liquid-phase catalytic systems, one problem encountered is the low concentration of the investigated metal under process relevant conditions. This can mean that it would take hours to measure the sample, therefore only the end state of the reaction could be probed. However, one approach to measuring liquid-phase catalytic systems is to use a combination of stopped flow and freeze quench systems. This has been used to investigate the mechanism of a Mo olefin oligomerisation catalyst.[107] The stopped-flow part of the set-up uses four syringes controlled *via* a computer so that precise volumes and injection rates can be used. This system was developed for the combined measurement of UV–Vis and XANES using an Energy Dispersive EXAFS (EDE) beamline for fast data acquisition.[108] An EDE beamline focuses a polychromatic beam on the sample and the resulting divergent outgoing beam is measured using a spatially sensitive detector. As the measurement does not require the stepwise movement of the monochromator through a range of energies, EDE could potentially be performed with sub-millisecond time resolution.[109] The stopped-flow system is attached to a Schlenk line so that the reaction can be performed under an inert atmosphere. The freeze-quench adaptation, Figure 5, consists of a Kapton® capillary mounted onto a specially designed holder allowing injection into the capillary under an argon atmosphere, the capillary is then sealed and immersed in liquid N_2, then measured whilst under a cryostream. Using the stopped flow different times during the reaction can be measured, with the freeze-quench attachment EXAFS measurements can be performed on the solution 5 s after mixing.[107] The system is versatile and can be used either as a stopped-flow device, when the element of interest is in high enough concentration for Quick-XAS or EDE measurements, or combined with the freeze-quench accessory for measurements requiring long acquisition times or fluorescence measurements. Fluorescence measurements are necessary when the element of interest is in low concentration or in a solution containing

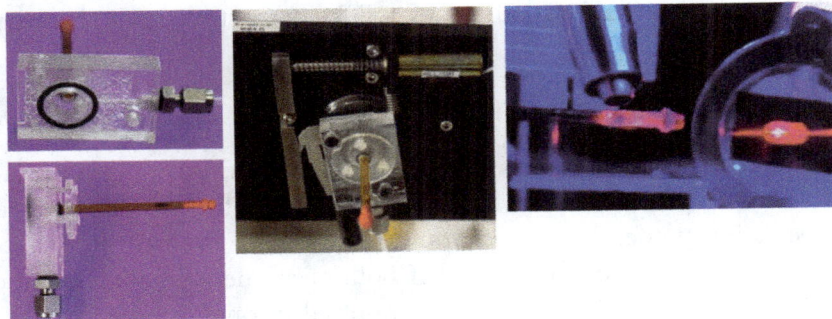

Figure 5:   Stopped flow freeze quench set up, Ref. [107].

highly X-ray absorbing elements or for elements with low X-ray absorption edge energies. An example of the latter is Cr, solutions of which have also been measured using the freeze-quench set-up described here.[110]

Another example of liquid-phase catalysis studied *in situ* by XAFS is the formation of $Cu_2(OH)_3Cl$ particles in the presence and absence of polyvinylpyrrolidone (PVP).[111] This experiment was performed in a dedicated reactor which has an adaptable cell thickness, enabling optimisation of the edge intensity depending on sample concentration.

To investigate the solid–liquid interface of liquid phase heterogeneous catalyst systems, Attenuated Total Reflection Infrared (ATR-IR) has proven to be highly informative. When combined with XAFS, information on the oxidation and coordination environment of the active metal and the molecular arrangement of the adsorbate layer can be simultaneously gathered. A combined ATR-IR/EXAFS cell with a vertical internal reflection element (IRE) made of ZnSe, has been designed by Ferri *et al.*[112] In this system, the catalyst which is deposited on the IRE is probed by X-rays from the side of the flowing solution, measuring XAFS spectra in fluorescence mode, whilst the IR light probes the catalyst from the side of the IRE.

Future improvements in cell design for liquid-phase catalytic reactions can be envisaged, such as the possibility to work under pressure and in stirred conditions similar to an autoclave. If this type of system were to be developed, three-phase catalytic reactions could be measured *in situ* without compromise to the operating conditions. However, data acquisition would have to be faster than the rate-limiting step of the reaction. With fast reaction rates, freeze-quench reactor systems could be used. The Catalysis Hub is currently working on a design for an autoclave type reactor containing a recirculating loop that could be optimised for different spectroscopic techniques.

3.3 *In situ* electrocatalysis followed by XAFS

XAFS can also be used for the *in situ* study of fuel cells, providing information on the electrode during reaction. To perform the measurement in transmission mode, the X-rays would have to pass through both electrodes, this would pose a problem if both electrodes contain the same element to be studied. This has been resolved in one instance, by Roth *et al.*, by removing a small part of the cathode which was in the beam.[113]

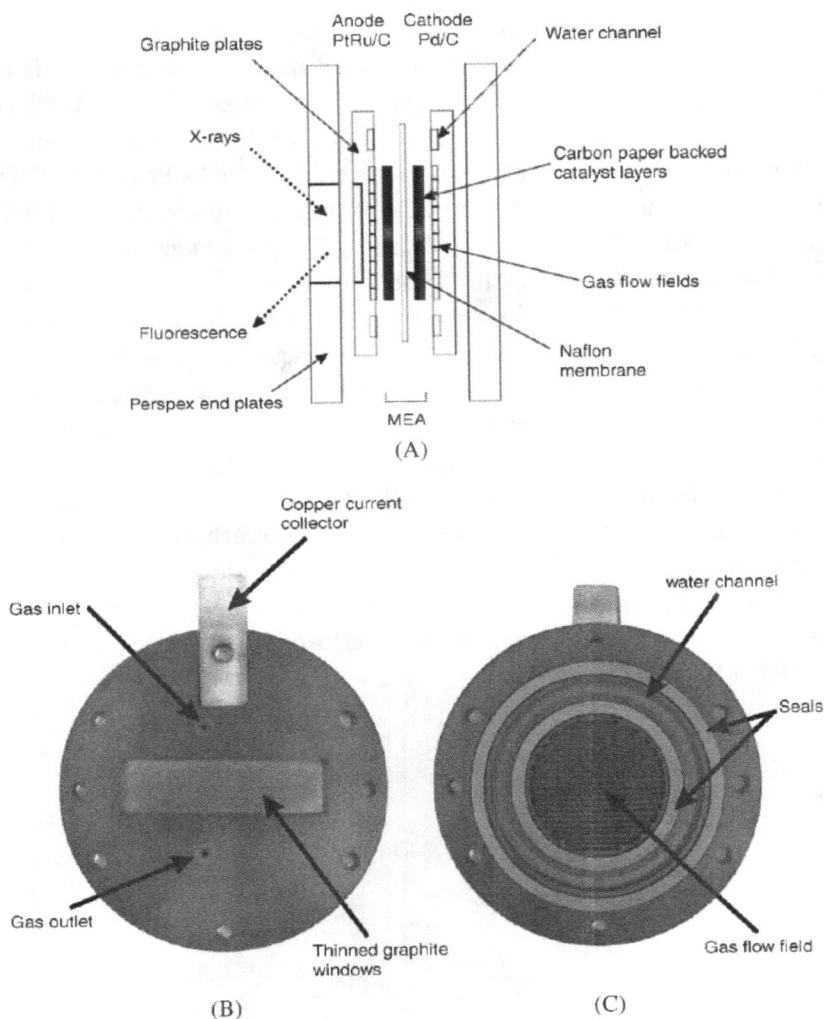

Figure 6: Schematic and cross section of the PEM fuel cell described in Ref. [114].

Another consideration for transmission mode experiments is the increase in metal loading required to achieve a good absorbance edge, increasing the loading well beyond that typically used in fuel cells.[114] One cell design which overcomes these problems measures the anode in fluorescence mode, Figure 6, was used to study a PtRu/C anode catalyst during operation.[114]

The formation of electrocatalysts can also be followed *in situ* using XAFS. One method of producing core shell nanoparticle catalysts is the galvanic displacement of an underpotentially deposited (upd) Cu layer. This first stage of the process, the formation of the Cu upd layer on Au nanoparticles supported on carbon, has been reported by Price *et al.*[115] The *in situ* EXAFS measurements were able to provide the structure of the Cu upd layer at each potential applied during the process and found that a complete Cu monolayer, required for the subsequent galvanic displacement, was not obtained. In a second study, the subsequent displacement of the Cu upd layer by Pd was investigated.[116] The Cu monolayer was achieved by holding the potential at −0.455 V for 30 min. Following the deposition of Cu, the Pd displacement was performed at the same potential after flushing the cell first with H_2SO_4 and then with the solution K_2PdCl_4 in H_2SO_4. The *in situ* electrochemical cell, Figure 7, consisted of a working electrode, containing the Au/C NP's painted onto carbon paper, a Pt wire counter electrode and a $Hg/HgSO_4$ reference electrode. A peristaltic pump was used to pump the electrolyte through the cell. The XAFS measurements on the working electrode were performed in fluorescence mode through a Kapton® window.

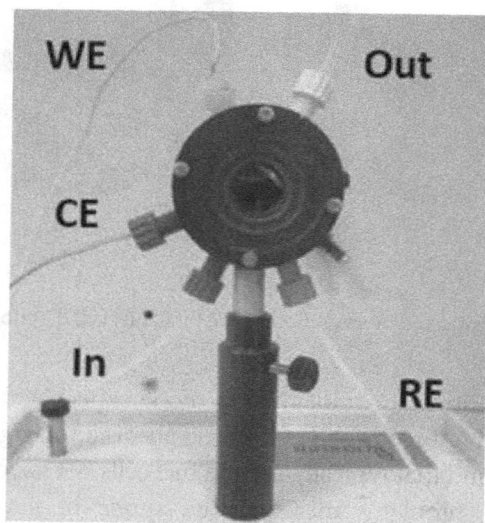

Figure 7:   Photograph of the *in situ* electrochemical cell used in Refs. [115] and [116], consisted of a working electrode (WE), counter electrode (CE) a reference electrode (RE).

3.4 FTIR using synchrotron radiation

Conventional infrared microscopy is a powerful tool for the characterisation of materials, but can offer a limited spatial resolution of ~20 μm. To increase spatial resolution, a smaller aperture is required, which decreases signal strength. The lower signal to noise ratio with smaller apertures therefore eventually results in an unacceptable spectrum.[117] Where a traditional Globar source uses a heated area to generate infrared light, the light generated at a synchrotron is effectively a point source, which allows tightly focused beams to be obtained. Synchrotron radiation infrared (SR-IR) sources are 100–1000 times brighter than a traditional infrared Globar and can provide spectra with a good signal to noise ratio with a resolution of ~5 μm,[102] where spatial resolution is not important as the greater flux provided by a Globar is preferred.[118]

SR-IR has been used to study reactions on large coffin-shaped H-ZSM-5 zeolite crystals ($100 \times 20 \times 20$ μm). The reaction was followed by measuring an area of 5×5 μm in the centre of the crystal during styrene oligomerisation.[119] The sample was subsequently mapped after reaction to determine the distribution of the product species. The *in situ* reaction was carried out in a commercial FTIR 600, Linkam Scientific Instruments cell. By connecting the outlet to a GC or mass spectrometer, this technique could be used under *operando* conditions.

Using SR-IR to map flow reactors can yield detailed kinetic information on the reaction performed. This has been achieved by Gross *et al.* using a flow microreactor made of two sandwiched CaF_2 windows with a spacer, allowing a 3-mm deep channel for the reactant and catalyst, Figure 8.[120] The reaction followed was the cascade dihydropropan formation over an Au/SiO_2 catalyst. Spectra were recorded with a resolution of 15 mm along the length of the reactor bed during steady state conditions. The reaction was reproduced in a stainless steel reactor and also using a quartz tube cell for XAFS with similar internal dimensions to ensure that the residence time of the reactants in each reactor was the same. In this way, the three results could be correlated, intermediate species in the reaction were observed from the IR measurements and the regions of high catalytic activity were found to match with areas of Au(III) as observed from the XAFS measurements.

Homogeneous catalytic flow reactions have also been monitored using micro-flow reactors and SR-IR microspectroscopy. Silicon microreactors

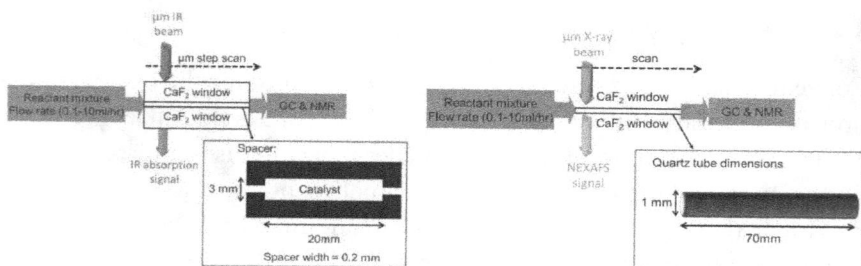

Figure 8: Scheme of the two reactors used in Ref. [120], (left), the CaF$_2$ flow reactor used for the micro-IR mapping experiments and (right) the micro-NEXAFS flow cell. Reprinted with permission from Ref. [120]. Copyright (2015) American Chemical Society.

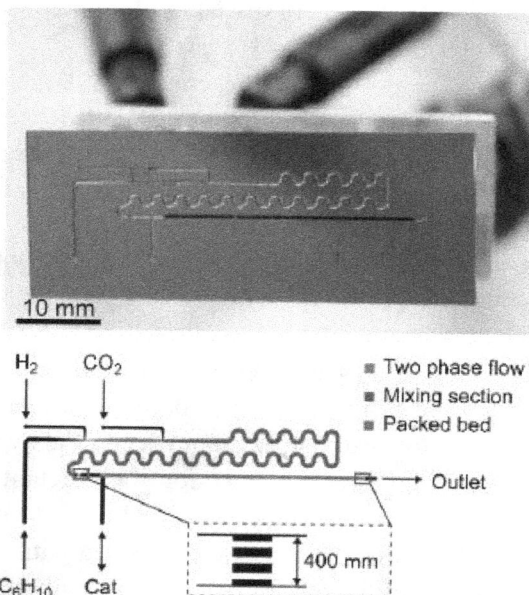

Figure 9: Photograph and schematic of a silicon/glass microreactor, Ref. [122].

can withstand high temperatures and pressures and can be etched with virtually any reactor design, they allow for good heat transfer and the smaller volumes used lower the risk factor of the experiments.[121] An example of a glass-silicon microreactor is shown in Figure 9.[122] The silicon is etched with the reactor design and then bonded to glass, the glass side can also be sputtered with Au of the same layout as the reactor channels, when the silicon and glass are sandwiched together the gold should match up with the channels, providing a reflective surface for infrared microscopy. An additional advantage is that microreactors can be used for both infrared and X-ray measurements, where the IR is performed in reflection mode and the XAFS measurements in fluorescence mode.

3.5 Combined XAFS/DRIFTS

XAFS is an averaging technique and so for, most catalyst systems is not surface sensitive. By combining with FTIR, the oxidation state and local coordination of the metal can be investigated whilst also measuring the surface adsorbed species. This combination of XAFS and DRIFTS first developed by Evans and Newton[123] has, for example, been used to study CO and NO cycling over Rh and Pd automative emission catalysts and methane over Pt/Al_2O_3.[124]

Adaptations to the original system have been developed to provide ease of alignment of the DRIFTS system on the X-ray beamline, namely a DaVinci arm allowing the IR beam to pass outside the spectrometer where a DRIFTS cell can be attached in line with the X-ray beam. The arm is also adjustable in x and y-axes allowing easy alignment of the X-ray beam with the 3 mm hole in the DRIFTS cell which allows transmission of X-rays through the sample.[125] The UK Catalysis Hub has purchased a similar system, supplied by Harrick, Figure 10. The sample cup is 3.17 mm in diameter and the holes allowing the transmission of X-rays through the sample are positioned 1.04 mm below the top surface of the sample. The distance from the top of the hole for the X-rays to the top surface of the sample is only 0.37 mm. As the infrared can penetrate up to 1–2 mm in non-absorbing materials such as Al_2O_3,[126] it is assumed that the IR is sampling the same volume of the catalyst as the X-rays. IR windows are ZnSe being compatible with most gas environments, the X-ray windows are made of glassy carbon.

The UK Catalysis Hub system has successfully studied the restructuring of AuPd nanoparticles during CO oxidation using the set-up described earlier.[127]

One disadvantage of the Harrick cell when used in transmission mode is the fixed X-ray path length, however, this can be easily fixed by inserting a

Figure 10: UK Catalysis Hub XAFS/DRIFTS set-up (A) Agilent 680 FTIR spectrometer with DaVinci arm and cell attached, (B) close-up of the Harrick XAFS/DRIFTS cell with ZnSe IR windows and glassy carbon X-ray transmission windows and (C) the sample cup inside the cell with holes for the X-rays visible.

smaller removable cup providing a smaller X-ray path length.[125] Another problem with the cell was the dependence on performing transmission XAFS measurements, limiting the choice of samples which can be measured to catalysts with supports containing only low Z-number elements such as Al_2O_3; CeO_2, for example, would be difficult. However, a recent adaptation of the dome to have a larger X-ray window has recently been used for the investigation of CO oxidation over CuO/CeO_2 using combined DRIFTS and fluorescence mode XAFS.[128] This new dome also permits XRD measurements and has also been used for combined DRIFTS and PDF measurements.[129]

A more difficult problem to address is the large dead volume of the cell of 14 mL. This has always been a problem with DRIFTS cells, and one that has previously been addressed by McDougall using a flat window design and filling some of the dead volume of the cell with a Macor gas guide.[130] Another common problem of DRIFTS cells is bypass of gas around the catalyst bed as the bed "shrinks" on heating or moves under gas flow, or in some DRIFTS cells where the sample holder meets the base of the cell. This latter problem has been successfully resolved by Meunier.[131] An adaptation of the original XAFS/DRIFTS cell[123] has been implemented along the lines of both Meunier and McDougall, limiting the dead volume to 2–2.5 cm³ whilst retaining a plug flow configuration.[132]

A recent design by D. Ferri *et al.* addresses the problem of dead volume in a different manner.[133] In this design, the catalyst is placed between two windows, window 2 (Figure 11(B)) is graphite allowing transmission of

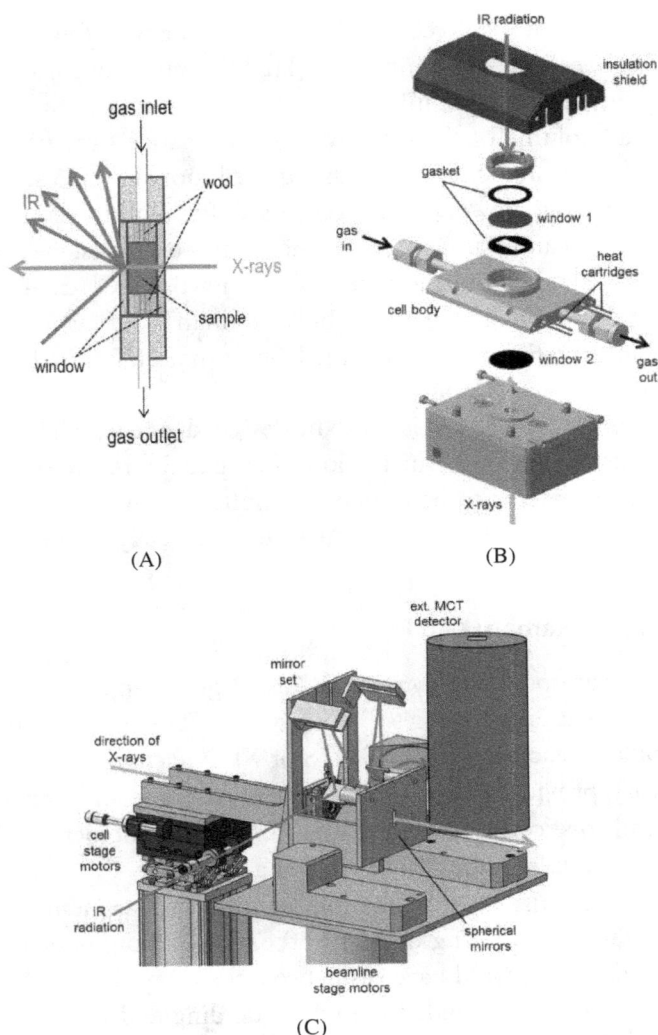

(A)

(B)

(C)

Figure 11: Cell design for XAS-DRIFTS measurements, from Ref. [133]. (A) cell geometry, (B) exploded view of cell and windows and (C) cell arrangement in the DRIFTS mirrors.

the X-rays, whilst the other has to transmit both X-rays and IR. A CaF_2 window was used for window 1 with a hole (0.5 mm dia.) drilled in the middle and filled with a high temperature carbon based glue. This hole allows for transmission of X-rays which would otherwise be absorbed and diffracted by the CaF_2. The cell is a plug flow reactor, with the catalyst reactor placed between the two windows and tightly kept in position by quartz wool plugs at both the gas inlet and outlet. The authors found that the reduced dead volume of the cell gave a gas exchange time of 5 s, much faster than the commercial cell which required more than 33 s.

A similar plug flow DRIFTS cell, used for simultaneous spatially resolved DRIFTS and Raman measurements, has been designed and used to study NO_x storage and reduction over a Pt–Ba/CeO$_2$ catalyst.[134] The advantage of this cell is that both bulk (Raman) and surface sensitive (DRIFTS) information can be obtained along the catalyst bed down to a resolution of less than 1 mm.

The optimal DRIFTS cell is yet to be designed, one which both resembles a tubular reactor and which allows for spatially resolved XAFS and DRIFTS measurements to be obtained simultaneously. One such idea is currently in the initial stages of design at the UK Catalysis Hub.

3.6 Combined Raman/XANES

Coupling Raman spectroscopy with XAFS can give complementary information on the catalyst species.[132] Combining Raman with traditional quartz capillary reactors used for XAFS or XRD *operando* experiments is relatively easy, providing no damage to the objective from sample heating occurs, usually overcome by using a long working distance lens.[132] The first Raman/XAFS coupled experiment studied a Mo/Al$_2$O$_3$–SiO$_2$ catalyst during room temperature post calcination hydration experiments.[135] Due to the strong Raman scattering of MoO_3, the small amounts of this oxide were detected which would have easily been missed using solely XAFS. In addition, the Raman data aided the understanding and fit of the EXAFS data, which otherwise would have missed the presence of the Mo–Al scattering path.

Recently, MoO_x shell–Fe$_2$O$_3$ core catalysts have been investigated using a combination of XAFS and Raman during calcination and reduction using the microreactor shown in Figure 12.[106] This set-up consists of a 6-mm OD quartz capillary with a wall thickness of 250 μm heated by a hot air blower and connected to a gas delivery system with the

Figure 12: *In situ* quartz capillary microreactor for XAFS studies, A — hot air blower, B — I_0, C — I_t and D — the fluorescence detector. This figure is from Ref. [106], published by The Royal Society of Chemistry.

outlet gas monitored by a mass spectrometer. In order to apply both techniques simultaneously, the capillary was placed at a 45° angle to the incoming X-rays and the XAFS measurements were performed in fluorescence mode. This limited the time resolution of the measurements to 1 min per scan. The Raman spectra were collected using 20 × 1 s scans using a Renishaw spectrometer with a 785 nm laser. The set-up can also be used for XAFS in transmission mode allowing the collection of spectra with higher time resolution, however, without the coupling with Raman spectroscopy.

This is by no means the only example of combined Raman and XAFS, for example, the SAMBA beamline at the SOLEIL synchrotron, Paris, has a cell which can be used in transmission or fluorescence modes and has been used to study the calcination of a hydrodesulphurisation Mo Ni bimetallic catalyst.[136] In this system, the catalyst is held between a graphite foil on one side and a graphite foil with a hole on the other side, the hole is then sealed with a mica window as shown in Figure 13. This cell geometry has the added advantage that transmission XAFS measurements can be performed simultaneously with Raman spectroscopy.

Further insight into catalyst mechanism can be obtained by combining a third spectroscopic tool, such as UV–Vis as has been employed by Beale

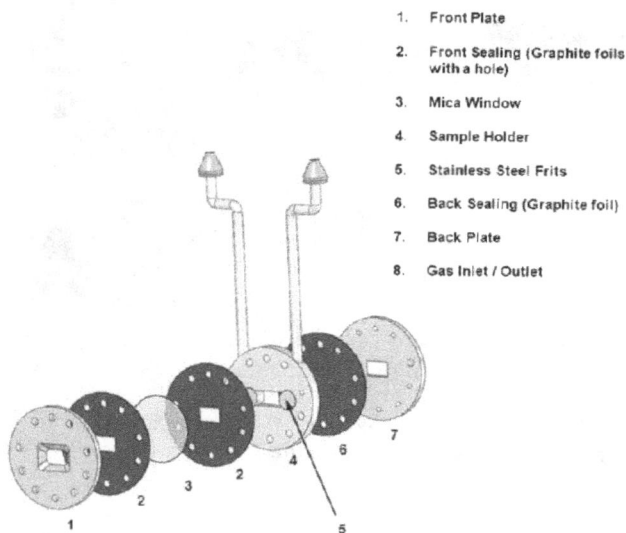

1. Front Plate

2. Front Sealing (Graphite foils with a hole)

3. Mica Window

4. Sample Holder

5. Stainless Steel Frits

6. Back Sealing (Graphite foil)

7. Back Plate

8. Gas Inlet / Outlet

Figure 13:   Exploded view of the sample holder assembly part of the transmission and fluorescence combined Raman cell at the SAMBA beamline of the SOLEIL synchrotron, Ref. [136].

et al. for the investigation of the deactivation mechanisms during propane dehydrogenation over molybdenum oxide catalysts.[137] With the combination of ED–XANES, Raman and UV–Vis the authors were able to propose that Mo^{4+} species are required for activity and that two deactivation processes occur. The first deactivation is due to coke formation, discerned from the Raman and UV–Vis results, and the second was due to an increased size of the MoO_3 species which were less easily reduced than the initial well dispersed Mo species. This latter, permanent deactivation, was evidenced from the ED–XANES and Raman data. The reactor used for this study was a modified 6 mm OD, 1 mm ID quartz tube. The quartz had been ground down on three sides to provide one 200 μm and two 100 μm windows for UV–Vis and X-ray measurements, respectively.

Many industrial reactions occur at high pressures and so reactors capable of withstanding these conditions are necessary to work under true *operando* conditions. One such cell has been recently published in which combined XAFS and Raman measurements can be performed up to

200 bar. To demonstrate the capabilities of the system, CO_2 hydrogenation to methanol over a $Cu/ZnO/Al_2O_3$ catalyst was performed.[138] This system, is also based on a capillary microreactor using a polyimide coated fused silica capillary (662 μm OD, 247 μm ID) attached to the gas lines using VICI bulkhead compression fittings with polyimide ferrules and heated using a hot air blower. An alternative heating system was attempted by placing the capillary inside a quartz tube with a heating wire coiled around it, providing uniform heating along the length of the catalyst bed. However, this system resulted in "unusual noise in the XAFS measurements".[138] This noise was found to be due to bending of the capillary when heated and so the capillary oscillated in and out of the X-ray beam. For Raman measurements, the polyimide coating on the capillary was removed. This resulted in a less flexible capillary but the authors state this did not limit the mechanical or thermal properties of the reactor, and was tested to 200 bar.

The simple design of the cell and the ability to easily adapt the length of the catalyst bed could allow this system to give spatially resolved spectroscopic information, if uniform heating along the bed can be resolved; some suggestions are given in Ref. [138].

3.7 XRD, Absorption-CT and XRD-CT and XAFS-CT?

There are numerous examples of powder XRD studies under *operando* conditions and also when combined with XAFS or XES measurements giving valuable information on many catalytic systems. However, as is also the case for the XAFS measurements (combined with FTIR or Raman) mentioned before, the catalyst studied is generally in powder form, and so not exactly as it would be in the industrial reactor. When used in industry, heterogeneous catalysts are usually in the form of mm to cm sized catalyst pellets, mainly to ensure limited pressure drop along the length of the reactor. When the catalyst is in the form of an extrudate or pellet, the active phase or active species may not be distributed uniformly through or around the surface of the catalyst body, it could, for example, be more concentrated at the surface, at the interior or as is referred to as an eggwhite distribution (Figure 14). Differences in species/phase distribution may occur during activation, reaction or during deactivation. To study

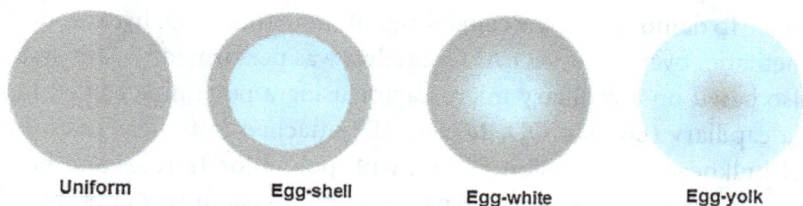

| Uniform | Egg-shell | Egg-white | Egg-yolk |

Figure 14: Illustration of the description of active species distribution in a catalyst pellet, shown when bisected. The grey represents the active species and the light blue the support material.

catalytic processes under true *operando* conditions not only the reaction conditions, temperature, pressure and gas composition, but also the catalyst form should be replicated as close to real industrial conditions as possible.

One technique that can study catalyst bodies *in situ* is XRD-Computed Tomography (XRD-CT). The highly collimated and high flux of X-rays available at third generation synchrotrons together with the recent advances in detectors has allowed 3D XRD-CT measurements on mm sized catalyst extrudates to be performed on minute timescales. These measurements usually use a 2–100 μm^2 sized pencil beam and record the diffraction pattern on a flat panel detector, whilst the sample is translated perpendicular to the beam.[101] The sample is then rotated and another translational map recorded, this is repeated until the sample has been rotated by a total of at least 180°. The powder ring diffraction data is then radially integrated for each measurement to give a "squeezed" pattern, and for each 2θ a sonogram is produced and back-projected to a square pixel image. Each pixel, resolution ≤ 100 μm, contains a full diffraction image, with the pixel representing a point inside the catalyst pellet. Individual peaks in the diffraction patterns can be selected and integrated over a slice of the pellet and their distribution imaged. This process is shown in Figure 15.

This technique has been used to study the calcination of a Ni supported γ-Al_2O_3 3 mm × 3 mm pellet which had been impregnated with $[Ni(en)(H_2O)_4]Cl_2$, where en is ethylenediamine. This study found that two crystalline phases form during the early stages of calcination, $Ni(en)_xCO_3$ and $Ni(en)(CO_3)_xCl_{2(1-x)}.yH_2O$. The former species showed an eggshell distribution and the latter an egg-white distribution. These two phases decompose at different temperatures, with $Ni(en)_xCO_3$ producing

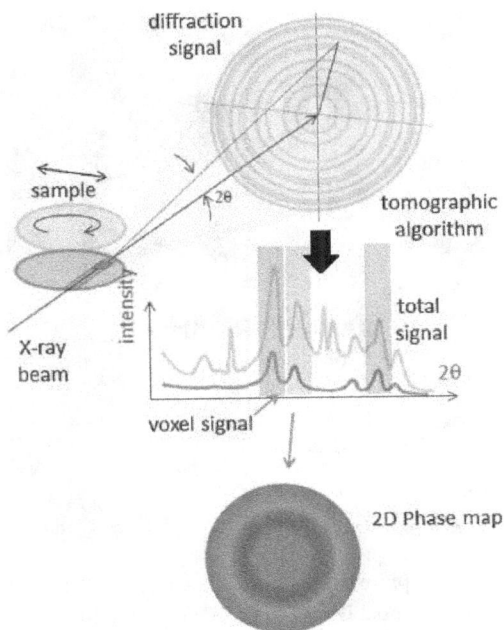

Figure 15: Illustration of the sampling methodology used to obtain 2D cross-sectional XRD-CT data. The process involves a data acquisition step (a few min per 2D slice), radial integration to give a 1D pattern (black arrow), feature selection/fitting (grey overlay rectangles are used to define different diagnostic features) and finally a sonogram construction and filtered back projection of these various features to yield their 2D distribution.[101]

fcc Ni and $Ni(en)(CO_3)_xCl_{2(1-x)} \cdot yH_2O$ initially decomposing to form hcp Ni which then forms fcc Ni above 400°C.[139] Although the technique is highly complex the reactor system was fairly simple, with the catalyst pellet placed on top of an alumina rod in a quartz reactor tube. The reactor was attached to a gas delivery stub which sat on a goniometer which itself was attached to a rotation stage. The whole system, including two heat blowers, was mounted on a translational stage, so that the heat guns would move with the sample. A similar tubular reactor type cell is described in detail in Ref. [140]. This cell, made of sapphire, can withstand up to 200 bar and can in theory be used to perform XRD-CT on catalyst pellets under reaction conditions. The cell is hung from a goniometer attached to a rotation and translation stage, inside a furnace, which can heat the sample to 1000°C, as shown in Figure 16.

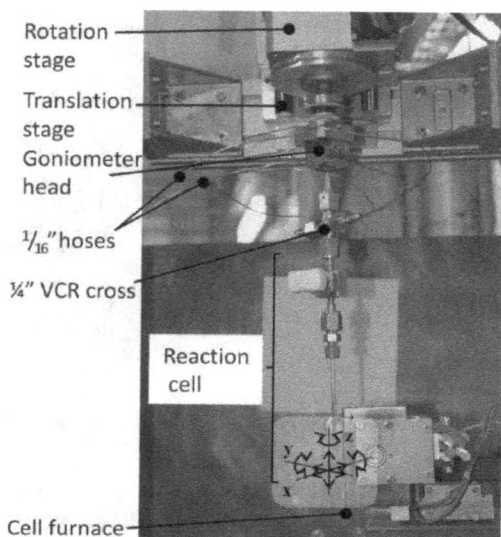

Figure 16:   Image of the sapphire cell alignment described in Ref. [140]. Reproduced with permission from the International Union of Crystallography, http://journals.iucr.org/.

This technique is not just limited to looking at catalyst pellets, but has recently been used to study a working catalytic membrane reactor (CMR) during the oxidative coupling of methane (OCM) to produce ethylene.[141] The CMR consisted of a Mn-Na-W/SiO$_2$ catalyst surrounded by an oxygen transport membrane (BaCo$_x$Fe$_y$Zr$_z$O$_{3-\delta}$), as shown on the top left of Figure 17(A). Figure 17(B) shows the evolution of five different crystalline phases identified in the catalyst during OCM. BaWO$_4$ was found to form in concentrated spots mostly at the catalyst/membrane interface, this phase is thought to form a layer at the membrane wall, limiting O$_2$ transfer to the catalyst.

XRD can be incredibly informative for catalysis, however it relies on the species of interest being crystalline with long-range order on the scale of >2–3 nm, which is not always the case. When the species of interest is on the shorter length scale, XAFS can be useful as it probes short-range order on the scale of ≤5 nm.[132] The combination of XRD and XAFS was first employed by Couves *et al.* for *in situ* measurements, and more recently for the first time under *operando* conditions by Grunwaldt *et al.*[142,143] However, no publications to date have appeared where the combination is used in a tomographic manner to provide 3D information of a

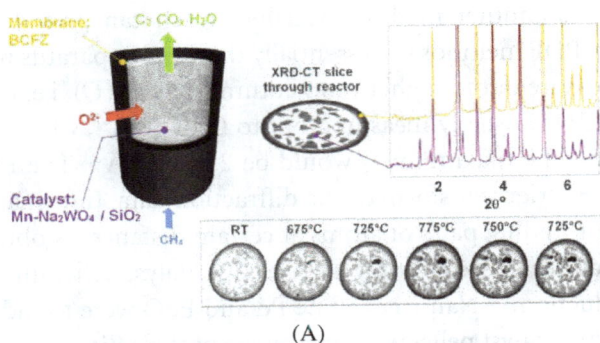

(A)

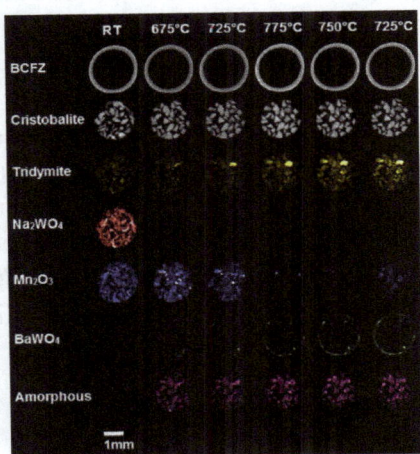

(B)

Figure 17: (A) The CMR reactor with an XRD-CT image, two selected diffraction patterns are shown from different pixels of the XRD-CT slice one from the membrane and another from the catalyst. XRD-CT images of one diffraction peak are shown during reaction. (B) Phase maps for five crystalline phases observed in the catalyst and their evolution during reaction. These images are from Ref. [141], published by The Royal Society of Chemistry.

catalyst. One example of μ-XANES-CT by Price *et al.* has been recently published, which describes the investigation of a Pt catalyst containing a Mo promoter supported on carbon.[144] The authors found a distribution of metallic Pt and Pt chloride species through the $30 \times 45 \times 80\ \mu$m particle as well as showing that no direct bimetallic interaction between Mo and Pt was present.

PDF-CT is another method available which can provide nanoscale information. PDF methods use essentially the same apparatus required for XRD but measure out to higher momentum transfer (Q), i.e. recording to higher 2θ. XRD is usually measured out to Q of 10 Å^{-1}, whereas, for PDF measurements the maximum Q would be 25 or 30 Å^{-1}. The PDF is then obtained by Fourier transform of the diffraction data. Information on the probability of finding pairs of atoms at certain distances is obtained. This technique has been used to study a Pd/Al$_2$O$_3$ catalyst pellet under calcination and reduction.[145] Nanocrystalline Pd and PdO were found to be present inside the catalyst pellet with an average particle diameter of ≤1.4 nm, which would have been invisible to XRD. This technique has an advantage over XRD-CT in that it can provide both crystalline and non-crystalline tomographic information in one measurement.

Advances in reactor design allowing measurements at high pressures encountered in industry will be needed to achieve tomographic, and in fact all spectroscopic investigations, under true *operando* conditions. In addition, increased data acquisition speeds will hopefully permit real industrial heterogeneous catalyst pellets to be investigated using either PDF-CT or XRD-CT in real time under transient conditions.

3.8 Possibilities with soft X-rays

X-rays have so far been used for catalysis under *operando* conditions for the investigation of metals. However, soft X-ray techniques (<3 keV) can provide information on, adsorption, desorption and surface reactions.[146] These techniques are limited in their ability to be performed under realistic conditions, for example the emitted photoelectron in XPS measurements would be highly attenuated by the surrounding gas or liquid phase molecules if performed *in situ*. However, recent advances in soft X-ray beamlines and the high brightness of synchrotron radiation have permitted near ambient pressure experiments to be performed.[147] Continued advances could possibly see soft X-rays being used to investigate *operando* style catalytic reactions.

4 Conclusions and Future Perspectives

The range of neutron and X-ray techniques for catalyst characterisation is broad, powerful and becoming more available. As these techniques become more routine, there is an increasing push to look at reactions in real time, under working conditions, and with multiple simultaneous techniques. This requires development of specialised cells, and complex engineering, but this allows linked techniques such as INS-Raman,[148] and XAFS-DRIFTS.[127]

In many cases, synchrotron and neutron sources are available on the same campus, (Diamond/ISIS, ESRF/ILL, SINQ/SLS and MAX4/ESS when the European Spallation Source (ESS) comes on line in ~2020) so it is somewhat surprising there are not many joint studies using neutrons and X-rays. Another drive in this field is to improve reactor engineering. For an accurate understanding of what is going on in an industrial reactor, it is important for our experiments to model the conditions or risk measuring the wrong things.

The history of *in situ* measurements at central facilities is illustrious. Whilst the sources and facilities continue to improve, the cutting edge extends, more difficult experiments become possible, and the future looks even brighter.

Figure 18: Photograph of the Rutherford Appleton Laboratory site, showing ISIS at the left, Diamond on the right, and the Research Complex (The physical hub of the Catalysis Hub project) in between. (credit STFC)

References

1. C. O. Arean, B. M. Weckhuysen and A. Zecchina, *Phys. Chem. Chem. Phys.*, 2012, **14**, 2125–2127.
2. S. Klotz, *Techniques in High Pressure Neutron Scattering*, CRC Press, 2012.
3. H.-W. Wang, V. R. Fanelli, H. M. Reiche, E. Larson, M. A. Taylor, H. Xu, J. Zhu, J. Siewenie and K. Page, *Rev. Sci. Instrum.*, 2014, **85**, 125116.
4. E. Polak, J. Munn, P. Barnes, S. E. Tarling and C. Ritter, *J. Appl. Crystal.*, 1990, **23**, 258–262.
5. J. F. C. Turner, S. E. McLain, T. H. Free, C. J. Benmore, K. W. Herwig and J. E. Siewenie, *Rev. Sci. Instrum.*, 2003, **74**, 4410–4417.
6. R. Haynes, S. T. Norberg, S. G. Eriksson, M. A. H. Chowdhury, C. M. Goodway, G. D. Howells, O. Kirichek and S. Hull, *J. Phys.: Conf. Ser.*, 2010, **251**, 012090.
7. J. F. C. Turner, R. Done, J. Dreyer, W. I. F. David and C. R. A. Catlow, *Rev. Sci. Instrum.*, 1999, **70**, 2325–2330.
8. T. Kandemir, D. Wallacher, T. Hansen, K.-D. Liss, R. Naumann d'Alnoncourt, R. Schlögl and M. Behrens, *Nucl. Instrum. Metho. Phys. Res. A*, 2012, **673**, 51–55.
9. K.-D. Liss, B. Hunter, M. Hagen, T. Noakes and S. Kennedy, *Phys. B: Condens. Matter*, 2006, **385–386, Part 2**, 1010–1012.
10. J. M. Nicol, *Spectrochimi. Acta: Mol. Spectrosc.*, 1992, **48**, 313–327.
11. I. P. Silverwood, N. G. Hamilton, A. McFarlane, R. M. Ormerod, T. Guidi, J. Bones, M. P. Dudman, C. M. Goodway, M. Kibble, S. F. Parker and D. Lennon, *Rev. Sci. Instrum.*, 2011, **82**, 034101.
12. A. R. McFarlane, I. P. Silverwood, R. Warringham, E. L. Norris, R. M. Ormerod, C. D. Frost, S. F. Parker and D. Lennon, *RSC Adv.*, 2013, **3**, 16577–16589.
13. A. R. McFarlane, I. P. Silverwood, E. L. Norris, R. M. Ormerod, C. D. Frost, S. F. Parker and D. Lennon, *Chem. Phys.*, 2013, **427**, 54–60.
14. N. G. Hamilton, R. Warringham, I. P. Silverwood, J. Kapitán, L. Hecht, P. B. Webb, R. P. Tooze, W. Zhou, C. D. Frost, S. F. Parker and D. Lennon, *J. Catal.*, 2014, **312**, 221–231.
15. D. Lennon, R. Warringham, T. Guidi and S. F. Parker, *Chem. Phys.*, 2013, **427**, 49–53.
16. E. Nowicka, J. P. Hofmann, S. F. Parker, M. Sankar, G. M. Lari, S. A. Kondrat, D. W. Knight, D. Bethell, B. M. Weckhuysen and G. J. Hutchings, *Phys. Chem. Chem. Phys.*, 2013, **15**, 12147–12155.
17. R. Warringham, D. Bellaire, S. F. Parker, J. Taylor, R. A. Ewings, C. M. Goodway, M. Kibble, S. R. Wakefield, M. Jura, M. P. Dudman, R. P. Tooze, P. B. Webb and D. Lennon, *J. Phys.: Conf. Ser.*, 2014, **554**, 012005.

18. M. O. Jones, A. D. Taylor and S. F. Parker, *Appl. Petrochem. Res.*, 2012, **2**, 97–104.

19. J. J. Pluth, J. V. Smith and Å. Kvick, *Zeolites*, 1985, **5**, 74–80.

20. N. Lock, M. Christensen, Y. Wu, V. K. Peterson, M. K. Thomsen, R. O. Piltz, A. J. Ramirez-Cuesta, G. J. McIntyre, K. Noren, R. Kutteh, C. J. Kepert, G. J. Kearley and B. B. Iversen, *Dalt. Transac.*, 2013, **42**, 1996–2007.

21. W. Baratta, S. Baldino, M. J. Calhorda, P. J. Costa, G. Esposito, E. Herdtweck, S. Magnolia, C. Mealli, A. Messaoudi, S. A. Mason and L. F. Veiros, *Chem. — A Eur. J.*, 2014, **20**, 13603–13617.

22. Q. Wan, B. C. Bennett, M. A. Wilson, A. Kovalevsky, P. Langan, E. E. Howell and C. Dealwis, *Proce. Nat. Acad. Sci.s*, 2014, **111**, 18225–18230.

23. S. Z. Fisher, M. Aggarwal, A. Y. Kovalevsky, D. N. Silverman and R. McKenna, *J. Am. Chem. Soc.*, 2012, **134**, 14726–14729.

24. J. Overgaard, B. Schiøtt, F. K. Larsen, A. J. Schultz, J. C. MacDonald and B. B. Iversen, *Angew. Chem. Inte. Edi.*, 1999, **38**, 1239–1242.

25. T. C. Hansen and H. Kohlmann, *Zeitschrift für anorganische und allgemeine Cheme*, 2014, **640**, 3044–3063.

26. A. K. Zbigniew, R. H. Jones, R. G. Bell, C. R. A. Catlow and J. M. Thomas, *Mole. Phys.*, 1996, **89**, 1345–1357.

27. T.-H. Bae, M. R. Hudson, J. A. Mason, W. L. Queen, J. J. Dutton, K. Sumida, K. J. Micklash, S. S. Kaye, C. M. Brown and J. R. Long, *Energ. Environ. Sci.*, 2013, **6**, 128–138.

28. L. M. Bull, A. K. Cheetham, B. M. Powell, J. A. Ripmeester and C. I. Ratcliffe, *J. Am. Chem. Soc.*, 1995, **117**, 4328–4332.

29. L. Smith, A. K. Cheetham, R. E. Morris, L. Marchese, J. M. Thomas, P. A. Wright and J. Chen, *Sci.*, 1996, **271**, 799–802.

30. H. Wu, J. M. Simmons, Y. Liu, C. M. Brown, X.-S. Wang, S. Ma, V. K. Peterson, P. D. Southon, C. J. Kepert, H.-C. Zhou, T. Yildirim and W. Zhou, *Chem. — A Eur. J.*, 2010, **16**, 5205–5214.

31. C. M. Brown, A. J. Ramirez-Cuesta, J.-H. Her, P. S. Wheatley and R. E. Morris, *Chem. Phys.*, 2013, **427**, 3–8.

32. J. Luo, H. Xu, Y. Liu, Y. Zhao, L. L. Daemen, C. Brown, T. V. Timofeeva, S. Ma and H.-C. Zhou, *J. Am. Chem. Soc.*, 2008, **130**, 9626–9627.

33. W. L. Queen, E. D. Bloch, C. M. Brown, M. R. Hudson, J. A. Mason, L. J. Murray, A. J. Ramirez-Cuesta, V. K. Peterson and J. R. Long, *Dalt. Transac.*, 2012, **41**, 4180–4187.

34. Y. Filinchuk, N. A. Tumanov, V. Ban, H. Ji, J. Wei, M. W. Swift, A. H. Nevidomskyy and D. Natelson, *J. Am. Chem. Soc.*, 2014, **136**, 8100–8109.

35. E. Mamontov, T. Egami, R. Brezny, M. Koranne and S. Tyagi, *J. Phys. Chem. B*, 2000, **104**, 11110–11116.

36. M. Ozawa and C. K. Loong, *Cataly. Today*, 1999, **50**, 329–342.
37. R. Martínez-Coronado, A. Aguadero, J. A. Alonso and M. T. Fernández-Díaz, *Solid State Sci.*, 2013, **18**, 64–70.
38. F. Basile, G. Fornasari, M. Gazzano and A. Vaccari, *Appl. Clay Sci.*, 2001, **18**, 51–57.
39. L. M. Plyasova, T. M. Yur'eva, I. Y. Molina, T. A. Kriger, A. M. Balagurov, L. P. Davydova, V. I. Zaikovskii, G. N. Kustova, V. V. Malakhov and L. S. Dovlitova, *Kinet. Cataly.*, 2000, **41**, 429–436.
40. G. N. Iles, F. Devred, P. F. Henry, G. Reinhart and T. C. Hansen, *J. Chem. Phys.*, 2014, **141**, 034201.
41. T. Kandemir, M. E. Schuster, A. Senyshyn, M. Behrens and R. Schlögl, *Angew. Chem. Int. Edi.*, 2013, **52**, 12723–12726.
42. T. Kandemir, F. Girgsdies, T. C. Hansen, K.-D. Liss, I. Kasatkin, E. L. Kunkes, G. Wowsnick, N. Jacobsen, R. Schlögl and M. Behrens, *Angew. Cheme Int. Ed.*, 2013, **52**, 5166–5170.
43. M. C. D. Mourad, M. Mokhtar, M. G. Tucker, E. R. Barney, R. I. Smith, A. O. Alyoubi, S. N. Basahel, M. S. P. Shaffer and N. T. Skipper, *J. Mater. Chem.*, 2011, **21**, 15479–15485.
44. H. Y. Playford, D. R. Modeshia, E. R. Barney, A. C. Hannon, C. S. Wright, J. M. Fisher, A. Amieiro-Fonseca, D. Thompsett, L. A. O'Dell, G. J. Rees, M. E. Smith, J. V. Hanna and R. I. Walton, *Chem. Mater.*, 2011, **23**, 5464–5473.
45. K. Page, J. Li, R. Savinelli, H. N. Szumila, J. Zhang, J. K. Stalick, T. Proffen, S. L. Scott and R. Seshadri, *Solid State Sci.*, 2008, **10**, 1499–1510.
46. D. Pickup, R. Moss and R. Newport, *J. Appl. Crystal.*, 2014, **47**, 1790–1796.
47. D. T. Bowron and S. Diaz Moreno, *Coordination Chem. Rev.*, 2014, **277–278**, 2–14.
48. J. F. C. Turner, C. J. Benmore, C. M. Barker, N. Kaltsoyannis, J. M. Thomas, W. I. F. David and C. R. A. Catlow, *J. Phys. Chem. B*, 2000, **104**, 7570–7573.
49. S. F. Parker, D. Bowron, S. Imberti, A. K. Soper, K. Refson, E. Lox, M. Lopez and P. Albers, *Chem. Commun.*, 2010, **46**, 2959–2961.
50. D. T. Bowron, E. C. Beret, E. Martin-Zamora, A. K. Soper and E. Sánchez Marcos, *J. Am. Chem. Soc.*, 2011, **134**, 962–967.
51. D. T. Bowron, M. Amboage, R. Boada, A. Freeman, S. Hayama and S. Diaz-Moreno, *RSC Adv.*, 2013, **3**, 17803–17812.
52. C. Hardacre, J. D. Holbrey, C. L. Mullan, T. G. A. Youngs and D. T. Bowron, *J. Chem. Phys.*, 2010, **133**, 074510.
53. C. L. Mullan, C. Hardacre, J. Holbrey, C. Lagunas, T. Youngs, D. Bowron, L. Gladden, M. Mantle and C. D'Agostino, *Meeting Abstracts*, 2010, **MA2010–02**, 2205.

54. V. Sears, *Neutron News*, 1992, **3**, 26.

55. M. A. da Silva, F. Bode, I. Grillo and C. A. Dreiss, *Biomacromolecules*, 2015, **16**, 1401–1409.

56. J. Milano-Brusco, S. Prevost, D. Lugo, M. Gradzielski and R. Schomacker, *New J. Chem.*, 2009, **33**, 1726–1735.

57. C. J. Glinka, L. C. Sander, S. A. Wise, M. L. Hunnicutt and C. H. Lochmuller, *Analy. Chem.*, 1985, **57**, 2079–2084.

58. O. Masakuni and L. Chun-Keung, *Phys. B: Condens. Matter*, 1997, **241–243**, 269–275.

59. A. V. Perdikaki, O. C. Vangeli, G. N. Karanikolos, K. L. Stefanopoulos, K. G. Beltsios, P. Alexandridis, N. K. Kanellopoulos and G. E. Romanos, *J. Phys. Chem. C*, 2012, **116**, 16398–16411.

60. C.-S. Tsao, M. Li, Y. Zhang, J. B. Leao, W.-S. Chiang, T.-Y. Chung, Y.-R. Tzeng, M.-S. Yu and S.-H. Chen, *J. Phys. Chem. C*, 2010, **114**, 19895–19900.

61. M. C. Diaz, P. J. Hall, C. E. Snape, S. D. Brown and R. Hughes, *Ind. Engine. Chem. Res.*, 2002, **41**, 6566–6571.

62. E. Valiev, S. Bogdanov, A. Pirogov, A. Teplykh, A. Ostrouschko and Y. Mogilnikov, *Phys. B: Condens. Matter.*, 2000, **276–278**, 854–855.

63. R. Winter, D. W. Hua, P. Thiyagarajan and J. Jonas, *J. Non-Cryst. Solids*, 1989, **108**, 137–142.

64. N. Miyamoto, Y. Inoue, S. Koizumi and T. Hashimoto, *J. Appl. Crystal.*, 2007, **40**, s568–s572.

65. T. Terashima, R. Motokawa, S. Koizumi, M. Sawamoto, M. Kamigaito, T. Ando and T. Hashimoto, *Macromolecules*, 2010, **43**, 8218–8232.

66. T. Youngs, H. Manyar, D. Bowron, L. Gladden and C. Hardacre, *Chem. Sci.*, 2013, **4**, 3484–3489.

67. S. Gautam, A. K. Tripathi, V. S. Kamble, S. Mitra and R. Mukhopadhyay, *Pramana — J. Phys.*, 2008, **71**, 1153–1157.

68. N. M. Gupta, D. Kumar, V. S. Kamble, S. Mitra, R. Mukhopadhyay and V. B. Kartha, *J. Phys. Chem. B*, 2006, **110**, 4815–4823.

69. H. Jobic and D. N. Theodorou, *Micropor. Mesopor. Mater.*, 2007, **102**, 21–50.

70. A. J. O'Malley and C. R. A. Catlow, *Phys. Chem. Chem. Phys.*, 2013, **15**, 19024–19030.

71. O.-E. Haas, J. M. Simon, S. Kjelstrup, A. L. Ramstad and P. Fouquet, *The J. Phys. Chem. C*, 2008, **112**, 3121–3125.

72. O.-E. Haas, J. M. Simon and S. Kjelstrup, *The J. Phys. Chem. C*, 2009, **113**, 20281–20289.

73. S. Adak, L.L. Daemen, T. Seydel, N. Strange, C. Sumner, J. Z. Larese, In *Inelastic Neutron Scattering (INS) Studies of Hydrogen Spillover on Pure and*

Pd Decorated Metal Oxides, 246th ACS National Meeting & Exposition, Indianapolis, IN, United States, *J. Am. Chem. Soc.*, 2013, pp CATL-110.

74. J. Z. Larese, In *Inelastic Neutron Scattering Studies of Hydrogen Spillover on Metal Oxide Surfaces*, 66th Southeast Regional Meeting of the American Chemical Society, Nashville, TN, United States, *J. Am. Chem. Soc.*, 2014; pp SERMACS-600.

75. J. Z. Larese, In *Inelastic Neutron Scattering Studies of Hydrogen Spillover on Metal Oxide Surfaces*, 248th ACS National Meeting & Exposition, San Francisco, CA, United States, *J. Am. Chem. Soc.*, 2014; pp CATL-228.

76. F. Bellezza, A. Cipiciani, S. Cinelli, A. Esposito, G. Onori and A. Paciaroni, *Chem. Phys. Lett.*, 2009, **478**, 260–265.

77. P. C. H. Mitchell, S. F. Parker, A. J. Ramirez-Cuesta and J. Tomkinson, *Vibrational Spectroscopy with Neutrons*, 1st edition World Scientific, 2005.

78. I. J. Braid, J. Howard and J. Tomkinson, *J. Chem. Soci., Faraday Transact. 2: Mol. Chem. Phys.*, 1983, **79**, 253–262.

79. A. R. McInroy, D. T. Lundie, J. M. Winfield, C. C. Dudman, P. Jones, S. F. Parker and D. Lennon, *Catal. Today.*, 2006, **114**, 403–411.

80. D. T. Lundie, A. R. McInroy, R. Marshall, J. M. Winfield, P. Jones, C. C. Dudman, S. F. Parker, C. Mitchell and D. Lennon, *J. Phys. Chem. B*, 2005, **109**, 11592–11601.

81. D. Lennon, D. T. Lundie, S. D. Jackson, G. J. Kelly and S. F. Parker, *Langmuir*, 2002, **18**, 4667–4673.

82. S. F. Parker, *Chem. Commun.*, 2011, **47**, 1988–1990.

83. N. G. Hamilton, I. P. Silverwood, R. Warringham, J. Kapitán, L. Hecht, P. Webb, R. Tooze, S. F. Parker and D. Lennon, *Angew. Chem. Int. Ed.*, 2013, **52**, 5608–5611.

84. P. Albers, R. Burmeister, K. Seibold, G. Prescher, S. F. Parker and D. K. Ross, *J. Catal.*, 1999, **181**, 145–154.

85. S. F. Parker, J. W. Taylor, P. Albers, M. Lopez, G. Sextl, D. Lennon, A. R. McInroy and I. W. Sutherland, *Vib. Spectrosc.*, 2004, **35**, 179–182.

86. P. W. Albers, W. Weber, K. Kunzmann, M. Lopez and S. F. Parker, *Surf. Sci.*, 2008, **602**, 3611–3616.

87. P. W. Albers, H. Klein, E. S. Lox, K. Seibold, G. Prescher and S. F. Parker, *Phys. Chem. Chem. Phys.*, 2000, **2**, 1051–1058.

88. I. P. Silverwood, N. G. Hamilton, J. Staniforth, C. Laycock, S. F. Parker, M. Ormerod and D. Lennon, *Catal. Today*, 2010, **155**, 319–325.

89. I. P. Silverwood, N. G. Hamilton, A. McFarlane, J. Kapitan, L. Hecht, E. Norris, R. Mark Ormerod, C. Frost, S. F. Parker and D. Lennon, *Phys. Chem. Chem. Phys.*, 2012, **14**, 15214–15225.

90. I. P. Silverwood, N. G. Hamilton, C. Laycock, J. Staniforth, M. Ormerod, C. Frost, S. F. Parker and D. Lennon, *Phys. Chem. Chem. Phys.*, 2010, **12**, 3102–3107.

91. R. Warringham, N. G. Hamilton, I. P. Silverwood, C. How, P. B. Webb, R. P. Tooze, W. Zhou, C. D. Frost, S. F. Parker and D. Lennon, *Appl. Catal., A*, 2015, **489**, 209–217.

92. C. R. Vanston, G. J. Kearley, A. J. Edwards, T. A. Darwish, N. R. de Souza, A. J. Ramirez-Cuesta and M. G. Gardiner, *Faraday Discuss.*, 2015, **177**, 99–109.

93. M. Trapp, M. Tehei, M. Trovaslet, F. Nachon, N. Martinez, M. M. Koza, M. Weik, P. Masson and J. Peters, *J. R. Soc. Interface*, 2014, **11**, 20140372/20140371–20140372/20140379.

94. R. Zboray and H. M. Prasser, *Exp Fluids*, 2013, **54**, 1–15.

95. P. B. Oberholzer, R. Siegrist, A. Kästner, E. H. Lehmann, G. G. Scherer, A. Wokaun, *Electrochem. Commun.*, 2012, **20**, 67–70.

96. A. Schröder, K. Wippermann, T. Arlt, T. Sanders, T. Baumhöfer, N. Kardjilov, J. Mergel, W. Lehnert, D. Stolten, J. Banhart and I. Manke, *Int. J. Hydrogen Energ.*, 2013, **38**, 2443–2454.

97. T. Kotaka, Y. Tabuchi, U. Pasaogullari and C.-Y. Wang, *Electrochim. Acta*, 2014, **146**, 618–629.

98. A. Byrne, V. Dakessian, R. Hughes, J. Santamaria-Ramiro and C. J. Wright, *J. Catal.*, 1985, **96**, 146–153.

99. H. Fritzsche, C. Ophus, C. T. Harrower, E. Luber and D. Mitlin, *Appl. Phys. Lett.*, 2009, **94**, 241901.

100. H. Fritzsche, W. P. Kalisvaart, B. Zahiri, R. Flacau and D. Mitlin, *Int. J. Hydrogen Energ.*, 2012, **37**, 3540–3547.

101. A. M. Beale, S. D. M. Jacques, E. K. Gibson and M. D. Michiel, *Coord. Chem. Rev.*, 2014, **277–278**, 208–223.

102. A. M. Beale, S. D. M. Jacques and B. M. Weckhuysen, *Chem. Soc. Rev.*, 2010, **39**, 4656–4672.

103. http://xafs.org/.

104. D. C. Koningsberger, B. L. Mojet, G. E. v. Dorssen and D. E. Ramaker, *Top. Catal.*, 2000, **10**, 143–155.

105. J. Evans, A. Puig-Molina and M. Tromp, *MRS BULLETIN*, 2007, **32**, 1038–1043.

106. C. Brookes, P. P. Wells, E. K. Gibson, D. Gianolio, K. M. H. Mohammed, S. Parry, S. M. Rogers, I. P. Silverwood and M. Bowker, *in review*, 2015.

107. S. A. Bartlett, P. P. Wells, M. Nachtegaal, A. J. Dent, G. Cibin, G. Reid, J. Evans and M. Tromp, *J. Catal.*, 2011, **284**, 247–258.

108. M. Tromp, J. R. A. Sietsma, J. A. v. Bokhoven, G. P. F. v. Strijdonck, R. J. v. Haaren, A. M. J. v. d. Eerden, P. W. N. M. v. Leeuwenb and D. C. Koningsberger, *Chem. Commun.*, 2003, 128–129.

109. M. A. Newton, A. J. Dent and J. Evans, *Chem. Soc. Rev.*, 2002, **31**, 83–95.
110. S. A. Bartlett, J. Moulin, M. Tromp, G. Reid, A. J. Dent, G. Cibin, D. S. McGuinness and J. Evans, *ACS Catal.*, 2014, **4**, 4201–4204.
111. J. Boita, M. d. Carmo, M. Alves and J. Morais, *J. Synchrotron Rad.*, 2014, **21**, 254–258.
112. M. A. Newton, M. Nachtegaal, O. Kröcher and D. Ferri, presented in part at the Operando V, Deauville, France, 2015.
113. C. Roth, N. Martz, T. Buhrmester, J. Scherer and H. Fuess, *Phys. Chem. Chem. Phys.*, 2002, **4**, 3555–3557.
114. R. J. K. Wiltshire, C. R. King, A. Rose, P. P. Wells, M. P. Hogarth, D. Thompsett and A. E. Russell, *Electrochim. Acta*, 2005, **50**, 5208–5217.
115. S. W. T. Price, J. D. Speed, P. Kannan and A. E. Russell, *J. Am. Chem. Soc.*, 2011, **133**, 19448–19458.
116. S. W. T. Price, J. M. Rhodes, L. Calvillo and A. E. Russell, *J. Phys. Chem. C*, 2013, **117**, 24858–24865.
117. E. Stavitski and B. M. Weckhuysen, *Chem. Soc. Rev.*, 2010, **39**, 4615–4625.
118. L. M. Miller and R. J. Smith, *Vib. Spectrosc.*, 2005, **38**, 237–240.
119. E. Stavitski, M. H. F. Kox, I. Swart, F. M. F. de Groot and B. M. Weckhuysen, *Angew. Cheme Int. Ed.*, 2008, **47**, 3543–3547.
120. E. Gross, X.-Z. Shu, S. Alayoglu, H. A. Bechtel, M. C. Martin, F. D. Toste and a. G. A. Somorjai, *J. Am. Chem. Soc.*, 2014, **136**, 3624–3629.
121. N. Al-Rifai, E. Cao, V. Dua and A. Gavriilidis, *Curr. Opin. Chem. Eng.*, 2013, **2**, 338–345.
122. F. Trachsel, B. Tidona, S. Desportes and P. R. v. Rohr, *J. Supercrit. Fluids*, 2009, **48**, 146–153.
123. M. A. Newton, B. Jyoti, A. J. Dent, S. G. Fiddy and J. Evans, *Chem. Commun.*, 2004, 2382–2383.
124. A. Kubacka, A. Martínez-Arias, M. Fernández-García, M. D. Michiel and M. A. Newton, *J. Catal.*, 2010, **270**, 275–284.
125. N. S. Marinkovic, Q. Wang and A. I. Frenkel, *J. Synchrotron Rad.*, 2011, **18**, 447–455.
126. J. G. Highfield, M. Prairie and A. Renken, *Cataly. Today*, 1991, **9**, 39–46.
127. E. K. Gibson, A. M. Beale, C. R. A. Catlow, A. Chutia, D. Gianolio, A. Gould, A. Kroner, K. M. H. Mohammed, M. Perdjon, S. M. Rogers and P. P. Wells, *Chem. Mater.*, 2015, **27**, 3714–3720.
128. S. Yao, K. Mudiyanselage, W. Xu, A. C. Johnston-Peck, J. C. Hanson, T. Wu, D. Stacchiola, J. A. Rodriguez, H. Zhao, K. A. Beyer, K. W. Chapman, P. J. Chupas, A. Martínez-Arias, R. Si, T. B. Bolin, W. Liu and S. D. Senanayak, *ACS Catal.*, 2014, **4**, 1650–1661.

129. K. A. Beyer, H. Zhao, O. J. Borkiewicz, M. A. Newton, P. J. Chupas and K. W. Chapman, *J. Appl. Crystal.*, 2014, **47**, 95–101.

130. M. Cavers, J. M. Davidson, I. R. Harkness, G. S. McDougall and L. V. C. Rees, *Reaction Kinetics and the Development of Catalytic Processes*, Elsevier, Amsterdam, 1999.

131. F. C. Meunier, A. Goguet, S. Shekhtman, D. Rooney and H. Daly, *Appl. Catal.*, 2008, **340**, 196–202.

132. M. A. Newton and W. v. Beek, *Chem. Soc. Rev.*,, 2010, **39**, 4845–4863.

133. G. L. Chiarello, M. Nachtegaal, V. Marchionni, L. Quaroni and D. Ferri, *Rev. Sci. Instrum.*, 2014, **85**, 074102.

134. A. Urakawa, N. Maeda and A. Baiker, *Angew. Chem. Int. Ed.*, 2008, **47**, 9256–9259.

135. V. Briois, S. Belin, F. Villain, F. Bouamrane, H. Lucas, R. Lescouezec, M. Julve, M. Verdaguer, M. S. Tokumoto, C. V. Santilli, S. H. Pulcinelli, X. Carrier, J. M. Krafft, C. Jubin and M. Che, *Phys. Scripta*, 2005, **T115**, 38–44.

136. C. La Fontaine, L. Barthe, A. Rochet and V. Briois, *Catal. Today*, 2013, **205**, 148–158.

137. A. M. Beale, A. M. J. v. d. Eerden, K. Kervinen, M. A. Newton and B. M. Weckhuysen, *Chem. Commun.*, 2005, 3015–3017.

138. A. Bansode, G. Guilera, V. Cuartero, L. Simonelli, M. Avila and A. Urakawa, *Rev. Sci. Instrum.*, 2014, **85**, 084105–084108.

139. S. D. M. Jacques, M. D. Michiel, A. M. Beale, T. Sochi, M. G. O. Brien, L. Espinosa-Alonso, B. M. Weckhuysen and P. Barnes, *Angew. Chem. Int. Ed.*, 2011, **50**, 10148–10152.

140. J. Andrieux, C. Chabert, A. Mauro, H. Vitoux, B. Gorges, T. Buslaps and V. Honkimaki, *J. Appl. Crystal.*, 2014, **47**, 245–255.

141. A. Vamvakeros, S. Jacques, V. Middelkoop, M. di Michiel, C. K. Egan, I. Ismagilov, G. Vaughan, F. Gallucci, M. van Sint Annaland, P. Shearing, R. J. Cernik and A. M. Beale, *Chem. Commun.*, 2015, **51**, 12752–12755.

142. J. W. Couves, J. M. Thomas, D. Waller, R. H. Jones, A. J. Dent, G. E. Derbyshire and G. N. Greaves, *Nature*, 1991, **354**, 465–468.

143. J.-D. Grunwaldt, N. v. Vegten and A. Baiker, *Chem. Commun.*, 2007, 4635–4637.

144. S. W. T. Price, K. Ignatyev, K. Geraki, M. Basham, J. Filik, N. T. Vo, P. T. Witte, A. M. Beale and J. F. W. Mosselmans, *Phys. Chem. Chem. Phys.*, 2015, **17**, 521–529.

145. S. D. M. Jacques, M. Di Michiel, S. A. J. Kimber, X. Yang, R. J. Cernik, A. M. Beale and S. J. L. Billinge, *Nat Commun.*, 2013, **4**.

146. R. Toyoshima and H. Kondoh, *J. Phys.: Condens. Metter.*, 2015, **27**, 083003–083017.
147. X. Liu, W. Yang and Z. Liu, *Adv. Mater.*, 2014, **26**, 7710–7729.
148. M. A. Adams, S. F. Parker, F. Fernandez-Alonso, D. J. Cutler, C. Hodges and A. King, *Appl. Spectrosc.*, 2009, **63**, 727–732.

Chapter 3

Outperforming Nature's Catalysts: Designing Metalloenzymes for Chemical Synthesis

Charlie Fehl[†], Amanda G. Jarvis*, Maria Palm-Espling[†], Ben Davis[†] and Paul C. J. Kamer*

* EaStCHEM, School of Chemistry, University of St Andrews, St Andrews, Fife, KY16 9ST, UK

[†] Department of Chemistry, University of Oxford, Oxford OX1 3TA, UK

The design of artificial metalloenzymes — protein scaffolds that have been adapted to introduce non-natural active sites and metals to enable chemistry beyond their evolutionary functions — is a rapidly expanding field in biocatalysis. The ability to control transition metal-promoted reactivity using the recognition and selectivity of enzymes has proven to be a powerful contribution to chemistry. This chapter aims to provide an overview of several key strategies used in the development of artificial metalloenzymes: re-engineering existing metalloenzymes, the creation of new metal sites within a protein scaffold, and the *de novo* construction of novel structural elements. The key methodologies we highlight include structural biology, the use of unnatural amino acids as metal ligands, and computational tools to design novel metalloenzymes targeting challenging problems in synthetic chemistry.

1 Introduction to Biocatalysis

As this book illustrates, developments in catalysis play a vital part in improving chemical synthesis. These innovations arise from a diverse array of sources. One area that has revealed powerful results for synthesis is only recently growing in popularity — the synthetic chemistry of biological systems.[1] This field, termed biocatalysis, focuses on advancing the reactions of protein catalysts beyond their evolved functions.[2] Enzymes are protein-based tools that have been honed by the forces of evolution to perform chemical reactions within a complex, folded active site. These active sites fit precisely around their substrates, and may also contain metal or organocatalytic cofactors to promote reactivity. Their benefits to catalysis spring from two major parameters: efficiency and selectivity, which form a balance between speed and chemical accuracy.[3] Several enzymes have evolved to achieve "kinetic perfection" — catalysts so fast that their rates are controlled by the diffusion limit of their substrates in solution.[4] On the other hand, the geometric constraints of protein catalytic sites evoke the size- and shape-selectivity of heterogeneous catalysts. The protein architecture fits the substrates and transition states of their reactions with exquisite shape complementarity.[3] These two aspects of biocatalysts can be tuned for use outside of their evolutionary settings to facilitate difficult chemical reactions at both large and small scales.[5] Applications are found particularly in pharmaceutical synthesis, where biocatalysts can reduce the cost[6] and environmental impact[7] of production, but these same factors drive biocatalyst development for a variety of fine and commodity chemicals including biofuels.[8]

This chapter is focused on the overlap between biocatalysis and transition metal-mediated catalysis, a field which has also contributed immensely to chemical synthesis.[9] Insights from transition metal catalyst development have been used to aid the design of biocatalytic systems that surpass or bypass natural functions. These are directed toward accessing improved combinations of organometallic reactivity and enzymatic selectivity.[10] In-depth reviews abound on the development of specific classes of biocatalysts,[11–13] so here we present general strategies for the re-design and creation of new metalloenzymes, or metal-utilising enzymes. Case studies for each of these strategies illustrate their power and limitations in the development of novel catalysts.

2 Overlap Between Chemical Transition Metal Catalysis and Biocatalysis

Catalytic chemical reactions have been driven in large part by the application of transition metals, which can promote reactivity in ways more diverse than the s- and p-block of the periodic table.[9] Several of these modes of activity and reactivity, such as catalytic oxidative addition/reductive elimination cycles, are uniquely accessible using these metals. Furthermore, their activities can be tuned using ligands to reach desired levels of selectivity.[14] Metalloenzymes could be thought of as a very special class of transition metal ligands, one with extremely complex reactivity and stereochemical directing abilities due to the large size and dynamic conformational flexibility of the protein scaffold. However, natural metalloenzymes use only a small subset of transition metals.[15] This suggests that additional metals can potentially be incorporated, given the right adjustments to the metalloenzyme architecture.

Though biocatalyst use in industry has grown tremendously in the past two decades, as reported in a 2012 white paper,[16] in practice traditional chemical reactions dominate industrial processes. However, there are both advantages and disadvantages for the use of transition metals in both traditional chemical synthesis and biocatalysis. Figure 1 illustrates this with a snapshot of challenging metal-catalysed reactions accessible to each approach. Though scope is often a challenge of biocatalytic reactions, their selectivity make them useful for large-scale applications.[17]

Though advantages and disadvantages will vary tremendously between the particular reactions to be catalysed, some generalisations can be made. These are summarised in Table 1. The essential nature of biocatalysts, precisely-shaped cavities around substrates and transition states that follow a defined reaction pathway, can provide excellent stereo- and regiochemical control and eliminate protection and deprotection steps from a synthetic route. Another key advantage that biocatalytic routes hold over many metal-catalysed chemical transformations is a lower environmental impact. Most enzyme reactions are performed in aqueous solutions at nearly ambient temperatures and pressures. These avoid intensive energy use and can reduce solvent usage and reagent waste.[5] However, a major drawback can be the significant intellectual and experimental investment that is often required to optimise biocatalysts into target parameters.[24]

(A)

cytochromes
P450 [17]

intermediate in industrial
glucocorticoid production;
70% of total steroids are
desired fermentation product

(B)

5% Ru(Cl)$_3$
KBrO$_3$ (3 equiv)

MeCN/H$_2$O/pyridine
60 °C

73%

Du Bois *et al.* [18]
challenges: only tertiary C–H oxidation
protecting groups required
limited scope

(C)

non-heme iron
halogenase [19]

challenges: scale
turnover number

(D)

10% Mn-porphyrin
NaOCl (3 equiv)

CH$_2$Cl$_2$, rt

42%

Groves *et al.* [20]
challenges: regiochemical selectivity
protecting groups required
scale
limited scope

(E)

adenosine
deaminase [21]

challenges: scale
limited scope

(F)

cat. RuH$_2$(CO)(PPh$_3$)$_3$

mesitylene, 160 °C

76%

Kakiuchi *et al.* [22]
challenges: chemoselecivity
forcing conditions

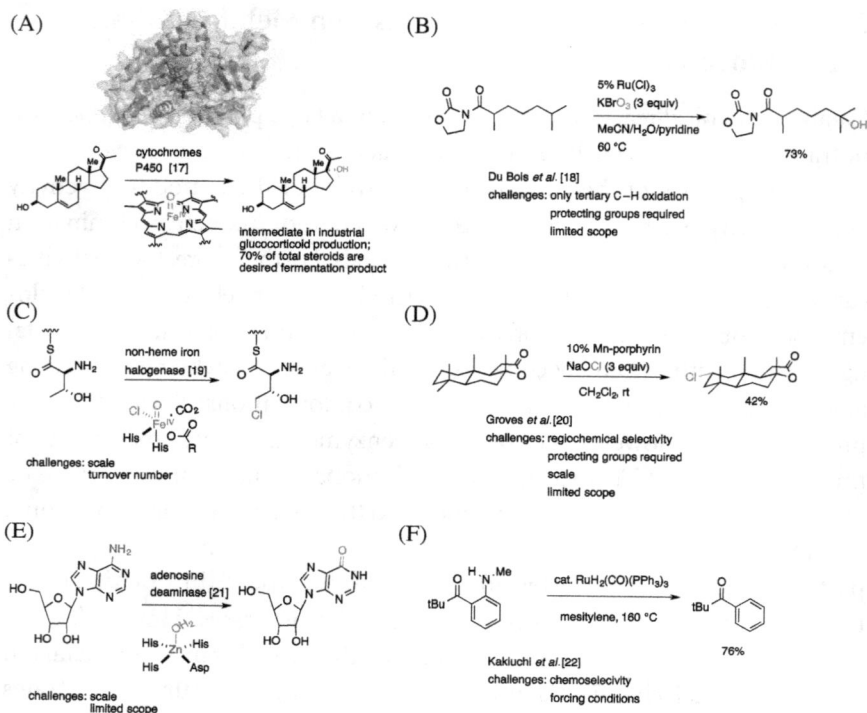

Figure 1:   Metalloenzyme transformations compared to transition metal chemical transformations. (A) Hydroxylation by cytochromes P450[17] vs. (B) C–H activation chemistry.[18]
(C) Halogenation by non-heme iron halogenases[19] vs. (D) C–H activation.[20] (E) Amine to
ketone functional group interconversion[21] vs. (F) hydrogenation.[22] The cytochrome P450
17A1 protein structure was constructed from Protein Data Bank accession number
3RUK.[23]

Though the same could be true of modifying ligands for chemical catalysts, the design stage can increase the time required for developing biocatalysts from an activity hit into their optimal final construct. Issues of
biocatalyst stability, especially within high organic solvent conditions, can
be an issue. However, these can also be addressed in the design stage, as
discussed later in this chapter and elsewhere.[5]

   In the design process of novel biocatalysts, the difficulties inherent to
adapting a protein for new chemistry present both a challenge and an
opportunity. The extremely broad range of transition metal-catalysed
reactions unknown in biology suggests that many useful reactions are able

Table 1: Comparison of design criteria for biocatalysis vs. metal-catalysed chemistry.

Parameter	Biocatalysis*	Chemical catalysis**
Chemistry		
Product concentration	>75 g L^{-1} (moderate)	Can be high
Conversion	>95% (high)	Very high
Catalyst activity (TOF/k_{cat})	10^3/h–10^7/s (moderate–exceptional)[25] (reactions typically complete <24 h)	50^{-2}–10^6/h (slow–very fast)[26]
Purity of product	High	High
Product separation	Often straightforward (phase separation), but can be challenging	Can be difficult; distillation, extraction, immobilisation helps
Catalyst loading	<2% (w/w) (moderate)	Low–high: 0.0001–20%
Enantioselectivity	>99% *ee* (very high)	>95% *ee* (high)
Regiochemistry	Often protecting group-free	Protected functionality often required
Environmental impact	Follows all 12 steps of "Green Chemistry" process design[27]	Often requires high energy (temperature, pressure) and/or organic solvents
Catalyst stability	Low; it may be necessary to add fresh biocatalysts to stalled reactions[28]	High, under the reaction conditions. Multi-ton scales are regularly utilised
Development speed	Can be slow: regularly extensive development is required; reactivity not always readily applicable to analogs	Fast: often readily adaptable directly from primary literature and condition screening

Note: *Based on target parameters as reported for Codexis' industrial process design criteria.[5]
**Generalised values based on common ranges in chemical literature.

to be redeveloped within protein frameworks.[15] The goals of artificial met-alloenzyme creation focus on combining the reactivity of metal-catalysed chemistry with the control of biocatalysis. Equally important goals include removing limitations such as the need for chemical protecting or directing groups, as well as conditions which are detrimental to the environment.

The following sections summarise the key strategies utilised to access new transition metal-based biocatalysts. Section 3 details the redesign of natural metalloenzymes to access novel, but often highly-related activity. The following Sec. 4, discusses the installation of alternative transition metals into protein cavities. This strategy could allow reactions wholly new to biology to take place, directed by a protein cavity. The combination of computational chemistry with bioinformatics towards the redesign of active sites to accommodate alternative transition states is discussed in Sec. 5. Finally, progress in the *de novo* design of novel structural features from first principles is addressed in Sec. 6. This approach can allow full control over the folding and shape of new biocatalysts. All of these strategies are complementary. Taken together, the construction of artificial metalloenzymes is an attainable goal that can enable powerful directions in catalysis.

3 Redevelopment of Natural Metalloenzymes for Alternative Chemistry

Metalloenzymes make up some of the most powerful enzymatic machinery in biology. By harnessing metal species that can be activated within the interior of highly specialised protein architectures, these enzymes accomplish carbon- and heteroatom-functionalisation reactions involving highly reactive species. The fact that the protein cavity can utilise these species in an ambient, aqueous environment is remarkable. Moreover, the reactions natural metalloenzymes catalyse are extremely specific, only recognising and reacting with a tiny subset of the complex mixtures found within cellular environments.[29]

The caveat to the selectivity of natural metalloenzymes is that they are often unsuitable for catalysing reactions on unnatural substrates, even those with only slight differences from natural motifs. To overcome this limitation, one approach is to screen genomic data to identify related proteins that have promiscuous activity with the desired substrate.[30] Another approach is to redesign the active catalytic site of a candidate protein to accommodate a non-natural reaction.

This redesign process begins with a biocatalyst known to perform similar chemistry to the desired transformation. Mutating the amino acid composition of the protein can reconfigure the active site shape toward novel substrates (Figure 2(A)). Additional mutations that can improve the activity of a biocatalyst include focusing on increasing the stability. This

(A)

(B)

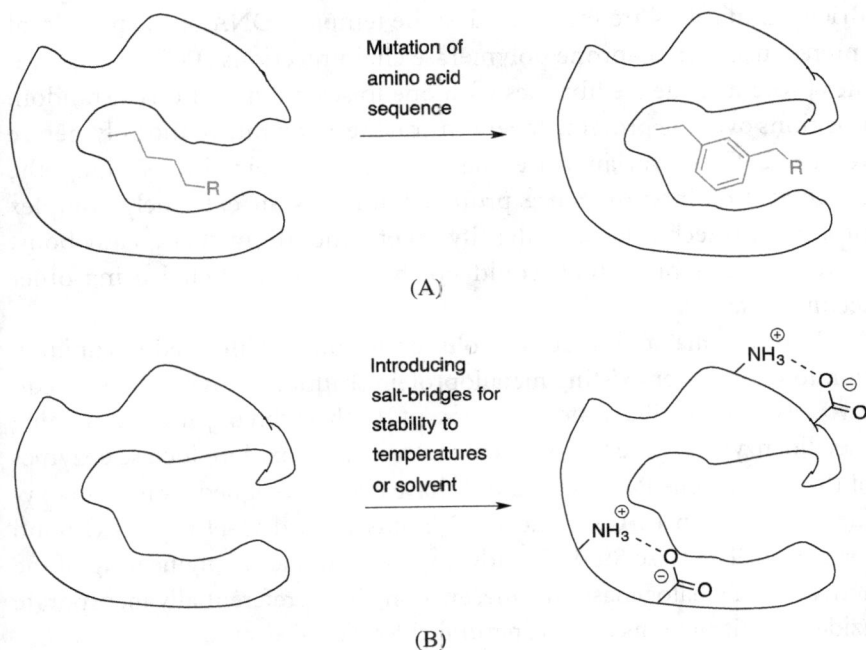

Figure 2: Improving activity for protein catalysts. (A) Active site mutations can expand substrate scope. (B) Surface or internal mutations can improve stability and solubility.

can be obtained through the addition of charged interactions[31] that form salt-bridges between protein units (shown in Figure 2(B)) or by improving the packing of a hydrophobic interior (not shown).[32]

These improvements can be incorporated using conformational information like the solved crystal or NMR structure of a protein, or from a homology model based on known structural information. These rational design strategies are used when available.[33]

Conversely, one of the most powerful techniques utilised for protein improvement do not rely on any *de novo* structural information, relying instead on random mutation pressure to phenotypically reconfigure a protein structure. The strategy of *directed evolution*[24] combines random mutagenesis with an unnatural evolutionary selection pressure such as yield, stability or enantioselectivity. Libraries of mutants are tested for improvements, and selected for their "fitness" along the target parameter.

Briefly, mutations are introduced to the template DNA of the protein of interest using error-prone polymerase chain reactions (PCRs). This creates protein sequence libraries with one to several mutations at random positions over the protein, depending on the error-rate of the polymerase reaction. These variants are then tested for their fitness along the parameter of interest. Since protein folding is an extremely complex process, this technique can identify cryptic mutations that greatly boost activity and stability that would not have been predicted using other techniques.[24]

Both rational and directed evolution techniques are used in conjunction to re-engineer existing metalloprotein scaffolds to fit synthetic needs. Two case studies illustrate this concept, both involving iron-containing metalloenzymes. Case Study 1 demonstrates how oxidoreductase enzymes of the cytochrome P450 superfamily have been redesigned away from oxy-functionalisation of C–H and C–C bonds toward C–N and C–C bond formation. The Case Study 2 study involves rational re-engineering of the active site of a halogenase enzyme, enabling it to preferentially incorporate azide and nitrate ions over its natural chloride substrates.

(A) Case Study 1: Cytochrome P450 re-engineering

One of the first families of metalloproteins targeted for biocatalyst reengineering is the superfamily of cytochromes P450.[34] These proteins share a structural motif that tightly associates heme, an iron-containing porphyrin prosthetic group.[35]

The solution chemistry of metalloporphyrins has been well studied. Before targeting P450 enzyme scaffolds for redesign, metalloporphyrin systems outside of enzymes were observed to have hydroxylation and epoxidation activities similar to those within P450 enzymatic frameworks (Figure 3(A)).[36] Furthermore, Breslow and Gellman showed that metalloporphyrins could extend beyond hydroxylation chemistry by demonstrating a metalloporphyin-catalysed intramolecular C–H amino-functionalisation (Figure 3(B)). This was a reactivity type that had not been observed with natural P450 enzymes.[37] Subsequently, they reconstituted this catalytic activity in a P450 scaffold, proving that P450 enzymes could be used for transformations beyond their natural functions albeit with low turnover numbers (Figure 3(C)).[38] Reactivity was only observed

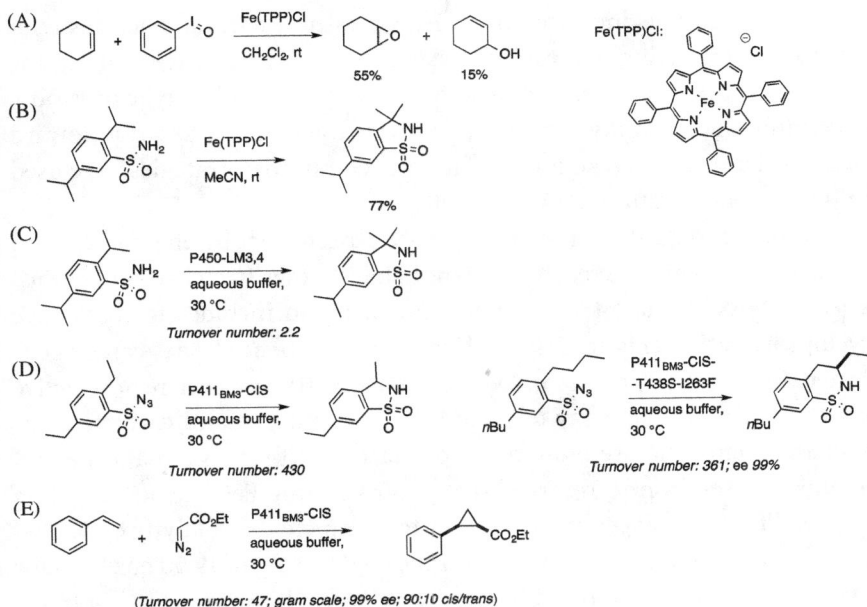

Figure 3: (A) Metalloporphyrin oxene[36] and (B) nitrene insertions.[37] (C) Nitrene insertions in a rabbit P450 scaffold.[38] (D) Enzyme evolution applied to P450-catalysed nitrene insertions.[11] (E) Carbene chemistry performed by evolved P450 enzymes.[39]

with one of several CYP isoforms from multiple organisms, suggesting that only a subset of promiscuous P450 enzymes could carry out this unnatural activity.

In the time since this discovery, cytochromes P450 have been extensively redesigned for higher activity using a combination of rational design and directed evolution.[11] Arnold and co-workers have made vast strides in improving turnover numbers, enantioinduction and preparative scale of P450 enzyme-catalysed reactions (Figure 3(D)). The workhorse scaffold is P450 BM3, a bacterial enzyme that incorporates a cytochrome P450 reductase domain and is self-sufficient in catalysis.[40] One of the key mutations for enabling these proteins to obtain higher turnover numbers is the cysteine 400 to serine mutation (mutations hereafter denoted as Cys400Ser, etc.), which pushes additional electron density into the heme and increases reactivity. Additionally, the Thr268Ala mutation removed a conserved hydrogen bond that activated C–H hydroxylation activity in

P450 enzymes. This increased the turnover of non-natural reactions relative to the competing natural hydroxylation chemistry.[41] Interestingly, the Ile263Phe mutation directed reactivity to the homobenzylic position, overriding the naturally higher reactivity of the benzylic position toward such C–H amination reactions.[41] Thus, varied rational strategies were used to improve non-natural P450 reactivity.

Noting a potential similarity between oxene or nitrene-based C–H insertions and carbene reactivity, Arnold has further stretched the boundaries of cytochrome P450-mediated chemistry to include stereoselective cyclopropanation (Figure 3(E)).[39] Here, heme-carbene species can be generated from diazo precursors, which react in stereo- and regioselective fashion with a broad array of olefins. In both nitrene and carbene engineered systems, the use of an azido or diazo motif serves a major benefit in allowing the iron to return directly to its resting Fe(II) state, avoiding the need for an endogenous reductant to complete the catalytic cycle.[11]

The intense work ongoing toward the engineering of P450 enzymes for use in synthesis has produced biocatalysts that are beginning to rival traditional chemical catalysts such as rhodium complexes for cyclopropanation.[39] The selectivity of P450 enzymes, both natural and evolved, can be highly relevant for the preparation of fine chemicals. For example, this technology was applied toward an enantioselective formal synthesis of the antipsychotic drug levomilnacipran from simple starting materials on preparative scale.[42]

(B) Case Study 2: Iron halogenase engineering

Iron is also the key species for a non-heme class of metalloproteins, the iron(II)- and α-ketoglutarate–dependent (Fe/αKG) hydroxylases and halogenases. Like most P450 enzymes, these enzymes use a high-valent iron-oxo species to abstract a hydrogen radical from unactivated aliphatic species.[43] Unlike the P450 enzymes, the active site iron of the Fe/αKG halogenases has an additional open coordination site suitable for halogen binding. This can react with the newly-generated aliphatic radical, leading to halogenation (Figure 4(A)). This system was targeted by Bollinger and co-workers to replace the halide with either nitro or azido ions.[19] When screening putative halogenase enzymes for this purpose, SyrB2 from the syringomycin biosynthetic pathway of the *Pseudomonas syringae* B301D strain was found to have low levels of nitration activity (Figure 4(B)).

(A)

(B)

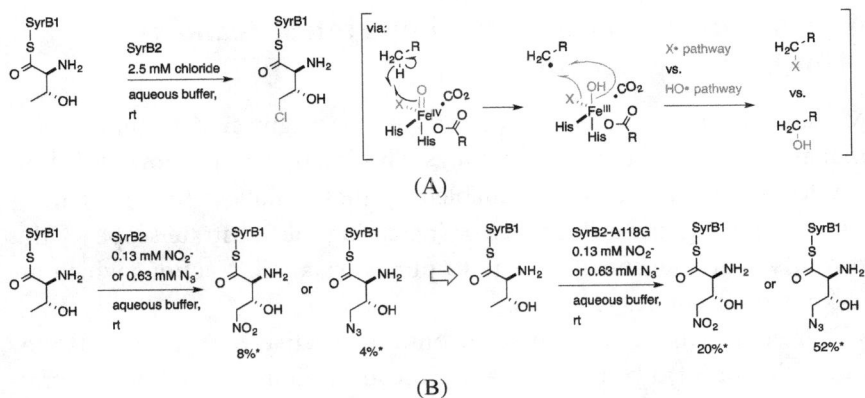

Figure 4: Halogenase activity expanded to azidation. (A) SyrB2 catalyses threonine chlorination, presented on the carrier protein SyrB1. The iron catalyst and reactive species are shown in brackets. (B) Expansion of activity to include amino incorporation. *Yield based on starting material consumption.[19]

The structure of SyrB2 had been previously solved by Drennan and Walsh, providing a basis for rational improvement of activity.[43] While chloride is a spherical ion, the azide ion is narrow and linear. Additional space to accommodate the unnatural ion was engineered by the Ala118Gly mutation, which expanded the active site in the suitable direction. SyrB2 Ala118Gly was found to have ca. 13-fold higher azidation activity compared to the natural, wild type enzyme.[19]

A significant limitation is the need for the substrate to be attached to the carrier protein SyrB1. Presumably this could be overcome by engineering a more accessible active site pocket. Additionally, the authors only report single-turnover conditions.[19] Despite these limitations, this type of amino-functionalisation reaction on unactivated C–H bonds is unprecedented in biology, and offers an exciting target for further enzyme engineering. The work on SyrB2 reveals that amine C–H functionalisation chemistry can be engineered from a naturally-occurring halogenase enzyme activity. This progress promises that further enzyme evolution work could overcome the current shortcomings of this system to provide valuable late-stage functionalisation reactions on biologically active molecules.

4 Incorporating Novel Metals into Protein Scaffolds for Expanded Activity

Metalloenzyme scaffolds can alternatively be engineered to bind metals that are not utilised in natural settings. This is a potentially powerful strategy to expand the chemistry enabled by these catalysts further than by simply allowing higher turnover or increasing the substrate scope — this comprises an additional strategy to pursue catalytic reactions which are unknown for natural enzymes.[10]

Proteins can be engineered to bind alternative transition metals by modifying or introducing compatible metal-binding motifs. For this strategy, both the metal and the biological framework in question need to be considered for mutual compatibility. Here we discuss both aspects, beginning with how new metals are incorporated within proteins. Throughout the design process, the protein scaffold's production capability, structural stability toward modifications toward metal binding, and ability to withstand putative reaction conditions must be evaluated.

Transition metal ions can be incorporated into the protein scaffold using several motifs, which can be categorised into three groups: direct covalent incorporation, a "tag-and-modify" approach, or through non-covalent interactions (Figure 5). The strategies used depend on the requirements for metal binding, i.e. if standard amino acids can form a suitable ligand framework, whether an unnatural motif must be introduced, or if the active site has evolved for binding a class of metal complexes.

In the first strategy, metals may be directly coordinated to the protein through backbone amino acid residues. These are either natural metal-binding amino acids such as cysteine or histidine residues, or may be unnatural amino acids incorporated by mutagenesis.[44] The geometry of the metal binding site is critical for affinity and selectivity between transition metals. When starting with existing metalloenzymes, the geometry of the metal-binding site can be retained, the metal removed, and an exogenous metal added.[45] An example is human carbonic anhydrase, which naturally binds $Zn(II)$. The zinc was removed with a chelating group and replaced with $Mn(II)$ or $Rh(I)$. These constructs were used to catalyse an enantioselective alkene epoxidation[46] or the regioselective hydroformylation

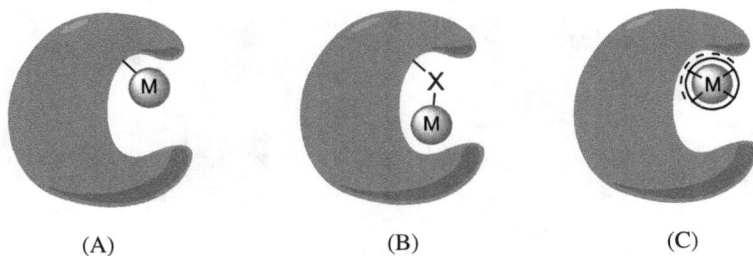

(A)   (B)   (C)

Figure 5: Different methods for metal insertion in artificial metalloenzymes. (A) The metal is directly covalently bound to amino acids in the protein. (B) The protein scaffold and the metal are "bridged" by a chemical modification step ("tag-and-modify"). (C) The metal ion or complex is bound by non-covalent interactions.

of styrene, respectively.[47] Often, however, the geometry must be re-optimised using unnatural amino acids or computational techniques (discussed in Sec. 5). The use of non-canonical amino acids can expand the repertoire of possible metal-coordinating groups.[44] These amino acid homologs include selenocysteine or a bipyridyl tyrosine mimic, and can be introduced using microbiological methods.[48]

The second strategy increases the complexity of the incorporation methodology, but expands the range of possible metal-binding motifs. A tag-and-modify sequence utilises the chemical reactivity or lability of a subset of amino acids. These are selectively modified in a chemical step following the biological production of the protein scaffold. These chemically-modified proteins are then incubated with the metal to be incorporated. An example of this approach with an unnatural amino acid is the incorporation of azidohomoalanine, the azide of which can be functionalised using azide-alkyne 1,3-dipolar cycloaddition chemistry ("click" chemistry).[49,50] Natural amino acids used in this approach are cysteine[50] or lysine[51] groups, which can be alkylated with various electrophiles. A key benefit to this approach is the ability to screen complex groups (Figure 6) that can be quickly exchanged, since this approach is modular. However, the scaffold must withstand the additional modification conditions utilised.

In the third approach, the metal ion, often with additional substituents such as a ligand framework, is coordinated to the scaffold non-covalently.

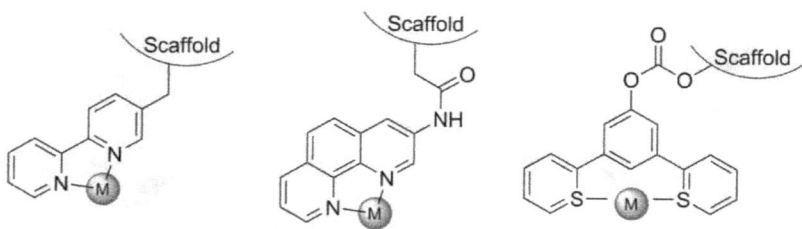

Figure 6: Example of metal-binding unnatural amino acid ligands, capable of binding a metal ion. Incorporated into the protein scaffold via mutagenesis and/or chemical modification.

Here, the use of covalent scaffold modification techniques is eliminated, but it is limited to proteins that already bind specific metal complexes. In the correct application, this can be less challenging than re-optimising amino acid-based metal binding geometries in a protein scaffold.[45,52] Heme proteins such as cytochromes P450 are often utilised for this purpose, since the large heme co-factor can be removed (to provide the apo-heme enzyme) and replaced, for instance with other metalloporphyins.[53] Another non-covalent interaction that has found use for this approach is the biotin-strepavidin ligand-receptor system.[54] Here, streptavidin is a non-catalytic receptor protein that can tightly bind chemical material attached to its ligand, biotin. Thus, a metal-chelating group can be docked to the streptavidin protein cavity. This approach was pioneered in 1978 by Whitesides and Wilson[55] and has since been used to successfully introduce a large number of complexes into protein scaffolds. Using this technology, Ward and co-workers have created a number of artificial hydrogenases, transfer hydrogenases, and artificial enzymes for allylic alkylation reactions.[54] As a powerful example, this system was used to create at Rhodium-based hydrogenase. Through further chemical and genetic optimisation, enantioselectivities of up to 94% were obtained.[56]

Considerations when choosing the protein scaffold involve the ease of expressing and purifying the protein, its stability toward the reaction conditions, and the possibility of retaining an active conformation after mutation and or chemical modification. In addition, the affinity of the metal complex must be sufficient and selective enough to avoid complex dissociation or multiple binding events. The natural reaction is of lower importance, since this activity is often abolished by the removal of the metal and modification to the backbone sequence. However, the numbers of available proteins are in practice limited for alternative metal incorporation. The active site needs to be large enough to fit both the metal-binding ligand complex and also its substrates. Therefore, this approach has been utilised mostly for hydrolases or proteases, β-barrel proteins and apo-heme proteins, which all can have large active site volumes.[57]

Though challenging, the use of alternative transition metals in a biological setting holds excellent potential promise for expanding the scope of metalloenzymatic reactions. To date, artificial metalloenzymes have been successfully utilised in a number of reaction classes; particularly oxidation (sulfoxidation, epoxidation) and reduction (hydrogenation) reactions,[54,58,59] alongside C–C bond forming reactions such as allylic alkylation and Diels–Alder reactions.[54,60]

There are expected to be some limitations on the reactions that can be conducted using artificial metalloenzymes, for example those reactions involving reagents incompatible with water. However, the solvent-shielded active sites of some protein scaffolds have allowed the pursuit of reactions in which the intermediates would be considered water sensitive. A recent example is the development of a dirhodium artificial metalloenzyme for selective olefin cyclopropanation.[61] The dirhodium-binding moiety was introduced to active sites using azide-alkyne "click chemistry." Notably, this artificial metalloenzyme was found to reduce the formation of by-products, including those resulting from the reaction of the metallocarbene intermediate with water. Thus, the protein scaffold improved the water tolerance of the reaction.

In many of these cases, metal environments similar to those used in traditional homogeneous chemistry were introduced to the protein environment via the methods described before. Optimisation using genetic tools such as directed evolution and rational design could then be used to obtain constructs which exhibit useful selectivities. Despite this possibility, only a few of the reported reactions were extensively subjected to scaffold optimisation, due to the substantial time investment and the challenge of manipulating covalently-bound metal complexes. Thus, this field has significant potential for improvement as methods toward metal incorporation progress.

Many other reactions still remain to be investigated in the context of artificial metalloenzymes. Whilst Rh and Ir-catalyzed reactions have been studied by a number of groups, the reactions of the neighbouring transition metals Pd and Pt have not yet been extensively studied. The following section discusses a potential strategy for developing a palladium-utilising metalloenzyme for cross-coupling reactions.

4.1 Opinion: C–C Cross-coupling with artificial palladium metalloenzymes

Palladium carbon–carbon cross–coupling reactions pose an example for which an artificial metalloenzyme system could be designed. Carbon–carbon bond formation via palladium-catalysed cross-coupling reactions such as the Heck-reaction[62] and the Suzuki–Miyaura-reaction[63] are essential tools in fine chemical production in the pharmaceutical and chemical industry. An artificial palladoenzyme could improve cross-coupling reactions by directing a reaction's stereo and regiochemistry. Palladium-catalysed reactions have been carried out in one-pot cascades alongside biocatalysts,[64] and palladium has been added to protein scaffolds both via the "tag-and-modify" approach[65] and using the biotin/streptavidin systems.[66] In the latter approach, the resulting artificial metalloenzymes were successful in catalysing asymmetric allylic alkylation. However, this metal has not been incorporated into a protein active site that retains catalytic activity for transition metal-mediated cross-coupling reactions.

In a parallel effort toward this goal, a palladium–protein cluster was synthesised through soaking free palladium into a large protein complex.[67] The cluster effectively catalysed both Heck and Suzuki coupling reactions, but activity as attributed to the bound palladium surface rather than within a designed active site. Similarly, the protein cage structure of apo-ferritin, a carrier protein for insoluble iron ions, was used as a scaffold to bind palladium ions.[68] This palladium–protein construct was also able to catalyse a Suzuki reaction. However, the role of the protein in both of these reactions was simply as a metal ion carrier rather than a complex active site, and did not take an active role in directing the catalytic reaction.

One strategy that is envisioned to accomplish the design of a true palladium-utilising artificial metalloenzyme is illustrated in Figure 7.

(A)

Bisubstrate localisation

(B)

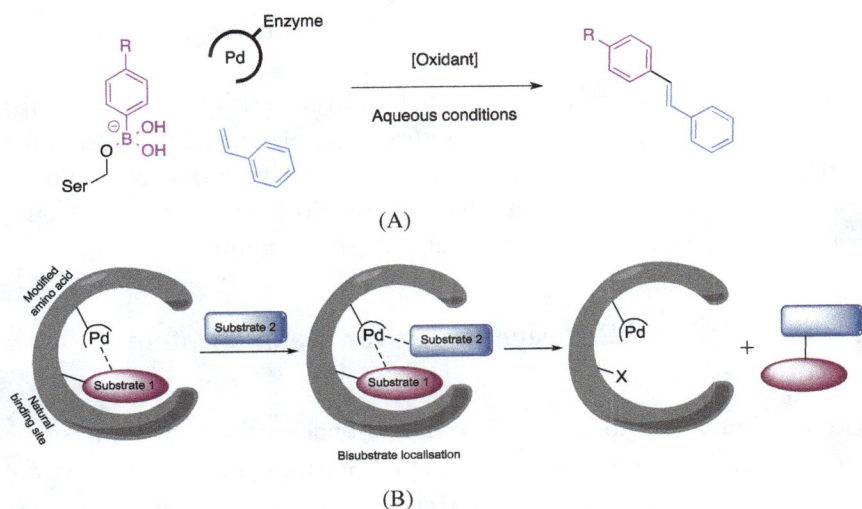

Figure 7:   A strategy for an artificial palladoenzyme. (A) The catalytic oxidative Heck-type reaction. (B) Schematic drawing of an artificial metalloenzyme incorporating the non-natural palladium ion. Substrate 1 binds in a reversible covalent manner, anchoring it in close proximity to the Pd metal. Diffusion of substrate 2 allows the formation of the cross-coupled product.

A potentially useful protein scaffold is the serine hydrolase family of enzymes. Though serine hydrolases are normally not metalloenzymes, they have two conserved features in their active sites that could be useful in directing a palladium-catalysed cross coupling reaction: the key serine residue and an adjacent oxyanion hole. The serine residue forms a charged tetrahedral intermediate with its carbonyl substrates, stabilised by the oxyanion hole, a small polar pocket in the active site that reversibly accommodates negatively-charged serine-bound adducts.[69] These two features are envisioned to bind boronic acids in a reversible covalent manner, localising them near the Pd metal for catalysis (Figure 7(B)).

To incorporate palladium into an active site that does not naturally contain a metal, a tag-and-modify approach is envisioned.[70] The scaffold can be mutated to incorporate a cysteine residue in the active site, enabling the attachment of a metal-coordinating ligand. This tag-and-modify approach has been previously applied to a serine hydrolase from *Bacillus lentus*.[71] This cysteine could be modified by covalently attaching a metal-binding ligand via nucleophilic attack on a halide-containing ligand (refer to Figure 5).

This type of design sequence illustrates the exciting possibilities for using and modifying protein architecture to enable powerful reactions for organic synthesis. We envision this type of biocatalyst redesign to be both feasible and useful to direct the stereo and regiochemistry of challenging reactions, in this case palladium-catalysed cross coupling reactions.

5 Computational Redesign Efforts for Metalloprotein Active Sites

The continued development of computational methods to understand protein structure and function is one of the most promising areas for improving enzyme activities.[33] The ability to align and compare sequences and structural information is widely used to identify potentially useful biocatalysts for given reactions, especially as improvements in DNA sequencing leads to larger biological datasets to draw from.[72] We focus here on strategies useful for the forward design of artificial metalloenzymes.

Structural biology is able to provide complete data about the architecture of proteins, which is compiled in the RCSB Protein Data Bank (PDB).[73] A subset of the PDB has been focused into MetalPDB, which simplifies the analysis of how metals are incorporated into biological sites.[74] Several innovative programs are able to utilise these structural resources to quickly identify useful scaffolds. One example is METAL-SEARCH, which probes scaffolds for possible metal-binding cavities and suggests affinity improvements by replacing natural amino acids with metal-coordinating residues such as histidine or cysteine.[75]

In reaction design, the multidisciplinary nature of biocatalysis benefits greatly from the application of diverse, complementary computational methods. Powerful results arise from the combination of density functional theory (DFT) and molecular modelling approaches to identify reactions' transition states with whole-protein conformation analysis efforts, in order to facilitate the formation of candidate transition states by designing active site complementarity.[76] Software suites such as RosettaDesign and ORBIT combine both molecular modelling and *de novo* protein structural prediction modules, and have led to improved workflows for the redesign of enzyme active sites for novel activities.[77] Further advancements in computational packages include the use of non-canonical amino acids, increasing the utility of Rosetta for artificial metalloenzyme development.[78] Intriguingly, this software has also been used to widen public participation in science through the use of Rosetta@Home, a Berkeley Open Infrastructure for Network Computing project,[79] which allows the idle computing time of home computers to contribute to protein folding computations.

In order to obtain the desired efficiency with novel enzymes, an iterative approach is typically utilised. These strategies cycle computational techniques such as active site redesign suites or molecular dynamics with the production and testing of candidate protein sequences. This approach can be applied to other biocatalyst systems, but is especially amenable for metalloenzyme redesign since modelling the metal ion-transition state complex formed the core of the approach. This reveals the power of computational methodologies such as RosettaDesign[80] for identifying initial starting points for designed enzyme activity. However, the need for non-rational improvement, based on random substitution of amino acids via directed evolution,[24] shows that our abilities for computational redesign face limitations when calculating the overall conformation and function of large protein architectures. Nevertheless, computational strategies are continually improving, and comprise an essential resource for biocatalyst development.[76]

Several recent reviews further describe the advances made in computational modelling of proteins and computational design.[81–83] To directly illustrate this concept, Case Study 3 details the re-engineering of a metalloenzyme for catalysis of an alternative reaction.

(C) Case Study 3: Zinc enzyme re-engineering toward an alternative reaction

Developing new catalytic activity toward inert substrates in a metalloprotein active site is conceptually more difficult than enhancing low levels of promiscuous enzyme activity as discussed in Sec. 3. As in the previous examples, analogies between enzyme and solution-phase chemistry can greatly enable reaching these aims. Zinc catalysts are used in phosphoester cleavage reactions, both in[84] and outside[85] of biological systems.

Introducing phosphohydrolase activity into a zinc-binding protein scaffold was a challenge undertaken by the Tawfik and Baker groups.[21] Zinc is a common metal in proteins and has diverse forms of reactivity, including Lewis acidity, general basicity, or hydroxide formation from water, depending on the environment.[86] Baker and Tawfik saw an opportunity to reengineer the substrate binding sites of zinc-containing proteins to accommodate aryl-substituted phosphates and catalyse their cleavage. Aryl phosphates are potent nerve toxins, and these enzymes were proposed to be useful in bioremediation or detoxification.

The approach centred on the structure of the transition state of this organophosphate hydrolysis reaction using the RosettaDesign program.[87,88] This analysis suggested the key structural requirements for substrate geometry, which facilitated the design of an artificial active site (Figure 8(A)). The hypothetical structure of this idealised active site pocket, called a "theozyme", was then aligned against the enzyme architectures of all zinc-binding proteins extracted from the PDB. This generated a series of putative redesigned proteins, incorporating as many as 15 simultaneous mutations in the active site region. These synthetic gene sequences encoding these scaffolds were then purchased and the proteins expressed for activity studies.

Only one of these scaffolds, termed PT3, proved active. This was based on a murine adenosine deaminase backbone.[89] In the natural adenosine

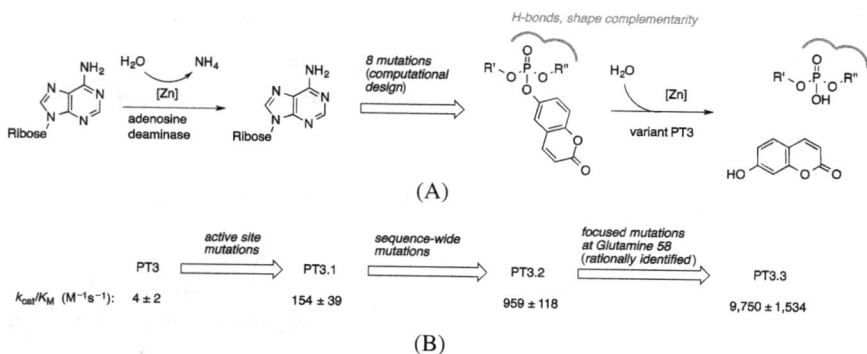

(A)

(B)

Figure 8: Redesign of adenosine deaminase toward organophosphate hydrolysis. (A) Natural and designed reactions achieved with active site redesign. (B) Directed evolution sequence to improve new enzyme activity.

deaminase, the zinc ion serves as both a Lewis acid toward the amine and an activator of a hydroxide nucleophile for the removal of the aniline moiety of adenosine. Molecular dynamics were used to redesign the active site into PT3.1, in which aniline deaminase activity was diminished 50,000-fold whereas organophosphate hydrolysis was enhanced 7-fold over the natural enzyme (Figure 8(A)). Further activity was obtained through two rounds of directed evolution aimed at increasing turnover.[90] The final variant, PT3.3, had organophosphate hydrolysis activity of 2,500-fold over baseline, and undetectable adenosine deaminase activity (Figure 8(B)).

6 New Protein Structural Motifs in Catalysis

Until now, we have discussed targeting natural protein sequences for redesign efforts. An alternative is the field of *de novo* structural design, a bottom-up approach to the synthesis of new protein structural domains from first principles. *De novo* design could enable precise control of peptide structures designed around specific metals, ligands, or properties such as stability. Furthermore, advances in this field can greatly expand knowledge of protein folding and conformational dynamics.[91] Researchers can thus effectively probe the role of each facet of the protein scaffold on catalysis, moving forward from biomimetic studies to protein folds that are designed specifically for abiotic chemical catalysis.

Since even small proteins, such as a 40 amino acids/~4400 Da peptide, are extremely conformationally complex, the computational power needed to screen libraries of potential sequences is expensive. Therefore, the most successful *de novo* approaches are based on reconfiguring natural backbone structures to provide a starting point for sequence exploration. To date, there is only a single protein, Top7, with a completely novel protein fold.[92] However, Top7 lacks functional capabilities, which simplified the modelling of its free-energy landscape, as allowing dynamic motions for catalysis or ligand binding were unnecessary in the design stage.

Thus, this approach is in practice still limited to the small subset of peptides whose folding properties have been characterised.[93,94] Typically, α-helical peptides such as α-helical coiled coils are utilised, but several examples focus on all β-sheet structures such as porin domains.[95] The majority of work has focused on mimicking heme and non-heme iron enzymes using α-helical single chain peptides or three- and four-helix bundles. The use of α-helices is attractive as the construction is well understood and can be achieved simply by using a repeating seven-residue pattern of polar and non-polar residues. Helix formation can be driven by providing a hydrophobic core, which can be designed computationally.

In the following case studies, we have focused on examples utilising α-helices, which hold higher promise for the catalytic chemistry community. Case Study 4 focuses on the ground-up design of redox catalysts based on heme systems, a discrete iron-binding cofactor commonly used in biology. The small size of these peptides places this class of catalysts into an interesting space between homogenous and biocatalysis.

Case Study 5 details the design of simplified protein mimics of complex dimetallic enzyme active sites (the Duo Ferri (DF) family), which show either phenol oxidation or *N*-hydroxylation activities. The efforts on (DF) protein design illustrate that complex biological motifs such as a dimetallic active site can be reconstructed in simplified form (further discussed later). Computational tools were again able to design and improve an artificial metalloenzyme active site for two different catalytic oxidation reactions. Furthermore, substrate activity could be controlled by the residues surrounding the di-iron site, suggesting that the substrate scope of these proteins can be controlled by design.

The work described in Case Studies 4 and 5 uses natural amino-acids and cofactors, but just as we saw in Sec. 4, unnatural co-factors and unnatural amino acids have also been used.[96,97] This results in artificial peptides with redox properties different to those of the natural cofactors, showing that it is possible to fine-tune the properties of the protein and possibly access different reactivity patterns than those observed in native systems.

Alongside the work on iron, α-helices have been used in the design of zinc and copper enzymes.[98,99] Notable examples include the development of the dizinc protein MID1-zinc[100] and the TRI family of peptides,[101] which both show esterase activity. The TRI peptides have been shown to be very versatile, also catalysing CO_2 hydration and the reduction of nitrite.[102–104] The *de novo* design of functional enzymes is not just limited to those for chemical synthesis, artificial enzymes for energy storage and transport, biochemical sensing, and for biological functions have been developed.[105–107] Dutton and co-workers have used alpha helices to create an O_2 transport protein.[108] Whilst Ghirlanda and co-workers described an artificial metalloenzyme which was able to interact with natural enzymes to effectively participate in redox activity.[109]

From the work presented, we can see that the design of novel protein structures is a strategy that can take full advantage of transition metal-promoted reactivity in both catalysis and catalysis related areas. The smaller size of these catalysts facilitate both the capacity to model and understand the interplay of their structure with function. Mainly due to limitations in our ability to fully understand peptide folding and conformational dynamics, these artificial catalysts do not yet approach the efficiencies of natural metalloenzymes. However, they allow access to novel protein sequence space, termed "post-evolution biology" by Baker.[91] This field is rich in possibilities for designing tailored peptide catalysts.

(D) Case Study 4: *De novo* mimics of heme proteins

As mentioned in Sec. 3, metalloporphyrin systems are active catalysts in both biological and abiological settings. Furthermore, their rigid structure can comprise a defined starting point for a small protein core. Some of the earliest *de novo* protein designs with catalytic activity were based on the creation of artificial heme active sites. In 1989, Sasaki and Kaiser described the synthesis and enzymatic activity of Helichrome, an artificial heme protein.[110] Helichrome was designed around a porphyrin scaffold, with amphiphilic α-helices attached at each corner. This provided a hydrophobic pocket for substrate binding. Catalytic studies showed that this construct was capable of acting as an aniline hydroxylase with similar activities to those observed in native heme proteins (Figure 9).

These systems were used as starting points for alternative *de novo* designed structures. The mimochromes from the groups of Lombardi and Pavone represent a more advanced example.[111,112] Early models consisted of a helix–heme–helix structure, in which the heme was sandwiched between histidines on each helix (Figure 10(A)). Consequently, no catalytic activity was possible in these systems due to the lack of a free coordination site for the substrate. In mimochrome VI (MC6), one of the His residues was removed to open up a free coordination site on iron (see Figures 10(B) and 10(C)).[113]

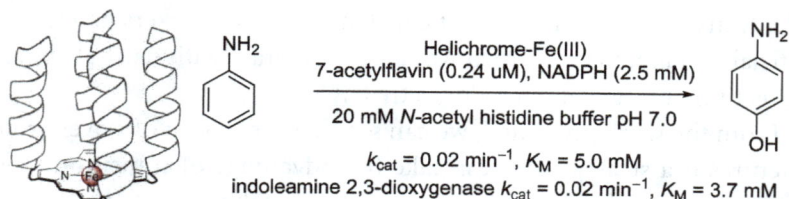

Figure 9: Depiction of Helichrome and its activity in the hydroxylation of aniline. Indoleamine-2,3-dioxygenase, a natural heme protein, is listed for comparison. Adapted with permission from Ref. [110]. Copyright 1989 American Chemical Society.

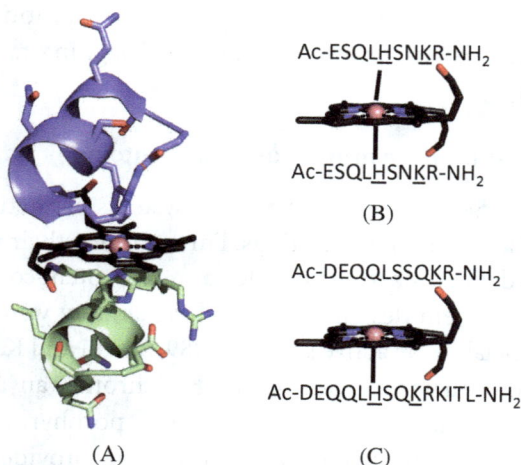

Figure 10: (A) Depiction of Co(III)-Mimochrome IV based on its crystal structure (PDB code 1PYZ).[116] (B) Schematic representation of mimochrome IV including amino acid sequence. (C) Schematic representation of mimochrome VI including amino acid sequence. Note the open coordination site for the metal, allowing catalysis. Key amino acid residues are underlined.

The catalytic activity of mimochrome VI was assessed in a colorimetric peroxidase assay, using H_2O_2 with 2,2′-azinodi(3-ethyl-benzothiazoline-6-sulfonic acid (ABTS). The presence of 2,2,2-trifluoroethanol (TFE) stabilised the mimochrome scaffold, resulting in turnover numbers of more than 4,000 for the oxidation of ABTS vs. around 150 in the absence of TFE. Mimochrome activity could be further improved by changing the N- and C-terminal residues on the tetradecapeptide chain to a hydrophobic

Table 2: Kinetic parameters for the oxidation of ABTS.

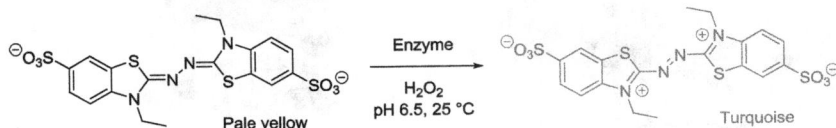

Enzyme	k_{cat} (s^{-1})	K_M (mM)	k_{cat}/K_m (mM^{-1}s^{-1})	TON
MC6[a,c]	370	0.084	4400	4000
E²L(TD)-MC6[a,c]	780	0.050	16000	5900
Fe(III)-MP3[a,d]	535	0.34	1570	>1000
HRP[b,c]	6200	1.07	5800	50000

Notes: (a) TFE added. (b) pH 4.6. (c) Data from Ref. [114]. (d) Data from Ref. [115].

leucine residue, resulting in turnovers of almost 6,000 (Table 2).[114] The catalytic efficiency for the activation of H$_2$O$_2$ ($k_{cat}/K_m = 25$ mM^{-1}s^{-1} for E²L(TD)-MC6), although still several orders of magnitude lower than that of Horse Radish Peroxidase (HRP, $k_{cat}/K_m = 7.3 \times 10^3$ mM^{-1}s^{-1}), is among one of the highest reported among synthetic mimics of peroxidase. It is also notable that the pH optimum for mimochrome VI is close to neutral pH (pH 6.5), showing that they can catalyse reactions that require higher pH conditions than HRP can provide. Besides peroxidase activity, mimochome VI has also been shown to catalyse the nitration of phenol using NaNO$_2$/H$_2$O$_2$ to give both 2- and 4-nitrophenol in about 15% yield.[113]

MiniPeroxidase 3 (MP3) was designed to mimic HRP, and has more functionality then those seen for the mimochromes.[115] Importantly, it was designed to mimic not just the axial residue but also its hydrogen bonded aspartate residue, which stabilises high-oxidation state intermediates, and the distal site arginine which is involved in O–O peroxide bond cleavage (Figures 11(B) and 11(C)). Once again, the catalytic activity was assessed in the oxidation of ABTS with H$_2$O$_2$. The Fe^{III}-MP3 shows a higher k_{cat} than mimochrome VI but a lower number of turnovers (Table 2).

(E) Case Study 5: Design of dimetallic α-helical bundles

α-Helices have also been used as a scaffold for another class of *de novo* designed metalloproteins, those containing a di-metal site. The α-helix-rich di-metal binding sites observed enzymes such as bacterioferritin

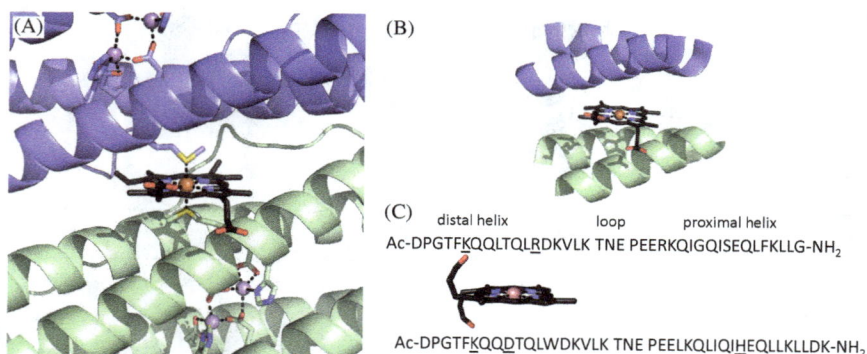

Figure 11: (A) Representation of the metal sites in the dimer sub-unit of bacterioferritin (PDB code 1BFR)[117] showing both the heme group and di-Mn site. (B) Starting point taken from BFR and used to design MP3. (C) Schematic representation of MP3 including amino acid sequence. Key amino acid residues are underlined. Images modified from Ref. [115]. Adapted with permission from WILEY-VCH Verlag GmbH & Co. KGaA, Weinheim, 2012.

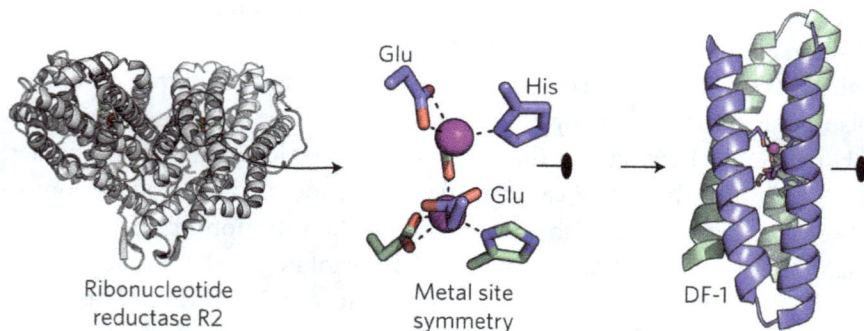

Figure 12: Design strategy of dinuclear metallopeptide DF-1, using natural metalloenzymes such as ribonucleotide reductase as inspiration. Reprinted by permission from Macmillan Publishers Ltd: *Nature Chemistry* (Ref. 81), copyright 2009.

and ribonucleotide reductase have been used as inspiration (Figure 12).[118] The so-called DF class of peptides were designed as mimics of natural di-iron metalloenzymes. The initial design was a peptide homodimer of helix-loop-helix structures, with each monomer containing two glutamate and one histidine residue for iron chelation.[119] Thus, where the dimers interfaced a Glu_4His_2 dimetallic binding site formed, as shown in Figure 12.[120]

This design was subsequently modified to improve solubility in aqueous solvents and increase substrate access to the active site. This was achieved by replacing the leucine residues at positions 7, 11, 21, 26 and 47 with more hydrophilic residues, and replacing a bulky set of leucine residues (Leu13 and its mirror across the dimer interface) with smaller alanine residues, giving DF2 and Leu13Ala-DF1.[121,122] This led to stabilisation of the protein with Zn(II), Co(II), Cu(II), Fe(II) and Mn(II). Di-Fe(II)-DF2 rapidly reacts with oxygen to give rise to the diferric-oxo units (Fe(III)-O-Fe(III)), a mimic of natural di-iron proteins. This complex was successfully probed with azide and acetate, indicating that small molecules were able to access the active site.

Using the knowledge gained from these initial studies into the structure and activity of the DF series of proteins, the authors were able to design artificial proteins showing phenol oxidase activity by increasing the cavity size for substrate binding. Initial catalytic variants were tetramers of the type referred to as DF_{tet} $A_aA_bB_2$ or DF_{tet} A_2B_2, which were identified using a combinatorial approach.[123] The most active variant was G_4-DF_{tet}, a symmetrical construct that contained the largest active site pocket of all proteins studied. This variant showed a 1000-fold rate enhancement in the oxidation of 4-aminophenol relative to the background reaction, giving rise to over 100 turnovers.

The authors then introduced the successful design features of G_4-DF_{tet} to the dimer DF1. This resulted in DF3, which was found to catalyse the oxidation of a number of substrates including the bulky 3,5-ditert-butyl-catechol, 4-aminophenol and *para*-phenylendiamine.[124] A similar strategy was applied to a single chain variant of the DF family, DFsc. Changes in the active site switched the activity of this diiron protein from aminophenol oxidation to arylamine *N*-hydroxylation.[125] The introduction of four smaller glycine residues improved substrate access.

7 Conclusion and Outlook

Biocatalysis with artificial metalloenzymes is a strategy that could potentially improve the selectivity, efficiency, and environmental impact for a wide range of chemical applications. Their challenging design has fostered the development of several key strategies that utilise theoretical and experimental techniques, bridging the biological and chemical interface.

These examples of catalysis-driven biological design illustrate that natural enzymes can be repurposed in a multitude of ways. Existing metal-binding active sites can be redesigned not only for higher activity, but also to access modes of reactivity not seen in nature, such as enantioselective carbene- and nitrene-based C–H functionalisation. Alternative metals can be incorporated into metalloproteins to change their chemical capabilities, and active sites can be computationally redesigned to accommodate novel transformations. These developments are excellent early steps towards the use of artificial metalloenzymes in a wide variety of settings.

Several emergent areas show future promise for novel biocatalytic reaction development. We suggest a possible strategy toward incorporating a wholly unknown metal in biology to create a protein biocatalyst for palladium-mediated cross coupling reactions. The validation of computational techniques to build new active sites around the transition state of a desired reaction is a powerful way to approach novel biocatalysts. Finally, *de novo* approaches to build structural features from first principles is an exciting way to devise biocatalyst-like entities with greater control over their shape and dynamics. Moving forward, we can expect that these approaches will be used together to build completely new biocatalysts, with *de novo* designed proteins being purpose-built around unnatural metal complexes. These constructs could then be optimised by the routes shown in Sec. 3, which are already used regularly in the biocatalytic industry.

The continuous development and application of these strategies show that these approaches are indeed useful for the design of new biocatalysts. Coupled with the recent expansion of biocatalysis in industry,[16] we expect a strong outlook for the catalysis-driven design of biological systems.

Acknowledgments

CF, ME and AGJ thank the UK Catalysis Hub for resources and support provided via our membership of the UK Catalysis Hub Consortium and funded by the EPSRC (portfolio grants EP/K014706/1, EP/K014668/1, EP/K014854/1, EP/K014714/1) AGJ also thanks the EPSRC through EP/J018138/1.

References

1. P. S. Coelho, F. H. Arnold and J. C. Lewis, in *Comprehensive Organic Synthesis II (2nd Edition)*, ed. P. Knochel, Elsevier, Amsterdam, 2014, pp. 390–420.
2. U. T. Bornscheuer, G. W. Huisman, R. J. Kazlauskas, S. Lutz, J. C. Moore and K. Robins, *Nature*, 2012, **485**, 185–194.
3. D. S. Tawfik, *Curr. Opin. Chem. Biol.*, 2014, **21**, 73–80.
4. K. C. Chou and G. P. Zhou, *J. Am. Chem. Soc.*, 1982, **104**, 1409–1413.
5. G. W. Huisman, in *Comprehensive Organic Synthesis II (2nd Edition)*, ed. P. Knochel, Elsevier, Amsterdam, 2014, pp. 421–437.
6. Q.-P. Zeng, L.-X. Zeng, W.-J. Lu, L.-L. Feng, R.-Y. Yang and F. Qiu, *Biocatal. Biotransform.*, 2012, **30**, 190–202.
7. S. K. Ma, J. Gruber, C. Davis, L. Newman, D. Gray, A. Wang, J. Grate, G. W. Huisman and R. A. Sheldon, *Green Chem.*, 2010, **12**, 81–86.
8. A. Robles-Medina, P. A. Gonzalez-Moreno, L. Esteban-Cerdan and E. Molina-Grima, *Biotechnol. Adv.*, 2009, **27**, 398–408.
9. R. Bates, *Organic Synthesis Using Transition Metals*, 2nd Edition, Wiley, 2012.
10. I. D. Petrik, J. Liu and Y. Lu, *Curr. Opin. Chem. Biol.*, 2014, **19**, 67–75.
11. H. Renata, Z. J. Wang and F. H. Arnold, *Angew. Chem. Int. Ed.*, 2015, **54**, 3351–3367.
12. J. B. Behrendorff, W. Huang and E. M. Gillam, *Biochem. J.*, 2015, **467**, 1–15.
13. V. Puchart, *Biotechnol. Adv.*, 2015, **33**, 261–276.
14. Z. Chen, B. Wang, J. Zhang, W. Yu, Z. Liu and Y. Zhang, *Org. Chem. Front.*, 2015, **2**, 1107–1295.
15. C. E. Valdez, Q. A. Smith, M. R. Nechay and A. N. Alexandrova, *Acc. Chem. Res.*, 2014, **47**, 3110–3117.
16. P. W. Sutton, J. P. Adams, I. Archer, D. Auriol, M. Avi, C. Branneby, A. J. Collis, B. Dumas, T. Eckrich, I. Fotheringham, R. ter Halle, S. Hanlon, M. Hansen, K. E. Holt-Tiffin, R. M. Howard, G. W. Huisman, H. Iding, K. Kiewel, M. Kittelmann, E. Kupfer, K. Laumen, F. Lefèvre, S. Luetz, D. P. Mangan, V. A. Martin, H.-P. Meyer, T. S. Moody, A. Osorio-Lozada, K. Robins, R. Snajdrova, M. D. Truppo, A. Wells, B. Wirz and J. W. Wong, in *Practical Methods for Biocatalysis and Biotransformations 2*, Hoboken John Wiley & Sons, Ltd., 2012, pp. 1–59.
17. F. M. Szczebara, C. Chandelier, C. Villeret, A. Masurel, S. Bourot, C. Duport, S. Blanchard, A. Groisillier, E. Testet, P. Costaglioli, G. Cauet, E. Degryse, D. Balbuena, J. Winter, T. Achstetter, R. Spagnoli, D. Pompon and B. Dumas, *Nat. Biotech*, 2003, **21**, 143–149.
18. E. McNeill and J. D. Bois, *J. Am. Chem. Soc.*, 2010, **132**, 10202–10204.

19. M. L. Matthews, W. C. Chang, A. P. Layne, L. A. Miles, C. Krebs and J. M. Bollinger, Jr., *Nat. Chem. Biol.*, 2014, **10**, 209–215.
20. W. Liu and J. T. Groves, *J. Am. Chem. Soc.*, 2010, **132**, 12847–12849.
21. S. D. Khare, Y. Kipnis, P. Greisen, Jr., R. Takeuchi, Y. Ashani, M. Goldsmith, Y. Song, J. L. Gallaher, I. Silman, H. Leader, J. L. Sussman, B. L. Stoddard, D. S. Tawfik and D. Baker, *Nat. Chem. Biol.*, 2012, **8**, 294–300.
22. T. K. Tetsuro Koreeda, Fumitoshi Kakiuchi, *J. Organomet. Chem.*, 2013, **741–742**, 148–152.
23. N. M. DeVore and E. E. Scott, *Nature*, 2012, **482**, 116–119.
24. N. J. Turner, *Nat. Chem. Biol.*, 2009, **5**, 567–573.
25. J. Hagen, in *Industrial Catalysis*, Wiley-VCH Verlag GmbH & Co. KGaA, 2006, DOI: 10.1002/3527607684.ch4, pp. 83–98.
26. J. Hagen, in *Industrial Catalysis*, Wiley-VCH Verlag GmbH & Co. KGaA, 2006, DOI: 10.1002/3527607684.ch3, pp. 59–82.
27. P. Anastas and N. Eghbali, *Chem. Soc. Rev.*, 2010, **39**, 301–312.
28. C. A. Martinez and S. G. Rupashinghe, *Curr. Top. Med. Chem.*, 2013, **13**, 1470–1490.
29. A. Messerschmidt, W. Bode and M. Cygler, *Handbook of Metalloproteins, Handbook of Metalloproteins*, Wiley, Hoboken, 2004.
30. M. E. Guazzaroni, R. Silva-Rocha and R. J. Ward, *Microb. Biotechn.*, 2015, **8**, 52–64.
31. T. Tu, H. Luo, K. Meng, Y. Cheng, R. Ma, P. Shi, H. Huang, Y. Bai, Y. Wang, L. Zhang and B. Yao, *Appl. Environ. Microbiol.*, 2015, **81**, 6938–6944.
32. T. Kawata and H. Ogino, *Biochem. Biophys. Res. Commun.*, 2010, **400**, 384–388.
33. H. J. Wijma and D. B. Janssen, *FEBS J.*, 2013, **280**, 2948–2960.
34. J. A. McIntosh, C. C. Farwell and F. H. Arnold, *Curr. Opin. Chem. Biol.*, 2014, **19**, 126–134.
35. *Cytochrome P450: Structure, Mechanism, and Biochemistry*, Kluwer Academic/ Plenum Publishers, New York, 3rd Edition, 2005.
36. J. T. Groves, T. E. Nemo and R. S. Myers, *J. Am. Chem. Soc.*, 1979, **101**, 1032–1033.
37. R. Breslow and S. H. Gellman, *J. Am. Chem. Soc.*, 1983, **105**, 6728–6729.
38. E. W. Svastits, J. H. Dawson, R. Breslow and S. H. Gellman, *J. Am. Chem. Soc.*, 1985, **107**, 6427–6428.
39. P. S. Coelho, Z. J. Wang, M. E. Ener, S. A. Baril, A. Kannan, F. H. Arnold and E. M. Brustad, *Nat. Chem. Biol.*, 2013, **9**, 485–487.
40. A. W. Munro, D. G. Leys, K. J. McLean, K. R. Marshall, T. W. Ost, S. Daff, C. S. Miles, S. K. Chapman, D. A. Lysek, C. C. Moser, C. C. Page and P. L. Dutton, *Trends Biochem. Sci.*, 2002, **27**, 250–257.

41. C. C. Farwell, J. A. McIntosh, T. K. Hyster, Z. J. Wang and F. H. Arnold, *J. Am. Chem. Soc.*, 2014, **136**, 8766–8771.

42. Z. J. Wang, H. Renata, N. E. Peck, C. C. Farwell, P. S. Coelho and F. H. Arnold, *Angew. Chem. Int. Ed. Engl.*, 2014, **53**, 6810–6813.

43. L. C. Blasiak, F. H. Vaillancourt, C. T. Walsh and C. L. Drennan, *Nature*, 2006, **440**, 368–371.

44. J. Xie, W. Liu and P. G. Schultz, *Angew. Chem. Int. Ed.*, 2007, **46**, 9239–9242.

45. J. C. Lewis, *ACS Catalysis*, 2013, **3**, 2954–2975.

46. K. Okrasa and R. J. Kazlauskas, *Chem. Eur. J.*, 2006, **12**, 1587–1596.

47. Q. Jing and R. J. Kazlauskas, *Chem. Cat. Chem.*, 2010, **2**, 953–957.

48. P. O'Donoghue, J. Ling, Y.-S. Wang and D. Soll, *Nat. Chem. Biol.*, 2013, **9**, 594–598.

49. Craig S. McKay and M. G. Finn, *Chem. Biol.*, **21**, 1075–1101.

50. J. M. Chalker, G. J. L. Bernardes, Y. A. Lin and B. G. Davis, *Chem. Asian. J.*, 2009, **4**, 630–640.

51. J. M. McFarland and M. B. Francis, *J. Am. Chem. Soc.*, 2005, **127**, 13490–13491.

52. I. D. Petrik, J. Liu and Y. Lu, *Curr. Opin. Chem. Biol.*, 2014, **19**, 67–75.

53. L. Fruk, C.-H. Kuo, E. Torres and C. M. Niemeyer, *Angew. Chem. Int. Ed.*, 2009, **48**, 1550–1574.

54. T. R. Ward, *Acc. Chem. Res.*, 2011, **44**, 47–57.

55. M. E. Wilson and G. M. Whitesides, *J. Am. Chem. Soc.*, 1978, **100**, 306–307.

56. M. Creus and T. R. Ward, *Org. Biomol. Chem.*, 2007, **5**, 1835–1844.

57. T. Ueno, S. Abe, Y. Watanable, *Top. Organomet. Chem.*, 2009, **25**, 25–43.

58. M. V. Doble, A. C. Ward, P. J. Deuss, A. G. Jarvis and P. C. Kamer, *Bioorg. Med. Chem.*, 2014, **22**, 5657–5677.

59. T. R. Ward, *Acc. Chem. Res.*, 2011, **44**, 47–57.

60. P. J. Deuss, G. Popa, A. M. Z. Slawin, W. Laan and P. C. J. Kamer, *Chem. Cat. Chem.*, 2013, **5**, 1184–1191.

61. P. Srivastava, H. Yang, K. Ellis-Guardiola and J. C. Lewis, *Nat. Commun.*, 2015, **6**.

62. A. Biffis, M. Zecca and M. Basato, *J. Mol. Catal. A: Chem.*, 2001, **173**, 249–274.

63. S. Kotha, K. Lahiri and D. Kashinath, *Tetrahedron*, 2002, **58**, 9633–9695.

64. C. A. Denard, J. F. Hartwig and H. Zhao, *ACS Catalysis*, 2013, **3**, 2856–2864.

65. W. Laan, B. K. Muñoz, R. den Heeten and P. C. J. Kamer, *Chembiochem*, 2010, **11**, 1236–1239.

66. J. Pierron, C. Malan, M. Creus, J. Gradinaru, I. Hafner, A. Ivanova, A. Sardo and T. R. Ward, *Angew. Chem. Int. Ed.*, 2008, **47**, 701–705.

67. M. Filice, M. Marciello, M. d. P. Morales and J. M. Palomo, *Chem. Commun.*, 2013, **49**, 6876–6878.

68. S. Abe, J. Niemeyer, M. Abe, Y. Takezawa, T. Ueno, T. Hikage, G. Erker and Y. Watanabe, *J. American Chem. Soc.*, 2008, **130**, 10512–10514.

69. R. A. Powers and B. K. Shoichet, *J. Med. Chem.*, 2002, **45**, 3222–3234.

70. J. M. Chalker, G. J. L. Bernardes and B. G. Davis, *Acc. Chem. Res.*, 2011, **44**, 730–741.

71. P. Kuhn, M. Knapp, S. M. Soltis, G. Ganshaw, M. Thoene and R. Bott, *Biochemistry*, 1998, **37**, 13446–13452.

72. J. Kennedy, J. Marchesi and A. Dobson, *Microb. Cell Fact.*, 2008, **7**, 27.

73. H. M. Berman, J. Westbrook, Z. Feng, G. Gilliland, T. N. Bhat, H. Weissig, I. N. Shindyalov and P. E. Bourne, *Nucleic. Acids. Res.*, 2000, **28**, 235–242.

74. C. Andreini, G. Cavallaro, S. Lorenzini and A. Rosato, *Nucleic. Acids. Res.*, 2013, **41**, D312–D319.

75. N. D. Clarke and S. M. Yuan, *Proteins*, 1995, **23**, 256–263.

76. J. Damborsky and J. Brezovsky, *Curr. Opin. Chem. Biol.*, 2014, **19**, 8–16.

77. S. D. Khare and S. J. Fleishman, *FEBS Lett.*, 2013, **587**, 1147–1154.

78. P. D. Renfrew, E. J. Choi, R. Bonneau and B. Kuhlman, *PLoS One*, 2012, **7**, e32637.

79. R. K. V. Lim and Q. Lin, *Chem. Commun.*, 2010, **46**.

80. Y. Liu and B. Kuhlman, *Nucleic. Acids. Res.*, 2006, **34**, W235–W238.

81. V. Nanda and R. L. Koder, *Nat. Chem.*, 2010, **2**, 15–24.

82. M. K. Tiwari, R. Singh, R. K. Singh, I. W. Kim and J. K. Lee, *Comput. Struct. Biotechnol. J.*, 2012, **2**, e201209002.

83. J. Damborsky and J. Brezovsky, *Curr. Opin. Chem. Biol.*, 2014, **19**, 8–16.

84. P.-C. Tsai, A. Bigley, Y. Li, E. Ghanem, C. L. Cadieux, S. A. Kasten, T. E. Reeves, D. M. Cerasoli and F. M. Raushel, *Biochemistry (Mosc.)*, 2010, **49**, 7978–7987.

85. L. V. Penkova, A. Maciag, E. V. Rybak-Akimova, M. Haukka, V. A. Pavlenko, T. S. Iskenderov, H. Kozłowski, F. Meyer and I. O. Fritsky, *Inorg. Chem.*, 2009, **48**, 6960–6971.

86. K. A. McCall, C.-C. Huang and C. A. Fierke, *J. Nutr.*, 2000, **130**, 1437S–1446S.

87. A. Zanghellini, L. Jiang, A. M. Wollacott, G. Cheng, J. Meiler, E. A. Althoff, D. Rothlisberger and D. Baker, *Protein Sci.*, 2006, **15**, 2785–2794.

88. B. Kuhlman and D. Baker, *Proc. Natl. Acad. Sci. U.S.A.*, 2000, **97**, 10383–10388.

89. Z. Wang and F. A. Quiocho, *Biochemistry (Mosc.)*, 1998, **37**, 8314–8324.

90. R. E. Cobb, N. Sun and H. Zhao, *Methods*, 2013, **60**, 81–90.

91. D. N. Woolfson, G. J. Bartlett, A. J. Burton, J. W. Heal, A. Niitsu, A. R. Thomson and C. W. Wood, *Curr. Opin. Struct. Biol.*, 2015, **33**, 16–26.
92. B. Kuhlman, G. Dantas, G. C. Ireton, G. Varani, B. L. Stoddard and D. Baker, *Science*, 2003, **302**, 1364–1368.
93. W. F. DeGrado, C. M. Summa, V. Pavone, F. Nastri and A. Lombardi, *Annu. Rev. Biochem.*, 1999, **68**, 779–819.
94. B. A. Smith and M. H. Hecht, *Curr. Opin. Chem. Biol.*, 2011, **15**, 421–426.
95. Y. Lu, N. Yeung, N. Sieracki and N. M. Marshall, *Nature*, 2009, **460**, 855–862.
96. H. K. Privett, C. J. Reedy, M. L. Kennedy and B. R. Gibney, *J. Am. Chem. Soc.*, 2002, **124**, 6828–6829.
97. F. V. Cochran, S. P. Wu, W. Wang, V. Nanda, J. G. Saven, M. J. Therien and W. F. DeGrado, *J. Am. Chem. Soc.*, 2005, **127**, 1346–1347.
98. A. F. Peacock, *Curr. Opin. Chem. Biol.*, 2013, **17**, 934–939.
99. F. Yu, V. M. Cangelosi, M. L. Zastrow, M. Tegoni, J. S. Plegaria, A. G. Tebo, C. S. Mocny, L. Ruckthong, H. Qayyum and V. L. Pecoraro, *Chem. Rev.*, 2014, **114**, 3495–3578.
100. B. S. Der, D. R. Edwards and B. Kuhlman, *Biochemistry*, 2012, **51**, 3933–3940.
101. M. L. Zastrow, A. F. Peacock, J. A. Stuckey and V. L. Pecoraro, *Nat. Chem.*, 2012, **4**, 118–123.
102. M. L. Zastrow and V. L. Pecoraro, *J. Am. Chem. Soc.*, 2013, **135**, 5895–5903.
103. V. M. Cangelosi, A. Deb, J. E. Penner-Hahn and V. L. Pecoraro, *Angew. Chem. Int. Ed.*, 2014, **53**, 7900–7903.
104. M. Tegoni, F. Yu, M. Bersellini, J. E. Penner-Hahn and V. L. Pecoraro, *Proc. Natl. Acad. Sci. USA.*, 2012, **109**, 21234–21239.
105. A. J. Boersma and G. Roelfes, *Nat. Chem.*, 2015, **7**, 277–279.
106. Z. T. Ball, *Acc. Chem. Res.*, 2013, **46**, 560–570.
107. F. Yu, V. M. Cangelosi, M. L. Zastrow, M. Tegoni, J. S. Plegaria, A. G. Tebo, C. S. Mocny, L. Ruckthong, H. Qayyum and V. L. Pecoraro, *Chem. Rev.*, 2014, **114**, 3495–3578.
108. R. L. Koder, J. L. Anderson, L. A. Solomon, K. S. Reddy, C. C. Moser and P. L. Dutton, *Nature*, 2009, **458**, 305–309.
109. A. Roy, D. J. Sommer, R. A. Schmitz, C. L. Brown, D. Gust, A. Astashkin and G. Ghirlanda, *J. Am. Chem. Soc.*, 2014, **136**, 17343–17349.
110. T. Sasaki and E. T. Kaiser, *J. Am. Chem. Soc.*, 1989, **111**, 380–381.
111. A. Ranieri, S. Monari, M. Sola, M. Borsari, G. Battistuzzi, P. Ringhieri, F. Nastri, V. Pavone and A. Lombardi, *Langmuir*, 2010, **26**, 17831–17835.
112. A. Lombardi, F. Nastri and V. Pavone, *Chem. Rev.*, 2001, **101**, 3165–3190.
113. F. Nastri, L. Lista, P. Ringhieri, R. Vitale, M. Faiella, C. Andreozzi, P. Travascio, O. Maglio, A. Lombardi and V. Pavone, *Chemistry*, 2011, **17**, 4444–4453.

114. R. Vitale, L. Lista, C. Cerrone, G. Caserta, M. Chino, O. Maglio, F. Nastri, V. Pavone and A. Lombardi, *Org. Biomol. Chem.*, 2015, **13**, 4859–4868.

115. M. Faiella, O. Maglio, F. Nastri, A. Lombardi, L. Lista, W. R. Hagen and V. Pavone, *Chemistry*, 2012, **18**, 15960–15971.

116. L. Di Costanzo, S. Geremia, L. Randaccio, F. Nastri, O. Maglio, A. Lombardi and V. Pavone, *JBIC J. Biol. Inorg. Chem.*, 2004, **9**, 1017–1027.

117. A. Dautant, J. B. Meyer, J. Yariv, G. Précigoux, R. M. Sweet, A. J. Kalb and F. Frolow, *Acta. Cryst. Sec. D Biolo. Cryst.*, 1998, **54**, 16–24.

118. M. K. Tiwari, R. Singh, R. K. Singh, I.-W. Kim and J.-K. Lee, *Comput. Struct. Biotechnol. J.*, 2012, **2**, e201209002.

119. A. Lombardi, C. M. Summa, S. Geremia, L. Randaccio, V. Pavone and W. F. DeGrado, *Proc. Natl. Acad. Sci. U. S. A.*, 2000, **97**, 6298–6305.

120. A. Lombardi, C. M. Summa, S. Geremia, L. Randaccio, V. Pavone and W. F. DeGrado, *Proc. Nat. Acad. Sci.*, 2000, **97**, 6298–6305.

121. A. Pasternak, J. Kaplan, J. D. Lear and W. F. Degrado, *Protein. Sci.*, 2001, **10**, 958–969.

122. L. Di Costanzo, H. Wade, S. Geremia, L. Randaccio, V. Pavone, W. F. DeGrado and A. Lombardi, *J. Am. Chem. Soc.*, 2001, **123**, 12749–12757.

123. J. Kaplan and W. F. DeGrado, *Proc. Natl. Acad. Sci. U.S.A.*, 2004, **101**, 11566–11570.

124. M. Faiella, C. Andreozzi, R. T. de Rosales, V. Pavone, O. Maglio, F. Nastri, W. F. DeGrado and A. Lombardi, *Nat. Chem. Biol.*, 2009, **5**, 882–884.

125. A. J. Reig, M. M. Pires, R. A. Snyder, Y. Wu, H. Jo, D. W. Kulp, S. E. Butch, J. R. Calhoun, T. Szyperski, E. I. Solomon and W. F. DeGrado, *Nat. Chem.*, 2012, **4**, 900–906.

Chapter 4

Immobilisation of Homogeneous Catalysts in Metal-Organic Frameworks: Methods and Selected Examples

Alexios Grigoropoulos

1 Introduction

1.1 Immobilisation of homogeneous catalysts

Homogeneous catalysts are predominantly well-defined highly selective and reactive transition metal (TM) complexes bearing tailored organic ligands, the design of which provides excellent control over the mechanism, the rate and the selectivity in catalytic reactions. Their solubility in common organic solvents or even water allows for the comprehensive characterisation of reaction intermediates and products by common spectroscopic techniques and kinetic analysis methods, thereby facilitating the elucidation of the overall catalytic mechanism. Mechanism can be further investigated at a theoretical level using computational chemistry methods which are now commonplace for discrete molecular systems. Despite these advantages, separation from the products and recycling of the homogenous catalyst after turnover is usually difficult and expensive. Moreover, TM catalysts are not in general stable under elevated temperature and pressure and are prone to deactivation, thus limiting their industrial use (Table 1).[1-3]

Table 1: Comparison of homogeneous and heterogeneous catalysts.

Property	Homogeneous	Heterogeneous
Form	Molecular metal complexes	Metals (supported) or metal oxides
Active site	Well-defined	Ill-defined
Temperature	Low (<250°C)	High (250–600°C)
Stability	Moderate	High
Product separation	Generally problematic	Facile
Catalyst recycling	Expensive	Cheap and simple
Selectivity	High	Low
Reaction Mechanism	Well understood	Poorly understood
Control and Tuning	Very good	Limited

On the contrary, heterogeneous catalysts are more chemically robust, can be easily separated from the reaction mixture and reused after turnover. They are however compositionally less sophisticated materials, therefore being less reactive and selective than homogenous catalysts, although progress has been accomplished in this field with the introduction of porous heterogeneous catalysts.[4–7] Moreover, limitations in both experimental and computational methods applicable for solid-state materials complicate the identification of the actual active site and the investigation of the corresponding catalytic mechanism. As a result, tuning of the reactivity and selectivity of heterogeneous catalysts based on the comprehensive understanding of the mechanism at work is also more challenging, compared to homogeneous catalysis (Table 1).

Considerable effort has thus been devoted to design new hybrid materials which combine the advantages of both genres. One particularly promising approach involves the immobilisation of homogeneous TM catalysts onto solid-state supports, a process also known as heterogenization.[8,9] Heterogenization can suppress common intermolecular deactivation pathways for TM catalysts such as dimerization or aggregation, provides better control of the chemical environment of the catalyst's active site for improved performance and permits the use of solvents that otherwise could not be employed because of the catalyst's poor solubility. Examples of solid-state supports include insoluble organic polymers, metal-oxide surfaces, nanoporous carbon, mesoporous aluminosilicates, zeolites and more recently Metal-Organic Frameworks (MOFs).[7,10–15]

Insoluble organic polymers and metal oxides such as silica, alumina and zirconia are the most extensively used solid-state supports up to date, however they do not show crystallinity and ordered porosity. Crystallinity is a highly desirable property in a solid-state support since the long-range atomic scale order allows for controlling the spatial location and the density of the immobilised catalytic sites, leading to consistent and reproducible catalytic activity. Porosity allows for the immobilisation of a TM catalyst not just on the surface but inside the pores of the support. In this way, the catalyst is better protected and the molecular sieve action of the pore itself can be exploited to regulate access of specific molecules to (or departure from) the catalytic sites to promote selectivity. Nanoporous carbon and mesoporous silicates are porous materials with controllable pore distribution and size but they lack atomic-scale periodicity. Zeolites and MOFs are both porous and crystalline materials. However, the number of available zeolites is rather limited (< 200 to date) and the number of zeolites with pores large enough to accommodate TM catalysts is even smaller.

MOFs are constructed by joining metal-based nodes, termed structural building units (SBUs), with multitopic organic linkers via strong coordination bonds to form crystalline permanently porous open frameworks. The large number of available building components (SBUs and organic linkers) and their almost infinite combinations has led to the synthesis of thousands of new MOFs during the last 15 years possessing a great variety of topologies and porosities compared to the traditional porous materials such as zeolites and mesoporous aluminosilicates.[16–22] MOFs are crystalline materials, providing a well-defined nanospace environment with pore sizes spanning from 3 Å– 0 Å (0.3– 3 nm). In addition, the shape, size and surface of the pores can be engineered, by simply considering the coordination geometry of the SBUs and the connectivity of the organic linkers. For example, introducing organic linkers with the same connectivity but different size affords isostructural frameworks of the same topology but with different pore size and surface (Figure 1). It should be noted that MOFs are not as stable as zeolites when exposed to organic solvents and particularly water, since solvent molecules can coordinate to the metal

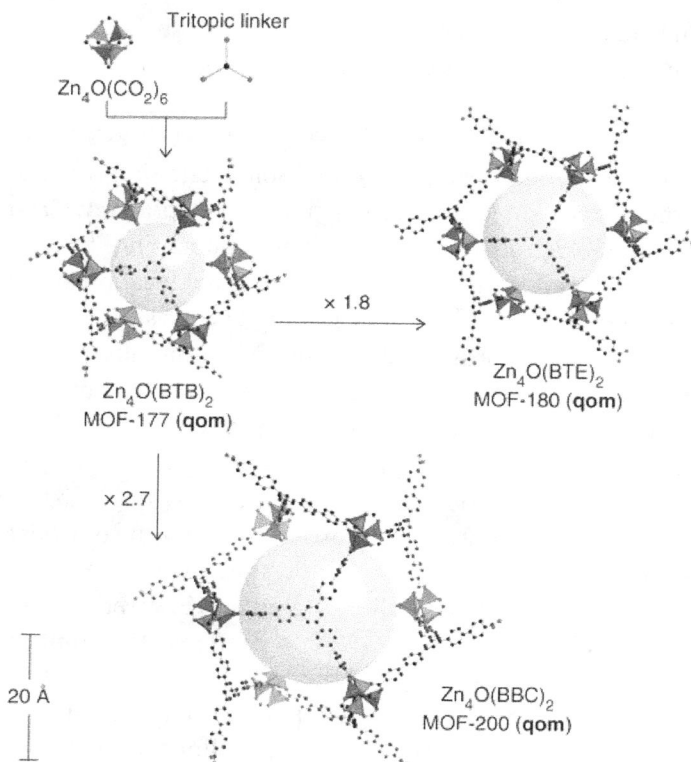

Figure 1: Isoreticular expansion of metal-organic frameworks by increasing the size of the organic linker whilst maintaining the same topology. In this example, MOFs are built from octahedral $[Zn_4O(CO_2)_6]$ clusters and tritopic linkers (BTB = benzene-tribenzoate (MOF-177), BTE = benzene-1,3,5-triyl-tris(ethyne-benzoate) (MOF-180), BBC = benzene-1,3,5-triyl-tris(benzene-benzoate) (MOF-200). Unit cell expansion factor is listed on top of each arrow. Reprinted with permission from Ref. [18].

ions and replace the linkers, thus inducing decomposition of the framework. Nevertheless, there are now several MOFs built from oxo-clusters of electropositive metal ions like Al(III), Cr(III) and Zr(IV) which are very robust and can survive treatment with even strong Brønsted acids or hydrothermal conditions for long periods of time. Therefore, MOFs are currently intensively investigated for a wide range of applications including gas storage, separation, molecular recognition, conductivity, nonlinear optics, drug-delivery, as well as catalysis.[23–31]

1.2 Metal organic frameworks as solid state supports

The fundamental design principle of MOFs requires the formation of strong coordinative bonds between metal ions and organic linkers. Consequently, metal-based nodes are usually coordinatively saturated and therefore not accessible to substrates for catalytic transformations. Nevertheless, there are MOFs in which the coordination sphere of the metal ions also contains coordinated solvent molecules which can be removed via mild heating under vacuum, a process known as activation. This in turn leads to the generation of coordinatively unsaturated metal sites (CUS) that can serve as Lewis acidic catalytic sites. It should be noted though that activation may result in the collapse of the structure due to the distortions produced upon changing the coordination environment of the SBUs and removing the guests from the pores. All the same, MOFs which can be directly used as heterogeneous catalysts even after activation represent only a small fraction. An excellent review on the field by Liu *et al.* in 2014 reported only 35 examples in which a MOF directly serves as a heterogeneous catalyst.[32] The full potential of MOFs can only be exploited via the incorporation of catalytically active guests inside their uniform pores. To that end, various MOFs have been explored as solid-state platforms (hosts) for the immobilisation of molecular TM catalysts, forming catalyst–MOF materials. Such hybrid systems could potentially outperform both homogeneous and heterogeneous catalysts since they contain a high proportion of spatially isolated but accessible, highly reactive and selective, well-defined catalytic sites and they can be easily recycled. The confined space may even dictate selectivity by controlling access of the substrates, stabilizing specific transition states or even assisting activation via auxiliary functionalities.

There are two major strategies for incorporating a homogeneous TM catalyst inside a MOF: (i) covalent bonding onto the linker or the SBU and (ii) direct encapsulation via non-covalent interactions (Figure 2). Covalent bonding, also known as grafting, usually takes place post-synthetically by attaching a TM catalyst onto an organic functionality of the linker. An interesting variation is the conversion of a TM catalyst to a metallolinker and employment of the latter to the direct synthesis of the targeted MOF. This obviously requires very robust TM catalysts which can survive the solvothermal process used for MOF synthesis.

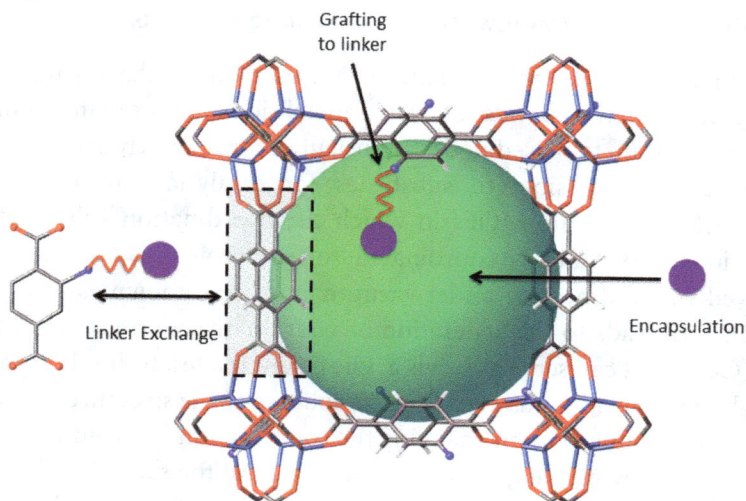

Figure 2:   Incorporation of a TM catalyst (purple spheres) inside a MOF: (i) covalent bonding to an available grafting site of the linker (depicted in blue); (ii) post-synthetic linker exchange (shaded in grey) and (iii) direct encapsulation via non-covalent interactions. Green sphere depicts the guest-accessible space.

Conversely, linkers containing suitable coordination or grafting sites can be post-synthetically exchanged into the framework under milder conditions. This process is known as post-synthetic exchange (PSE)[33-35] and has been widely explored. However, it requires multiple synthetically demanding steps typically involving deprotection of the grafting sites and modification of the coordination sphere of the TM catalyst which may significantly change its electronic and structural properties and as a result its overall catalytic activity.

On the contrary, simple encapsulation does not perturb the first coordination sphere of the catalyst and in principle allows for the direct transfer of solution-based chemistry to the solid state. However, although simple in concept, encapsulation is synthetically challenging since the catalyst must be firmly held inside the pores of the host via non-covalent interactions in order to avoid leaching in the liquid phase upon turnover. Therefore, examples are still rather limited.

The scope of this work is to introduce the reader to the fascinating chemistry of MOFs and their potential as solid-state platforms for the immobilisation of homogeneous TM catalysts. To this end, representative examples of the aforementioned methodologies are selected and discussed involving some of the more well-documented MOFs like UiO-66, MIL-100 and HKUST-1. These specific examples may not necessarily report remarkable catalytic activity, however they present in a comprehensive manner the potential advantages as well as problems of each methodology and demonstrate the benefits of MOFs compared to other solid-state supports.

2 Covalent Bonding

This approach requires the presence of suitable sites on the linkers where TM catalysts can be covalently bonded (grafted). The respective linkers can be either employed directly in the synthesis of the host MOFs or they can be incorporated into the framework post-synthetically. The former approach often requires protection (and subsequent deprotection) of the grafting sites during MOF synthesis under solvothermal conditions. On the contrary, post-synthetic exchange of linkers can be performed under significantly milder conditions.

2.1 Post synthetic exchange

A well documented and investigated family of particularly stable MOFs is the UiO-series (UiO stands for University of Oslo). UiO-MOFs are built from the interconnection of Zr(IV)-oxoclusters with carboxylate-based linkers. The prototypical member of the series is UiO-66, built from cuboctahedral $[Zr_6(\mu_3\text{-}O)_4(\mu_3\text{-}OH)_4(COO)_{12}]$ SBUs and linear ditopic BDC linkers (BDC = 1,4-benzene dicarboxylate). The SBU consists of an inner octahedral Zr_6-cluster whose triangular faces are capped by μ_3-O and μ_3-OH groups in an alternating fashion to account for charge balance (Figure 3). All 12 octahedral edges are bridged by carboxylate groups originating from different BDC linkers giving rise to a porous MOF with an overall $[Zr_6(\mu_3\text{-}O)_4(\mu_3\text{-}OH)_4(BDC)_6]$ formula and an augmented face-centred cubic array of 12-fold interconnected cuboctahedral SBUs (fcu-a topology).[36] UiO-66 contains both octahedral and tetrahedral pores with a diameter of 8.4 Å and 7.6 Å, respectively, after taking into account van der Walls radii. It is remarkably stable due to its high connectivity between the

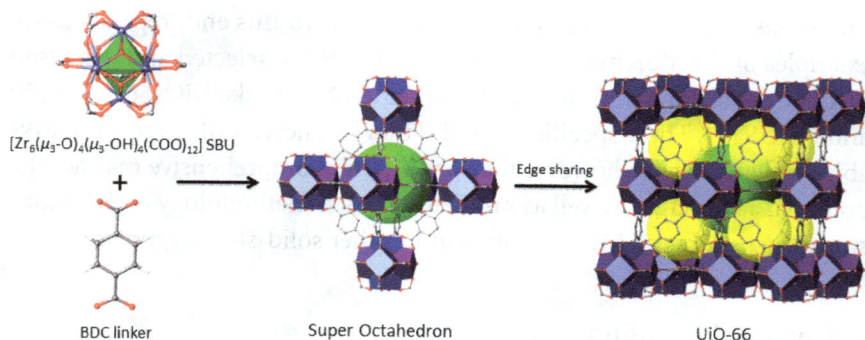

Figure 3: Crystal structure of UiO-66 built from Zr$_6$-based cuboctahedral SBUs and linear ditopic BDC linkers. Six SBUs occupy the vertices of a super octahedron interconnected with BDC linkers. The 12-fold connected SBUs form an augmented face centred cubic array (fcu-a topology) with octahedral (green) and tetrahedral (yellow) pores.

SBUs and the intrinsic bond strength of the coordination bonds between carboxylate groups and highly electropositive Zr(IV)-oxoclusters. UiO-66 retains its crystallinity upon heating to 400°C or treatment with water, commonly used organic solvents and strong Brønsted acids or bases.[37] Replacement of BDC with longer linear ditopic linkers such as 4,4-biphenyl dicarboxylate (BPDC) and terphenyl dicarboxylate (TPDC) leads to the synthesis of isostructural analogues, namely UiO-67 and UiO-68. These MOFs contain even larger pores and have been vastly explored as solid-state supports, although they are relatively less stable than UiO-66.[38]

The excellent chemical and structural stability of UiO-MOFs allows for partial PSE of the linkers. PSE is a powerful and facile method for functionalisation of the framework under mild conditions, leading to materials which otherwise would be inaccessible.[33–35] For example, Cohen *et al.* reported the successful introduction of catecholate groups into UiO-66 via the exchange of BDC with catecholate-containing linkers, such as dixydroxybenzene dicarboxylate (DHBDC), forming UiO-66-CAT (Scheme 1).[39] Subsequent metallation of the catecholate groups with Cr(III) afforded UiO-66-CAT(Cr). This MOF contains spatially isolated Cr-monocatecholate species and is therefore a very good and recyclable heterogeneous catalyst for the oxidation of secondary alcohols to ketones. Attempts to synthesize UiO-66-CAT directly from its components under

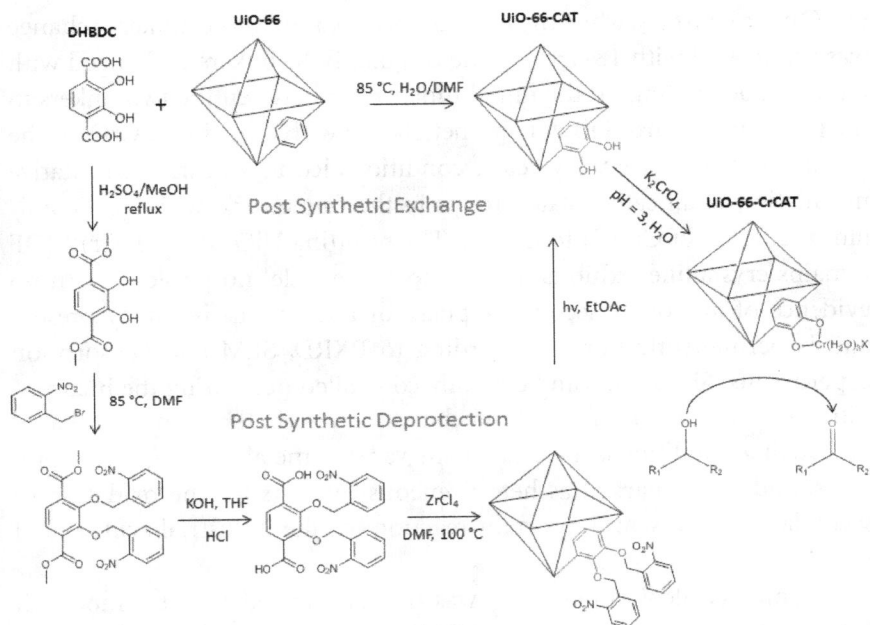

Scheme 1: Synthesis of UiO-66-CAT via post synthetic exchange or post-synthetic deprotection. Post-metallation of the final material leads to UiO-66-CAT(Cr) which serves as a heterogenised catalyst for the oxidation of secondary alcohols to ketones. Adapted with permission from Ref. [39].

solvothermal conditions were not successful, presumably due to the presence of the hydroxyl groups which can also coordinate to the Zr_6-based SBUs and disrupt the MOF synthesis. This demonstrates the value and applicability of the PSE method. Direct synthesis was only feasible if both hydroxyl groups were initially protected via coupling to *ortho*-nitrobenzyl. In this case, the targeted material was obtained under solvothermal conditions employing $ZrCl_4$ as the metal source and adding both linkers in the reaction mixture. The photocleavable protecting groups were post-synthetically removed via photolysis (Scheme 1). Clearly this route, known as post-synthetic deprotection, involves multiple and synthetically challenging steps. Moreover, the degree of MOF functionalisation is usually limited (in this case <35%) due to the large size of the protecting groups.

On the contrary, when the PSE approach was followed, linker exchange was modulated with 18–75% of the original BDC linkers exchanged with DHBDC, depending on the initial molar ratio between the two linkers in the reaction mixture. Subsequent metallation with Cr, using K_2CrO_4 as the metal source under weakly acidic conditions led to a nearly quantitative metallation of all catecholato groups in the framework, with a concomitant reduction of Cr(VI) to Cr(III). The resulting UiO-66-CAT(Cr) MOF remains crystalline, exhibits homogeneous particle morphology, with no evidence of any other secondary phase or amorphous impurity present and is permanently porous according to PXRD, SEM and N_2 sorption experiments. Since PSE can be readily controlled permitting the incorporation of a different number of catecholato groups in the framework based on starting conditions, materials with various metal loadings were prepared and benchmarked as heterogeneous catalysts for the oxidation of secondary alcohols, an important reaction for the industrial synthesis of ketones.[40,41]

Secondary alcohol oxidation was initially carried out at 1 mol% Cr loading in the form of UiO-66-CAT(Cr) with t-butyl hydroperoxide (TBHP) as the oxidant in chlorobenzene at 70°C. Nine secondary aliphatic, alicyclic or benzylic alcohols were used as substrates. The majority of them gave nearly quantitative yields after 8–24 h (entries labelled as A in Table 2). Performing the catalysis in a "greener" solvent-free environment increased the TOF under otherwise identical conditions (entries labelled as B in Table 2). Control experiments using the same amount of either the parent UiO-66 MOF or the chromium-free UiO-66-CAT MOF gave yields below 15%, suggesting a minor contribution of the host to the overall yield. The heterogeneous nature of the reaction was confirmed by a hot filtration test. The supernatant was physically separated from the solid-state catalyst via simple filtration after 2 h of reaction using 2-heptanol as the substrate. Negligible additional product yield was observed in the supernatant after additional 22 h, verifying that no catalytically active species were released into the liquid phase. Moreover, ICP-OES analysis confirmed that UiO-66-CAT(Cr) is a true heterogeneous catalyst since only traces of Cr were detected in the supernatant after turnover.

UiO-66-CAT(Cr) outperforms a known Cr-exchanged zeolite in terms of yield and TOF. In addition, significant Cr leaching was observed for the Cr-exchanged zeolite, a problem often encountered for zeolite-based

Table 2: Catalytic oxidation of secondary alcohols to ketones by UiO-66-CAT(Cr).[a]

	Product	Solvent	Oxidant	Time (h)	Yield (%)	TOF (h⁻¹)
1A		C_6H_5Cl	TBHP	24	99	4.20
1B		Solvent free	TBHP	8	99	12.48
1C		CH_3CN	H_2O_2	24	39	1.63
2A		C_6H_5Cl	TBHP	24	92	3.84
2B		Solvent free	TBHP	24	99	4.20
3A		C_6H_5Cl	TBHP	24	70	2.94
3B		Solvent free	TBHP	24	72	3.00
4A		C_6H_5Cl	TBHP	24	95	4.02
4B		Solvent free	TBHP	24	99	4.20
5A		C_6H_5Cl	TBHP	12	99	8.34
5B		Solvent free	TBHP	24	99	4.20
5C		CH_3CN	H_2O_2	24	95	4.02
6A		C_6H_5Cl	TBHP	12	99	8.34
6C		CH_3CN	H_2O_2	24	99	4.20
7A		C_6H_5Cl	TBHP	8	99	12.48
7C		CH_3CN	H_2O_2	24	99	4.20

Note: [a]Reaction conditions: loading = 1 mol%, $T = 70°C$, [sec-alcohol] = 1.0 M (or neat), [THBP] = 1.1 mmol (or 1.3 mmol), $[H_2O_2] = 2.0$ M.

catalysts.[42,43] The homogenous reaction using $Cr(acac)_3$ or K_2CrO_4/catechol gave significantly lower yields (<40%) under identical conditions. Moreover, UiO-66-CAT(Cr) shows comparable catalytic efficiency with other more efficient homogeneous Cu-based catalysts which may use air or O_2 as the oxidant but also require higher metal loadings (5–10 mol%) and toxic or expensive co-catalysts.[44–46] UiO-66-CAT(Cr) can be easily separated from the reaction mixture by centrifugation and recycled for at least five times without any obvious decrease of its catalytic activity. Oxidation of secondary alcohols was also carried out using H_2O_2 as the oxidant an environmentally benign alternative which produces water as the sole by-product. Excellent yields (>95%) were achieved when benzylic alcohols were used as substrates (entries labelled as C in Table 2). However, the oxidation of 2-heptanol and other aliphatic alcohols produced poor results.

This selected example clearly demonstrates the benefits of using robust MOFs for the immobilisation of homogeneous catalysts inside their uniform pores. It also highlights the value of PSE as a relatively straightforward method for immobilisation under mild conditions, leading to materials which cannot be otherwise obtained. Various linkers have been also exchanged into UiO-type MOFs, establishing the generality of the PSE method. The same group has employed thiocatecholato-based linkers containing softer dimercapto groups which can be post-synthetically metallated. The degree of linker exchange was again stoichiometrically controlled, varying between 40% and 71% depending on the initial molar ratio between BDC and its dimercapto-derivative. Subsequent metallation with Pd resulted in a highly efficient, reusable and selective heterogeneous catalyst for the oxidative functionalization of aromatic C–H bonds.[47]

2.2 Direct synthesis

Although solvothermal synthesis of MOFs using linkers with available grafting sites most often requires pre-synthetic protection of the latter, there are a few examples in which this lengthy and often laborious procedure is not necessary. For example, Lin *et al.* published the synthesis of a UiO-type MOF using 2,2′-bipyridine-5,5′-dicarboxylate (BPyDC) as the organic linker. This MOF, abbreviated as UiO-BPyDC, is isostructural with UiO-67 (Figure 4) with an overall $[Zr_6(\mu_3\text{-}O)_4(\mu_3\text{-}OH)_4(BPyDC)_6]$ formula and an augmented fcu topology.[48] It contains considerably larger octahedral and tetrahedral pores than UiO-66 with a pore diameter of 12.4 Å and 9.6 Å, respectively, including van der Walls radii, and a higher BET surface area of 2277 m^2 g^{-1} (*c.f.* 1580 m^2 g^{-1} for UiO-66). Moreover, it features uncoordinated 2,2′-bipyridine (bpy) groups which can be readily post-synthetically metallated with various TMs, forming UiO-BPyDC(M) hybrid materials (M = Fe, Co, Cu, Ir, Pd).[49] For example, treatment of UiO-BPyDC with $[Ir(COD)(OMe)]_2$ (COD = 1,5-cyclooctadiene) afforded UiO-BPyDC(Ir) in which approximately 25% of the bpy groups were metallated. UiO-BPyDC(Ir) shows a significantly reduced BET surface area (365 m^2 g^{-1}) compared to the parent UiO-BPyDC MOF due to the presence of Ir and associated ligands inside the pores.

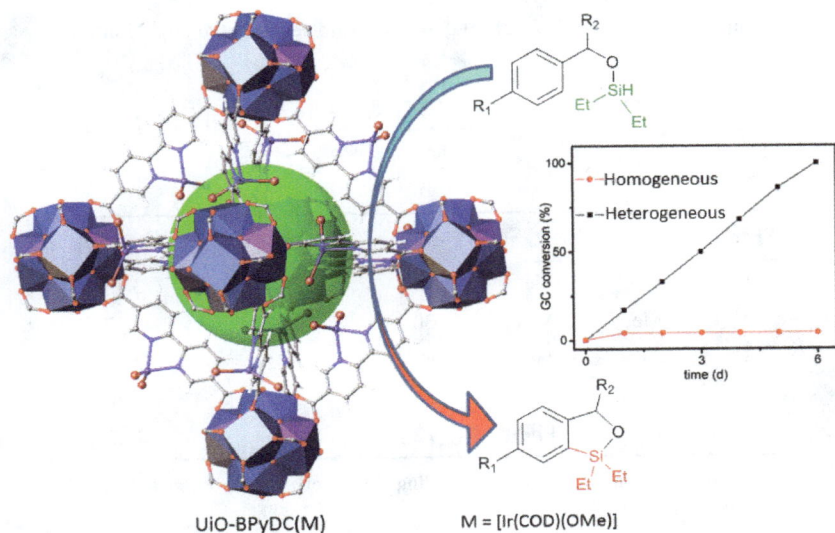

Figure 4: Crystal structure of UiO-BPyDC(Ir) built from Zr_6-based cuboctahedral SBUs and linear ditopic BPyDC linkers. Post-synthetic metallation with $[Ir(COD)(OMe)]_2$ forms a heterogeneous catalyst for *ortho*-silylation which is 1,250 times more active than its homogeneous counterpart (inset). Adapted with permission from Ref. [48].

UiO-BPyDC(Ir) exhibited excellent catalytic activity in intramolecular *ortho*-silylation of benzylicsilyl ethers to give benzoxasiloles which are important intermediates in organic synthesis.[50] The homogeneous reaction using $[\{bpy-(COOMe)_2\}Ir(COD)(OMe)]$ at 5.0 mol% loading under otherwise identical conditions afforded only 4% conversion in 24 h, after which the reaction essentially stops. In contrast, when UiO-BPyDC(Ir) was used as the catalyst even at a lower loading of 0.1 mol% (based on Ir), the yield increased linearly with time until completion after six days (Figure 4, inset). Therefore, UiO-BPyDC(Ir) is at least 1,250 times more active than its homogeneous counterpart. When Ir loading was raised to 1.0 mol%, in the form of UiO-BPyDC(Ir), good yields (83–94%) were observed for various substrates without adding any H-acceptor, also a significant improvement over the corresponding homogeneous reaction in terms of atom efficiency (Table 3). Lower TOF was observed when larger substrates carrying Ph groups were used, presumably because of the slower substrate diffusion through the MOF channels.

Table 3:   UiO-BPyDC(Ir) catalysed intramolecular *ortho*-silylation[a]

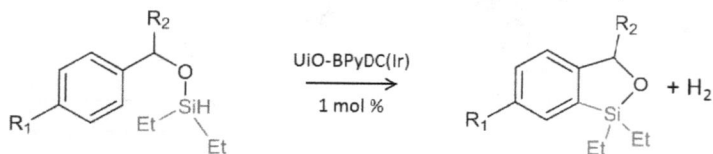

Entry	R_1	R_2	Time (h)	Yield (%)	TOF (h^{-1})
1	H	Me	30	94	3.13
2	Me	Me	30	92	3.07
3	Cl	Me	44	92	2.09
4	OMe	Me	50	85	1.70
5	H	Ph	72	83	1.15

Note: [a]Reaction conditions: : catalyst loading = 1.0 mol% (based on Ir), solvent = *n*-heptane, $T = 100°C$.

Various other examples of post-synthetic metallation of UiO-BPyDC have been reported in the last two years, clearly demonstrating the beneficial effect of active site isolation via the immobilisation of TM catalysts inside the pores of this robust MOF, leading to reusable single-site solid catalysts for important organic transformations. Employment of $[Pd(CH_3CN)_4][BF_4]_2$ as a TM precursor resulted in an active heterogeneous catalyst for the dehydrogenation of substituted cyclohexenones to phenols under an atmosphere of O2.[48] Li *et al.* used the same MOF as a solid-state support for grafting $PdCl_2$ which was subsequently reduced to Pd(0) forming embedded Pd nanoparticles. Excellent catalytic efficiency and shape-selectivity was observed in olefin hydrogenation and aerobic oxidation of alcohols. Moreover, stability was considerably improved compared to materials synthesised via traditional impregnation methods.[51]

In the quest for solid-state supports with even larger pores, UiO-type MOFs with even longer linear linkers have been recently synthesised containing different coordination sites like bpy,[52] 2,2′-bis(diphenylphosphino)-binaphthyl (BINAP),[53,54] norbornadiene[55] or mercapto groups.[56] One particularly interesting case involves sal-MOF(M) (M = Fe, Co) which is isostructural to UiO-68 with the middle benzene ring of the terphenyl dicarboxylate linker bearing a salicylaldimine group (Figure 5).[57]

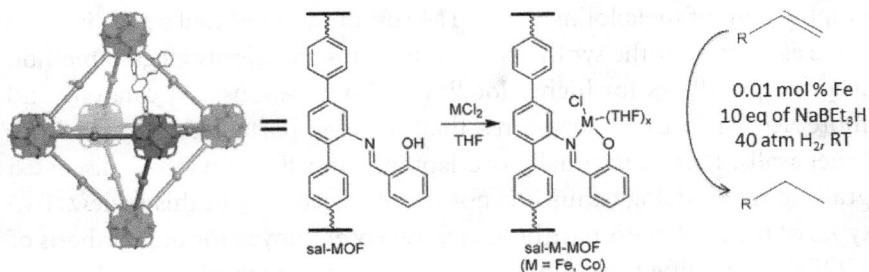

Figure 5: Post-synthetic metallation of sal-MOF to form a heterogeneous catalysts for olefin hydrogenation. Reprinted with permission from Ref. [57].

Post-synthetic metallation with iron or cobalt affords robust, highly active and recyclable heterogeneous catalysts for olefin hydrogenation. The crystallinity of the sal-MOF was maintained upon metallation with approximately 60–70% of the salicylaldimine sites being metallated. Even after metallation, the BET surface area was still higher than 3,000 m^2 g^{-1}, indicating the benefits of isoreticular chemistry in MOFs in order to increase pore aperture and channel dimensions without disturbing the overall topology. The material obtained after post-synthetic metallation and pretreatment with NaBEt$_3$H displayed excellent catalytic activity in the hydrogenation of a wide range of aliphatic and aromatic terminal alkenes and dialkenes. Even at 0.05–0.01 mol% Fe loading, very high turnover numbers up to 1.45×10^5 were observed, whereas the catalyst was recycled and reused nine times without and significant drop in reactivity. Even after partial deactivation, the catalyst was successfully regenerated upon treatment with NaBEt$_3$H. Notably, the synthesis of the corresponding homogeneous catalyst [(sal)Fe(THD)$_x$Cl] was not possible, consistently leading to tetranuclear clusters. This suggests that such low coordination number Fe species are not stable in solution and form oligomers instead. This process is obviously prohibited upon immobilisation, demonstrating how MOFs may also provide a platform for obtaining new TM catalysts.[57]

2.3 Metallolinkers

Linkers containing coordination or grafting sites can be incorporated into a MOF either directly or via PSE. A third attractive option involves employment of metallolinkers, i.e. TM complexes modified accordingly to serve as linkers for the synthesis of MOFs. This chemically elegant method in principle allows for higher loading of homogenous TM catalysts and more accessible catalytic centres that are now part of the framework's inner walls. It is conceptually overlapping with the previously discussed grafting methodology, thus it is not treated separately in this review. Two types of metallolinkers have been extensively employed for the synthesis of MOFs, i.e. modified metallosalens[58] and metalloporphyrins.[59,60] The most characteristic example of a metallolinker is *meso*-tetrakis(4-carboxyphenyl)porphyrin (H_4TCPP-H_2) which can be employed as a linker either in its deprotonated metal-free (TCPP-H_2) or metallated form (TCPP-M).

Yaghi *et al.* reported the synthesis of MOF-525 which consists of typical 12-fold connected cuboctahedral $[Zr_6(\mu_3\text{-}O)_4(\mu_3\text{-}OH)_4(COO)_{12}]$ SBUs,[61] also found in the UiO-series of MOFs,[36] and deprotonated metal-free TCPP-H_2 linkers. Each SBU is situated on the vertex of a cube and each tetratopic TCPP-H_2 linker is capping one face giving rise to a 3D porous framework with an overall $[Zr_6(\mu_3\text{-}O)_4(\mu_3\text{-}OH)_4(TCPP\text{-}H_2)_3]$ formula and an augmented ftw topology (Figure 6, top). If the metal to linker stoichiometric ratio is increased, then PCN-222(M) is formed instead (M = Fe, Mn, Co, Ni, Cu, Zn).[61–63] In PCN-222(M), only eight edges of the Zr_6-based SBUs are bridged by TCPP-M linkers, in contrast to MOF-525, while the remaining four edges are occupied by terminal water and hydroxyl groups (Figure 6, bottom). Consequently, PCN-222(M) is built from $[Zr_6(\mu_3\text{-}O)_4(\mu_3\text{-}OH)_4(t\text{-}OH)_4(H_2O)_4(COO)_8]$ SBUs serving as 8-fold connecting nodes and giving rise to a 3D structure of augmented csq topology and an overall $[Zr_6O_4(OH)_8(H_2O)_4(TCPP\text{-}M)_2]$ formula. PCN-222(M) contains very large hexagonal and triangular open channels with a diameter of 31.4 Å and 11.0 Å respectively, accounting for van der Waals radii, and a BET surface area of approximately 2200 $m^2\ g^{-1}$. Moreover, PCN-222(M) shows exceptional stability even in concentrated hydrochloric acid.[62]

PCN-222 was benchmarked as a biomimetic analogue of heme proteins like cytochromes and peroxidases which are capable of performing oxidation of organic substrates at remarkable rates using H_2O_2 as the oxidant.[64] Direct application of a metalloporphyrin as an oxidation catalyst in aqueous solutions is usually hampered by the formation of dimers and decomposition of the catalyst in the presence of oxidizing media.

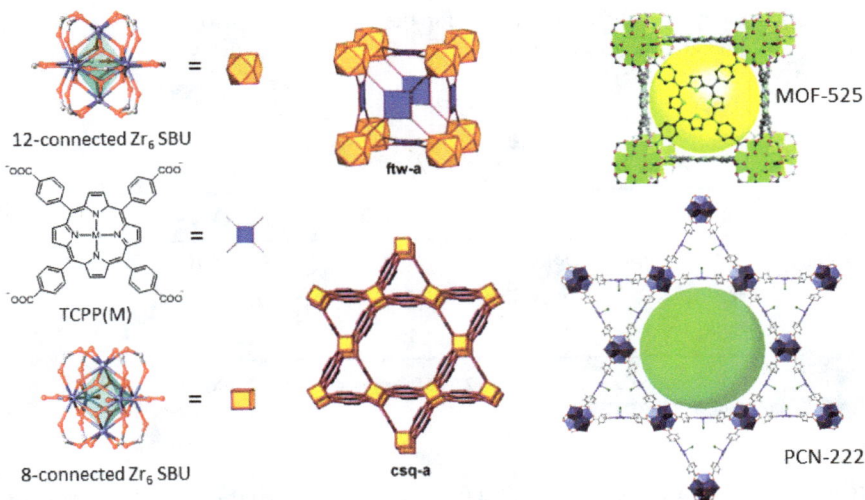

Figure 6: Crystal structure and topology of different Zr-based MOFs bearing TCPP linkers. MOF-525 is built from 12-fold connected SBUs (top) whereas PCN-222 is built from 8-fold connected SBUs (bottom) giving rise to different topologies. Cubic pores for MOF-525 and hexagonal channels for PCN-222 are illustrated with yellow and green spheres respectively. Adapted with permission from Ref. [61].

Therefore, bulky ligands are attached to the porphyrin backbone to protect the active sites, a process involving multiple synthetically demanding steps. An alternative approach is immobilisation of the metalloporphyrin on a solid-state support, a practice which inevitably dilutes the density of active sites. Employing TCPP(M) as a metallolinker for the construction of PCN-222(M) immobilizes the catalytic sites, maximises their density and renders them accessible to substrates, since they are situated on the walls of quite large channels.

To test the catalytic activity of PCN-222(M) pyrogallol was chosen as a substrate, commonly used in standard assays to characterize the catalytic performance of heme-like biomimetic analogues. Only an activated sample of PCN-222(Fe) exhibited peroxidase-like catalytic activity. The peroxidation of pyrogallol follows the conventional enzymatic dynamic regulation of the Michaelis–Menten equation. A comparison of the k_{cat} values of the natural enzyme horseradish peroxidase (HRP) and PCN-222(Fe) reveals that the native enzyme is approximately two orders of magnitude more active, as expected (Table 4). Nevertheless, the Michaelis

Table 4:   Kinetic parameters for the oxidation of pyrogallol by different catalysts in water.[a]

Catalyst	K_m (mM)	k_{cat} (min^{-1})	k_{cat}/K_m (M^{-1} min^{-1})
PCN-222(Fe)	0.33	16.1	4.85×10^4
Hemin	Not available	2.4	Not available
HRP	0.81	1.8×10^3	2.20×10^4

Note: [a]Reaction conditions: [Fe] = 2.5 μM, [pyrogallol] = 0.2–2 mM, [H$_2$O$_2$] = 50 mM, $T = 25$°C.

constant (K_m) value for PCN-222(Fe) is lower than that of HRP, indicating a higher affinity of the substrate to PCN-222(Fe). Therefore, it is the slower diffusion of the substrate into the pores of PCN-222(Fe) that slows down the catalytic reaction. Moreover, the obtained k_{cat} value of PCN-222(Fe) is seven times larger than the one of hemin which is the Fe-metallated protoporphyrin IX (Table 4). Oxidation of other substrates verified that PCN-222(Fe) shows superior catalytic activity compared to its homogeneous analogue hemin, with the obtained k_{cat} values often measured two orders of magnitude higher.[62,63] Such an excellent catalytic performance can be attributed to the construction of stable 3D structures with large open channels in which dimerization of porphyrin centres is prevented and diffusion of substrates is not a rate limiting factor. Compared with other solid-state supports used for the immobilization of metalloporphyrins, PCN-222 shows a higher density of catalytic centres and a faster diffusion rate of substrates. Other metalloporphyrins and metallosalen complexes have been employed as metallolinkers for the synthesis of MOFs, showing in most cases increased catalytic efficiency compared to analogous homogeneous systems.[65–76]

3 Encapsulation

This approach involves the encapsulation of a TM catalyst inside the pores of a MOF via non-covalent interactions. It does not require any modification of the catalyst's coordination environment and allows for the direct application of solution-based chemistry to the solid state. However, there are few examples up to date reporting the successful encapsulation of a TM catalyst inside a MOF, mainly due to the inherently weaker nature of non-covalent interactions. These examples include (i) assembling the catalyst from its components *in situ* inside the confined environment of the MOF, a methodology known as "ship in a bottle"; (ii) assembling the MOF around the catalyst, a methodology known as "bottle around ship" or template synthesis and (iii) direct cation exchange of positively charged TM catalyst inside the pores of an anionic MOF.

3.1 Ship in a bottle

The ship in a bottle methodology represents a general and versatile method for the encapsulation of TM catalysts inside the pores of MOFs (Figure 7). By definition, it involves multiple synthetic steps whereas systematic optimisation of the assembly reaction is necessary in order to avoid side reactions leading to multiple species present alongside the active catalyst.

This elegant but hard to realise strategy was originally employed using zeolites as hosts[77–84] and was very recently demonstrated using MOFs. Ma *et al.* selected an anionic MOF, $[Me_2NH_2]_2[Zn_8(\mu_4\text{-}O)(ad)_4(BPDC)_6]$ (BioMOF-1) as the host (ad = adeninate, BPDC = biphenyldicarboxylate).[85] Anionic MOFs are mainly built from negatively charged SBUs, giving rise to an anionic framework whose charge is balanced by small organic cations trapped inside the cavities.

In particular, BioMOF-1 consists of infinite zinc-adeninate columns with an overall formula of $[Zn_8(\mu_4\text{-}O)(ad)_4(COO)_{12}]^{2-}$ which are interconnected via ditopic BPDC linkers, giving rise to an anionic framework with large channels running parallel to the *c*-axis.[86] The negative charge is balanced by $[Me_2NH_2]^+$ cations, produced *in situ* upon DMF solvent decomposition and residing in the channels along with solvent molecules (Figure 8(A)). Exchange of these cations with Co(II) leads to Co–BioMOF-1 (Figure 8(B)) with a concomitant colour change from white (colour of the parent host) to pink due to Co(II) encapsulation.

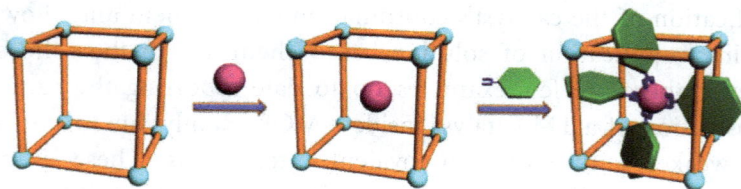

Figure 7: Ship in a bottle methodology. Key: light blue sphere = metal-based SBU; orange rod = organic linker; pink sphere = metal cation encapsulated inside the pores of a MOF (cube); green hexagon = component of guest molecule. Reprinted with permission from Ref. [85].

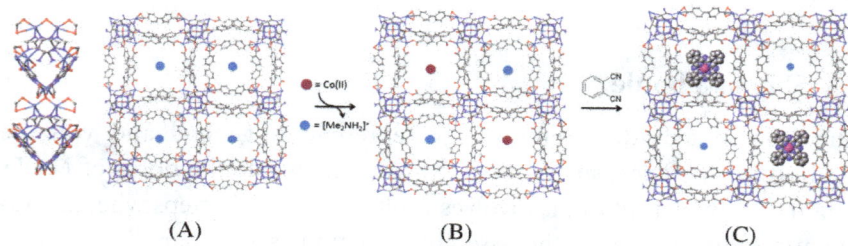

(A)   (B)   (C)

Figure 8: (A) Structure of BioMOF-1 consisting of zinc-adeninate columns interconnected via ditopic BPDC linkers. The negative charge is balanced by $[Me_2NH_2]^+$ cations (blue spheres) generated *in situ* upon DMF decomposition. Step-wise incorporation of Co(II) (magenta spheres) (B) and phthalonitrile (C) inside the pores leads to the formation of the CoPC phthalocyanine complex which is too large to exit from the pores.

Subsequent treatment with a formamide solution of phthalonitrile (1,2-dicyanobenzene) at 190°C for 10 min triggers the formation of the phthalocyanine (Pc) ligand inside the pores and subsequent coordination to Co(II), producing CoPC–BioMOF-1 via a typical condensation-complexation reaction (Figure 8(C)). The molecular dimensions of CoPC (1.3 × 1.3 nm) are larger than the pore windows of BioMOF-1 (1 × 1 nm), hence CoPC is trapped inside the pores after assembly. For the same reason, attempts to encapsulate pre-assembled CoPC inside the pores of BioMOF-1 were not successful, verifying the validity of the approach.

CoPC–BioMOF-1 was benchmarked as a heterogeneous catalyst in styrene epoxidation using TBHP as the terminal oxidant (Table 5). It can

Table 5: CoPC–BioMOF-1 catalytic epoxidation of olefins to epoxides.[a]

Entry	Catalyst	Substrate	Conversion (%)	Selectivity (%)
1	BioMOF-1	Styrene	8	61
2	CoPC	Styrene	38	60
3	CoPC–BioMOF-1	Styrene	72	65
4	After 3 cycles	Styrene	64	63
5	CoPC–BioMOF-1	Cis-stilbene	14	68

[a] Reaction conditions: 1.5 mol% Co loading, solvent = MeCN, T = 60°C, t = 16 h, [styrene] = 0.125 M, [TBHP] = 0.188 M.

efficiently catalyse the epoxidation of styrene to styrene oxide with a conversion of 72% (1.5 mol% loading for 16 h at 60°C) and a 65% selectivity towards styrene oxide (entry 3). In comparison, the homogeneous counterpart under the same reaction conditions exhibits a much lower conversion of 38% whereas selectivity is approximately the same (entry 2), suggesting that encapsulation suppresses formation of aggregates via π–π stacking of CoPC molecules, a well-established deactivation process in solution.[87] The parent host has a negligible contribution to the overall yield (entry 1) and CoPC–BioMOF-1 can be recycled at least three times without any significant decrease in its catalytic activity (entry 4). Moreover, leaching of the catalytically active species in the liquid phase is not observed. The heterogeneous nature of the catalytic reaction was further established by using the bulkier cis-stilbene as a substrate which showed much lower conversion of 14% (entry 5). The larger substrate cannot readily access the catalytic sites, suggesting that the reaction takes place inside the pores of the host and not on its surface.

It is not necessary to use an anionic MOF as the host for this methodology. Very recently, Li *et al.* reported the successful ship in a bottle encapsulation of [Cu(phen)$_3$]Cl$_2$ (phen = 1,10-phenanthroline) inside the pores of MIL-100(Al) and its use as a catalyst for the oxidation of cycloalkanes.[88] MIL-100(M) MOFs {M = Fe(III), Al(III), Cr(III)} were initially synthesised by Férey and co-workers (MIL stands for Materials Institute Lavoisier). They are constructed from [M$_3$(μ_3-O)(X)(COO)$_6$] SBUs (X = halide or hydroxide) and tritopic BTC linkers with a [M$_3$O(X)(H$_2$O)$_2$

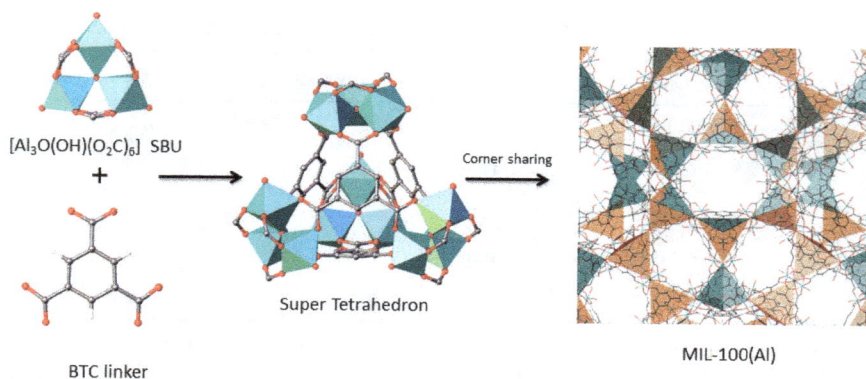

Figure 9: Crystal structure of MIL-100(Al) built from $[Al_3O(OH)(COO)_6]$ SBUs and tritopic BTC linkers. Four SBUs form a super-tetrahedron with BTC linkers capping all six faces. Corner sharing between the super-tetrahedra gives rise to MIL-100(Al) which exhibits two types of mesopores with pentagonal and hexagonal windows.

$(BTC)_2]\cdot nH_2O$ overall formula (BTC = benzene-1,3,5-tricarboxylate).[89–91] MIL-100(M) MOFs are chemically robust and contain two types of mesopores with a diameter of 25 Å and 29 Å (Figure 9) which are accessible via considerably smaller windows (5.5 Å and 8.7 Å, respectively). Therefore, they are ideal supports for the ship in a bottle methodology. Li *et al.* selected MIL-100(Al) as the host and incorporated $CuCl_2$ as the TM precursor and phen as the ligand inside the pores, leading to the *in situ* formation of $[Cu(phen)_3]Cl_2$. The assembled TM catalyst was firmly trapped inside the mesopores of MIL-100(Al) since it is too large to pass through the pore windows and no leaching was observed even after carrying out a soxhlet extraction for three days.

The catalytic performance of the hybrid material $[Cu(phen)_3]$–MIL-100(Al) was investigated in the cyclohexane oxidation towards cyclohexanol and cyclohexanone, using H_2O_2 as the oxidant. As shown in Table 6, parent MIL-100(Al) is catalytically inactive (entry 1), whereas the homogeneous counterpart using $[Cu(phen)_3]Cl_2$ as the catalyst leads to a TON of 71 with 75% selectivity towards cyclohexanol (entry 2). On the other hand, the heterogeneous reaction using the hybrid $[Cu(phen)_3]$–MIL-100(Al) material as the catalyst shows a six-fold increase in reactivity with a TON of 458 without any significant decrease in selectivity (entry 5),

Table 6: MIL-100(Al) catalytic oxidation of cyclohexane to cyclohexanol and cyclohexanone.[a]

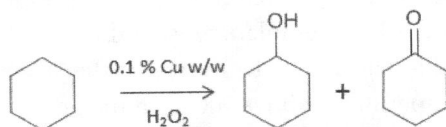

Entry	Catalyst	Conversion (%)	Selectivity (%)[b]	TON
1	MIL-100(Al)	—	—	—
2	[Cu(phen)$_3$]Cl$_2$	12	75:25	71
3	Cu–MIL-100(Al)	8	67:33	93
4	[Cu(phen)$_3$]–Zeolite-Y	12	73:27	195
5	[Cu(phen)$_3$]–MIL-100(Al)	21	69:31	458
6	After three cycles	18	69:31	390

[a] Reaction conditions: loading = 0.1 wt.%, solvent = MeCN, T = 70°C, t = 3 h, [cyclohexane] = 0.95 M.
[b] Cyclohexanol to cyclohexanone ratio.

suggesting that degradation of the catalytic species in the presence of H$_2$O$_2$ is inhibited due to site-isolation.[92] [Cu(phen)$_3$]–MIL-100(Al) is also more reactive than Cu–MIL-100(Al) (entry 3), indicating that phen successfully replaces the halide anions in the Cu(II) coordination sphere and promotes reactivity. Notably, when [Cu(phen)$_3$]–Zeolite-Y was investigated as a heterogeneous catalyst, it also synthesized via the ship in a bottle method,[82] a lower TON of 195 was obtained (entry 4). The difference in catalytic activity obtained was attributed to the larger cavities and windows present in MIL-100(Al) compared to zeolite Y (pore diameter of 13 Å and window of 7.4 Å) which facilitate the diffusion of the substrates and the generation of active transition-states,[93] demonstrating again the potential benefits of MOFs as hosts compared to zeolites.

3.2 Bottle around ship

The bottle around ship or template synthesis approach requires very robust catalysts which serve as structural directing agents (templates) around which the framework is built. Overall, template synthesis has been extensively used in the field of zeolites[94] and mesoporous silicates.[95] By default, the catalyst used as a template must endure conditions used for the synthesis of the solid-state support. Given that the vast majority of MOFs is synthesized at temperatures above 100°C in the presence of water, examples where a TM catalyst can fulfil such a role are rather scarce, currently restricted to polyoxometalates (POMs) and metalloporphyrins.[96–100] For example, Su *et al.* reported the successful incorporation of a series of Keggin-type POMs with an overall formula $[XM_{12}O_{40}]^{n-}$ (M = Mo or W, X = Si, Ge, P, As) into the HKUST-1 MOF via a one-step bottle around ship synthesis.[99]

HKUST-1 is one of the most systematically investigated MOFs. It is built from $[Cu_2(COO)_4(H_2O)_2]$ SBUs with a paddle-wheel geometry linked with tritopic BTC linkers.[101] Each dinuclear Cu_2-cluster is bridged by four carboxylate groups from four different BTC linkers, thus serving as a 4-fold connecting node. HKUST-1 is electrically neutral and diamagnetic with a covalent single bond formed between the two Cu(II) cations of each SBU and two weakly bound H_2O molecules occupying the *trans* apical positions. The fundamental component in HKUST-1 is a super-octahedron formed by six paddle-wheel SBUs situated at the vertices with four BTC linkers capping half of the triangular faces (Figure 10). Corner-sharing between these super-octahedra leads to a framework with a face-centered cubic unit cell and an overall $[Cu_3(BTC)_2(H_2O)_3]$ formula. HKUST-1 contains two types of octahedral pores with a diameter of 8.0 Å and 10.7 Å, interconnected via 4.7 Å windows (van der Walls radii included). The pores do not have the same diameter since half of them also contain H_2O molecules coordinated to the paddle-wheel SBUs. These can be readily removed under vacuum or replaced with other σ-donor ligands or coordinating solvents.

When a Keggin-type $[XM_{12}O_{40}]^{n-}$ POM is added in the reaction mixture as a template along with a small tetramethylammonium cation for charge balancing purposes, a new series of MOFs is obtained, known as NENU-MOFs, which are isostructural to HKUST-1 (NENU stands for Northeast Normal University). The POM is encapsulated and firmly trapped inside the larger pores of HKUST-1 since it is too bulky to exit through the relatively small 4.7 Å windows. Systematic investigation of the

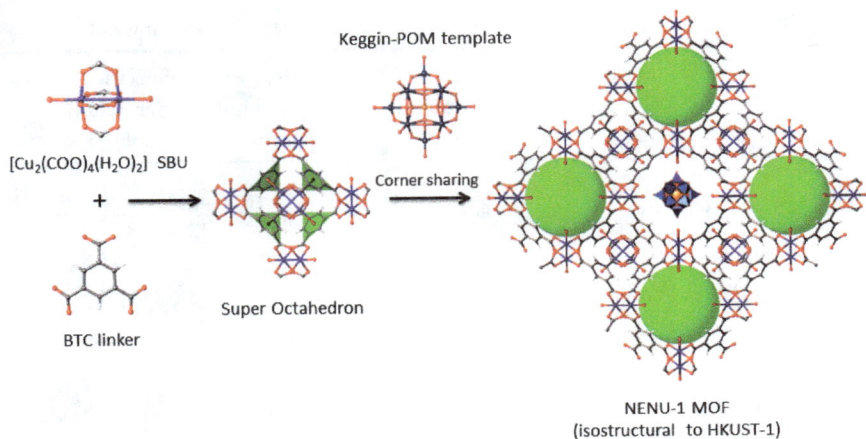

Figure 10: Crystal structure of NENU-1 with Keggin-POMs as templates. NENU-1 is isostructural to HKUST-1 and is built from 4-fold connecting Cu_2-based paddle-wheel SBUs and tritopic BTC linkers (left). Six paddle-wheel SBUs form a super-octahedron with four BTC linkers capping four of the eight faces (center). Corner-sharing leads to NENU-1 with the Keggin-POM encapsulated inside the larger size pores (right). Green spheres depict guest-accessible space

stability of the as synthesized NENU-MOFs reveals that they are stable up to 250°C with the encapsulated POM and the framework remaining intact. Moreover, they do not show any framework decomposition, loss of crystallinity or POM leaching in aqueous solutions (pH = 2–12) and common organic solvents even at elevated temperatures up to 80°C.[99]

The Keggin-type POMs are widely used as Brønsted acidic catalysts in solution, however catalytic applications of pure bulk POMs in the solid-state are limited due to their small BET surface area ($<10 \ m^2 \ g^{-1}$) rendering the active sites inaccessible to substrates.[102] Various supports with a higher surface area have been employed for the immobilisation and dispersion of POMs with limited success. To this end, NENU-3 containing the $[PW_{12}O_{40}]^{3-}$ POM was benchmarked as a heterogenised catalyst in the hydrolysis of ethyl acetate before and after thermal activation. When NENU-3 was degassed under high vacuum at 200°C all the solvent molecules and organic cations were removed from the pores. In order to maintain charge neutrality, the encapsulated POM was fully protonated,

Table 7: Activity of various catalysts for ethyl acetate hydrolysis in water.

Entry	Catalyst	Acid Amount (mmol g⁻¹)	Activity per Weight (μmol g_{cat}^{-1} min⁻¹)	Specific Activity per Acid (mmol mol_{acid}^{-1} min⁻¹)
1	HKUST-1	—	—	—
2	NENU-3[a]	0.18[b]	49.3	273.9
3	NENU-3act[a]	0.57[b]	178.9	313.8
4	$H_2PW_{12}O_{40}^{1-}$/SBA-15	0.091	25.1	275
5	$Cs_{2.5}H_{0.5}PW_{12}O_{40}$[c]	0.15	30.1	200.6
6	$Cs_3PW_{12}O_{40}$[c]	—	—	—
7	SO_4^{2-}/ZrO_2[c]	0.20	25.5	127.5
8	H-ZSM-5[c]	0.39	27.6	70.8
9	Nb_2O_5[c]	0.31	4.0	12.9
10	HY zeolite[c]	2.60	0.0	0.0
11	Nafion-H resin[d]	0.80	161.9	202.3
12	Amberlyst-15[d]	4.70	193.6	41.2
13	$H_3PW_{12}O_{40}$[e]	1.0	70.4	70.4
14	H_2SO_4[e]	19.8	911.9	46.1

[a] Reaction conditions: loading = 5.0 wt.%, solvent = H_2O, T = 60°C, t = 2 h, [AcOEt] = 0.56 M.
[b] Determined by acid-base titration.
[c] inorganic solid acids.
[d] organic solid acids.
[e] liquid acids.

forming [$H_3PW_{12}O_{40}$]–HKUST-1 (NENU-3act). The higher number of acidic sites after activation was verified with an acid-base titration.

Table 7 compares the catalytic activity in ethyl acetate hydrolysis of NENU-3act in excess water to that of other Brønsted acidic catalysts under the same reaction conditions. The activity of NENU-3act per catalyst weight (entry 3) was far superior to most inorganic solid acids (entries 5–10) and comparable to organic solid acids (entries 11–12). Importantly, when specific activity was measured, *i.e.* activity per number of Brønsted acidic sites of each catalyst (column 5), then NENU-3act outperforms all the other acids tested, including $H_3PW_{12}O_{40}$ immobilised on SBA-15 (a mesoporous silica) which was previously considered as the best catalyst (entry 4).[103] In addition, the yield of ethyl acetate hydrolysis was gradually increased for longer reaction periods, reaching a maximum of approximately 95% after 7 h. Control experiments confirmed the heterogeneous nature of the reaction since no reaction took place in the absence of NENU-3act or in the presence of only HKUST-1 (host MOF without an encapsulated POM). Even when $H_3PW_{12}O_{40}$ was simply mixed with HKUST-1, very low conversion was observed which can be attributed to the activity of $H_3PW_{12}O_{40}$ as a homogeneous catalyst (entry 13). This also shows that the POM anion can only be encapsulated via the bottle around ship approach and more importantly that encapsulation significantly increases catalytic activity. Excess water did not have any effect on the Brønsted acidic sites and NENU-3act was readily separated from the reaction mixture by simple filtration and recycled at least 15 times without losing its reactivity and selectivity. The encapsulated POM anion did not leach into the liquid phase even after thorough washing and drying at 200°C.

These results confirm that NENU-type MOFs possess excellent catalytic properties, comparable to the more conventional POM-immobilised catalysts and suggest that the encapsulation of robust TM catalysts inside the pores of MOFs via the bottle around ship approach could lead to hybrid material with unique catalytic properties. However, it should be reminded that the vast majority of MOFs is synthesized under rather harsh conditions. Therefore, the number of TM catalysts which can be used as templates is currently limited.

3.3 Direct cation exchange

An interesting alternative for the encapsulation of homogeneous TM catalysts inside the pores of MOFs is direct cation exchange during which a positively charged TM catalyst can be encapsulated as a whole in one step inside the pores of an anionic MOF through complementary electrostatic interactions to form a [catalyst]⁺–[MOF]⁻ material. Anionic MOFs are predominantly built from negatively charged SBUs interconnected by polytopic carboxylate-based linkers giving rise to anionic frameworks. The negative charge is balanced by organic cations trapped inside the pores which are either generated *in situ* upon solvent decomposition during the solvothermal synthesis or added in the reaction mixture as templates. These organic cations can be readily exchanged with cationic organic dyes,

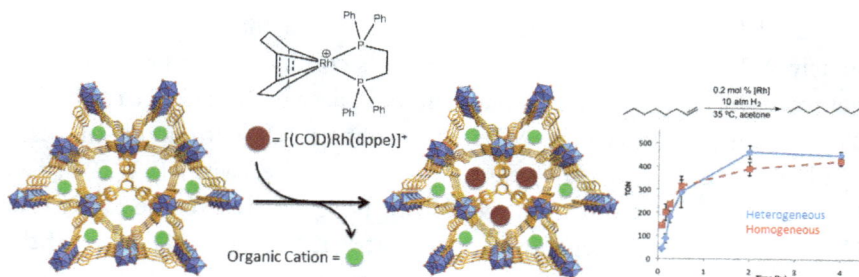

Figure 11: Encapsulation of [(COD)Rh(dppe)]$^+$ in ZJU-28 MOF via cation exchange. The immobilised catalyst compares favourably to its homogeneous counterpart (inset). Adapted from Ref. [115].

drugs or metal cations like Li$^+$ and Na$^+$.[104–114] Sanford *et al.* reported for the first time the heterogenisation of a series of cationic TM catalysts via a one-step cation exchange inside the pores of ZJU-28, an anionic indium-based MOF.[115] ZJU-28 consists of distorted tetrahedral [In^{III}(COO)$_4$]$^-$ SBUs interconnected via tritopic BTB linkers (BTB = benzene tribenzoate) forming an anionic [In$_3$(BTB)$_4$]$^{3-}$ framework. The negative charge is balanced by [Me$_2$NH$_2$]$^+$ cations generated *in situ* upon decomposition of DMF solvent used in synthesis.[116] ZJU-28 contains two types of channels running parallel to the long axis of the hexagonal unit cell (Figure 11) of approximately 8.0 Å, accounting for van der Waals radii.

[Rh(dppe)(COD)]$^+$ was successfully exchanged inside the pores of ZJU-28 via cation exchange [dppe = bis(diphenylphosphino)ethane; COD = 1,5-cyclooctadiene]. The degree of cation exchange was stoichiometrically controlled and 7–40% of the [Me$_2$NH$_2$]$^+$ cations were replaced by [Rh(dppe)(COD)]$^+$ depending on the nominal stoichiometric ratio between the endogenous and exogenous cations. Notably, encapsulation did not take place at all when neutral TM complexes or considerably larger cationic metalloporphyrins were examined as potential guests, verifying that the process is indeed driven by electrostatic interactions but limited by the size of the pore windows.

The catalytic activity of [Rh(dppe)(COD)]$^+$–ZJU-28 was investigated in the hydrogenation of 1-octene to *n*-octane (0.2 mol% Rh loading at 35°C under 10 atm H$_2$ in acetone). As shown in Figure 11, both the homogeneous [Rh(dppe)(COD)][BF$_4$] and heterogeneous [Rh(dppe)(COD)]$^+$–ZJU-28 catalysts afforded approximately 450 turnovers after 4 h corresponding to 90% conversion of 1-octene with the latter being slower only at the initial stages of the hydrogenation reaction. Moreover, the encapsulated catalyst was recycled four times under these conditions with minimal loss of activity. On the contrary, when the homogeneous reaction was recharged with additional substrate, only 200 additional turnovers were observed after further 4 h. Therefore, the immobilized catalyst exhibits superior catalytic performance relative to its homogeneous counterpart.

Direct cation exchange of a well-defined TM catalyst inside the pores of a solid state support requires relatively large pore windows, usually larger than 1 nm. Therefore, it is not considered as a viable option using zeolites as hosts and it has been only recently demonstrated with mesoporous silicates which however are not crystalline materials.[117] Following Sanford's work, a few more examples have been reported in which TM catalysts are immobilized inside the pores of MOFs via electrostatic interactions.[118-120] A number of other well-established positively charged TM catalysts with excellent catalytic activity in asymmetric hydrogenation, dehydrogenation and olefin metathesis could be used as guests. However, the number of anionic MOFs which contain large enough pore windows and are chemically and structurally stable under turnover conditions is currently rather small, thus restricting the applicability of this approach.

4 Conclusion and Perspective

The objective of this short review was to present comprehensive and representative examples in which MOFs serve as solid-state supports for the immobilisation of homogeneous TM catalysts. The most extensively used method for heterogenisation is the covalent bonding (grafting) of a TM catalyst onto the framework. This approach permits by far more flexibility regarding the selection of the TM catalyst for immobilization and the MOF as the host. On the other hand, grafting requires multiple synthetic steps involving amongst others the modification of the coordination environment of the catalyst which may eventually affect its overall catalytic properties.

By contrast, encapsulation does not perturb the catalyst's coordination environment and allows for a direct and more accurate comparison between the heterogeneous and the homogeneous systems. However, encapsulation requires that the host MOF must possess large enough pores to accommodate relatively bulky TM catalysts but narrow pore windows to prevent leaching of the encapsulated catalyst into the liquid phase. Taking also into account that the pore windows dictate circulation of the substrates and the products to and from the catalytic sites, it becomes clear that this approach calls for a very specific host–guest combination. Three different variations have been examined: (i) ship in a bottle; (ii) template synthesis also known as bottle around ship and (iii) direct cation exchange. Ship in a bottle is a very elegant approach, however it still requires multiple synthetic steps and most importantly the assembly reaction must be fast and produce a single product to suppress undesired side reactions. Template synthesis by default requires very robust catalysts that must survive the solvothermal conditions used for MOF synthesis. If successful, it results in ideally dispersed and firmly trapped catalytic species inside a porous host. In general, if the catalyst can be made easily in a small number of steps, then ship in a bottle is preferred. If the catalyst is difficult to make but is stable under solvothermal conditions, then bottle around ship is preferred. An interesting alternative is direct cation-exchange which in principle is the most facile method of encapsulation. It is based on the exchange of small organic or inorganic cations trapped inside the pores of anionic MOFs with cationic TM catalysts. Therefore, it is limited to cationic catalysts and most importantly anionic MOFs which are usually not as stable as their neutral analogues.

All the examples presented herein also demonstrate the benefits of immobilization of a TM catalyst inside a MOF, compared to other more traditional porous materials. MOFs are crystalline materials allowing for tailoring the density of the catalytic sites and for assessing structure-activity relationships. They show permanent porosity which can be pre-engineered via the proper selection of the principal structural components (linkers and SBUs). This provides a chemically and spatially tunable confined environment which can be crucial for advanced catalytic properties. The main disadvantage of MOFs is their instability compared to zeolites or mesoporous silicates. However, more and more thermally and chemically stable MOFs have been prepared in the last five years. This will allow for employing MOFs in more sophisticated areas of catalysis like asymmetric catalysis and one-pot multi-component coupling or tandem reactions where the superior tunability of MOFs can be exploited without being hampered by stability issues.

Acknowledgements

UK Catalysis Hub is kindly thanked for resources and support provided via our membership of the UK Catalysis Hub Consortium and funded by EPSRC (grants EP/K014706/1, EP/K014668/1, EP/K014854/1, EP/K014714/1 and EP/M013219/1). The author would like to acknowledge Professors Matthew J. Rosseinsky and Andrew S. Weller for helpful advice on the general outline of this chapter.

References

1. P. W. N. M. Leeuwen and J. C. Chadwick, *Homogeneous Catalysts: Activity, Stability, Deactivation,* Weinheim: Wiley-VCH, 2011.
2. R. H. Crabtree, *Chem. Rev.,* 2012, **112**, 1536.
3. A. Kobayashi, Y. Suzuki, T. Ohba, T. Ogawa, T. Matsumoto, S.-i. Noro, H.-C. Chang and M. Kato, *Inorg. Chem.,* 2015, doi: 10.1021/ic5021302.
4. T. M. McDonald, W. R. Lee, J. A. Mason, B. M. Wiers, C. S. Hong and J. R. Long, *J. Am. Chem. Soc.,* 2012, **134**, 7056.
5. J. M. Thomas and W. J. Thomas, *Principles and Practice of Hetergoeneous Catalysis,* VCH, Weinheim, 1997.
6. N. Mizuno and M. Misono, *Chem. Rev.,* 1998, **98**, 199.
7. A. Corma and H. García, *Chem. Rev.,* 2002, **102**, 3837.
8. U. Diaz, D. Brunel and A. Corma, *Chem. Soc. Rev.,* 2013, **42**, 4083.
9. C. E. Song and S.-g. Lee, *Chem. Rev.,* 2002, **102**, 3495.
10. D. E. De Vos, M. Dams, B. F. Sels and P. A. Jacobs, *Chem. Rev.,* 2002, **102**, 3615.
11. Q.-H. Fan, Y.-M. Li and A. S. C. Chan, *Chem. Rev.,* 2002, **102**, 3385.
12. N. E. Leadbeater and M. Marco, *Chem. Rev.,* 2002, **102**, 3217.
13. C. A. McNamara, M. J. Dixon and M. Bradley, *Chem. Rev.,* 2002, **102**, 3275.
14. P. McMorn and G. J. Hutchings, *Chem. Soc. Rev.,* 2004, **33**, 108.
15. J. M. Fraile, J. I. García and J. A. Mayoral, *Chem. Rev.,* 2009, **109**, 360.
16. G. Ferey, *Chem. Soc. Rev.,* 2008, **37**, 191.

17. M. Yoon, R. Srirambalaji and K. Kim, *Chem. Rev.*, 2012, **112**, 1196.
18. H. Furukawa, K. E. Cordova, M. O'Keeffe and O. M. Yaghi, *Science*, 2013, **341**, 1230444.
19. C. Wang, D. Liu and W. Lin, *J. Am. Chem. Soc.*, 2013, **135**, 13222.
20. Z.-J. Lin, J. Lu, M. Hong and R. Cao, *Chem. Soc. Rev.*, 2014, **43**, 5867.
21. A. Schneemann, V. Bon, I. Schwedler, I. Senkovska, S. Kaskel and R. A. Fischer, *Chem. Soc. Rev.*, 2014, **43**, 6062.
22. G. Férey, *Chem. Mater.*, 2001, **13**, 3084.
23. J.-R. Li, R. J. Kuppler and H.-C. Zhou, *Chem. Soc. Rev.*, 2009, **38**, 1477.
24. L. J. Murray, M. Dinca and J. R. Long, *Chem. Soc. Rev.*, 2009, **38**, 1294.
25. M. Kurmoo, *Chem. Soc. Rev.*, 2009, **38**, 1353.
26. P. Horcajada, T. Chalati, C. Serre, B. Gillet, C. Sebrie, T. Baati, J. F. Eubank, D. Heurtaux, P. Clayette, C. Kreuz, J.-S. Chang, Y. K. Hwang, V. Marsaud, P.-N. Bories, L. Cynober, S. Gil, G. Ferey, P. Couvreur and R. Gref, *Nat. Mat.*, 2010, **9**, 172.
27. A. Corma, H. García and F. X. Llabrés i Xamena, *Chem. Rev.*, 2010, **110**, 4606.
28. P. Horcajada, R. Gref, T. Baati, P. K. Allan, G. Maurin, P. Couvreur, G. Ferey, R. E. Morris and C. Serre, *Chem. Rev.*, 2012, **112**, 1232.
29. P. Garcia–Garcia, M. Muller and A. Corma, *Chem. Sci.*, 2014, **5**, 2979.
30. T. Zhang and W. Lin, *Chem. Soc. Rev.*, 2014, **43**, 5982.
31. C. He, D. Liu and W. Lin, *Chem. Rev.*, 2015, **115**, 11079.
32. J. Liu, L. Chen, H. Cui, J. Zhang, L. Zhang and C.-Y. Su, *Chem. Soc. Rev.*, 2014, **43**, 6011.
33. Z. Wang and S. M. Cohen, *Chem. Soc. Rev.*, 2009, **38**, 1315.
34. K. K. Tanabe and S. M. Cohen, *Chem. Soc. Rev.*, 2011, **40**, 498.
35. S. M. Cohen, *Chem. Rev.*, 2012, **112**, 970.
36. J. H. Cavka, S. Jakobsen, U. Olsbye, N. Guillou, C. Lamberti, S. Bordiga and K. P. Lillerud, *J. Am. Chem. Soc.*, 2008, **130**, 13850.
37. L. Valenzano, B. Civalleri, S. Chavan, S. Bordiga, M. H. Nilsen, S. Jakobsen, K. P. Lillerud and C. Lamberti, *Chem. Mater.*, 2011, **23**, 1700.
38. J. B. DeCoste, G. W. Peterson, H. Jasuja, T. G. Glover, Y.-g. Huang and K. S. Walton, *J. Math. Chem. A*, 2013, **1**, 5642.
39. H. Fei, J. Shin, Y. S. Meng, M. Adelhardt, J. Sutter, K. Meyer and S. M. Cohen, *J. Am. Chem. Soc.*, 2014, **136**, 4965.
40. T. Mallat and A. Baiker, *Chem. Rev.*, 2004, **104**, 3037.
41. C. Parmeggiani and F. Cardona, *Green Chemistry*, 2012, **14**, 547.
42. H. E. B. Lempers and R. A. Sheldon, *J. Catal.*, 1998, **175**, 62.
43. Z. Lounis, A. Riahi, F. Djafri and J. Muzart, *Appl. Catal. A: Gen.*, 2006, **309**, 270.

44. P. J. Figiel, A. Sibaouih, J. U. Ahmad, M. Nieger, M. T. Räisänen, M. Leskelä and T. Repo, *Adv. Synth. Catal.*, 2009, **351**, 2625.
45. M. Davi and H. Lebel, *Org. Lett.*, 2009, **11**, 41.
46. J. M. Hoover and S. S. Stahl, *J. Am. Chem. Soc.*, 2011, **133**, 16901.
47. H. Fei and S. M. Cohen, *J. Am. Chem. Soc.*, 2015, **137**, 2191.
48. K. Manna, T. Zhang and W. Lin, *J. Am. Chem. Soc.*, 2014, **136**, 6566.
49. M. I. Gonzalez, E. D. Bloch, J. A. Mason, S. J. Teat and J. R. Long, *Inorg. Chem.*, 2015, **54**, 2995.
50. C. Cheng and J. F. Hartwig, *Chem. Rev.*, 2015, **115**, 8946.
51. L. Chen, H. Chen, R. Luque and Y. Li, *Chem. Sci.*, 2014, **5**, 3708.
52. K. Manna, T. Zhang, F. X. Greene and W. Lin, *J. Am. Chem. Soc.*, 2015, **137**, 2665.
53. J. M. Falkowski, T. Sawano, T. Zhang, G. Tsun, Y. Chen, J. V. Lockard and W. Lin, *J. Am. Chem. Soc.*, 2014, **136**, 5213.
54. T. Sawano, N. C. Thacker, Z. Lin, A. R. McIsaac and W. Lin, *J. Am. Chem. Soc.*, 2015, **137**, 12241.
55. T. Sawano, P. Ji, A. McIsaac, Z. Lin, C. Abney and W. Lin, *Chem. Sci.*, 2015, doi: 10.1039/C5SC02100F.
56. B. Gui, K. K. Yee, Y. L. Wong, S. M. Yiu, M. Zeller, C. Wang and Z. Xu, *Chem. Commun.*, 2015, **51**, 6917.
57. K. Manna, T. Zhang, M. Carboni, C. W. Abney and W. Lin, *J. Am. Chem. Soc.*, 2014, **136**, 13182.
58. M. C. Das, S. Xiang, Z. Zhang and B. Chen, *Angew. Chem. Int. Ed.*, 2011, **50**, 10510.
59. W.-Y. Gao, M. Chrzanowski and S. Ma, *Chem. Soc. Rev.*, 2014, **43**, 5841.
60. M. Zhao, S. Ou and C.-D. Wu, *Acc. Chem. Res.*, 2014, **47**, 1199.
61. W. Morris, B. Volosskiy, S. Demir, F. Gándara, P. L. McGrier, H. Furukawa, D. Cascio, J. F. Stoddart and O. M. Yaghi, *Inorg. Chem.*, 2012, **51**, 6443.
62. D. Feng, Z.-Y. Gu, J.-R. Li, H.-L. Jiang, Z. Wei and H.-C. Zhou, *Angew. Chem. Int. Ed.*, 2012, **51**, 10307.
63. Y. Chen, T. Hoang and S. Ma, *Inorg. Chem.*, 2012, **51**, 12600.
64. H. Lu and X. P. Zhang, *Chem. Soc. Rev.*, 2011, **40**, 1899.
65. C.-D. Wu, A. Hu, L. Zhang and W. Lin, *J. Am. Chem. Soc.*, 2005, **127**, 8940.
66. S.-i. Noro, S. Kitagawa, M. Yamashita and T. Wada, *Chem. Commun.*, 2002, doi: 10.1039/b108695b, 222.
67. S. Kitagawa, S.-i. Noro and T. Nakamura, *Chem. Commun.*, 2006, doi: 10.1039/B511728C, 701.
68. M. Frisch and C. L. Cahill, *Dalton Trans.*, 2006, DOI: 10.1039/B608187H, 4679.
69. C. Wang, Z. Xie, K. E. deKrafft and W. Lin, *J. Am. Chem. Soc.*, 2011, **133**, 13445.

70. C. Wang, J.-L. Wang and W. Lin, *J. Am. Chem. Soc.*, 2012, **134**, 19895.

71. D. Feng, W.-C. Chung, Z. Wei, Z.-Y. Gu, H.-L. Jiang, Y.-P. Chen, D. J. Darensbourg and H.-C. Zhou, *J. Am. Chem. Soc.*, 2013, **135**, 17105.

72. D. Feng, Z.-Y. Gu, Y.-P. Chen, J. Park, Z. Wei, Y. Sun, M. Bosch, S. Yuan and H.-C. Zhou, *J. Am. Chem. Soc.*, 2014, **136**, 17714.

73. K. Wang, D. Feng, T.-F. Liu, J. Su, S. Yuan, Y.-P. Chen, M. Bosch, X. Zou and H.-C. Zhou, *J. Am. Chem. Soc.*, 2014, **136**, 13983.

74. O. K. Farha, A. M. Shultz, A. A. Sarjeant, S. T. Nguyen and J. T. Hupp, *J. Am. Chem. Soc.*, 2011, **133**, 5652.

75. X.-L. Yang and C.-D. Wu, *Inorg. Chem.*, 2014, **53**, 4797.

76. W. Xi, Y. Liu, Q. Xia, Z. Li and Y. Cui, *Chem. Eur. J.*, 2015, **21**, 12581.

77. A. Kozlov, A. Kozlova, K. Asakura and Y. Iwasawa, *J. Mol. Catal. A: Chem.*, 1999, **137**, 223.

78. N. A. Caplan, F. E. Hancock, P. C. Bulman Page and G. J. Hutchings, *Angew. Chem. Int. Ed.*, 2004, **43**, 1685.

79. K. Kervinen, P. C. A. Bruijnincx, A. M. Beale, J. G. Mesu, G. van Koten, R. J. M. Klein Gebbink and B. M. Weckhuysen, *J. Am. Chem. Soc.*, 2006, **128**, 3208.

80. J. Połtowicz, K. Pamin, E. Tabor, J. Haber, A. Adamski and Z. Sojka, *Appl. Catal., A*, 2006, **299**, 235.

81. R. J. Corrêa, G. C. Salomão, M. H. N. Olsen, L. C. Filho, V. Drago, C. Fernandes and O. A. C. Antunes, *Applied Catalysis A* 2008, **336**, 35.

82. K. K. Bania and R. C. Deka, *The Journal of Physical Chemistry C*, 2012, **116**, 14295.

83. K. K. Bania, G. V. Karunakar, K. Goutham and R. C. Deka, *Inorg. Chem.*, 2013, **52**, 8017.

84. X. Shi, B. Fan, H. Li, R. Li and W. Fan, *Microporous Mesoporous Mater.*, 2014, **196**, 277.

85. B. Li, Y. Zhang, D. Ma, T. Ma, Z. Shi and S. Ma, *J. Am. Chem. Soc.*, 2014, **136**, 1202.

86. J. An, S. J. Geib and N. L. Rosi, *J. Am. Chem. Soc.*, 2009, **131**, 8376.

87. Á. Zsigmond, F. Notheisz and J.-E. Bäckvall, *Catal. Lett.*, 2000, **65**, 135.

88. Y. Chen, B. Fan, N. Lu and R. Li, *Catal. Commun.*, 2015, **64**, 91.

89. G. Férey, C. Serre, C. Mellot-Draznieks, F. Millange, S. Surblé, J. Dutour and I. Margiolaki, *Angew. Chem. Int. Ed.*, 2004, **43**, 6296.

90. P. Horcajada, S. Surble, C. Serre, D.-Y. Hong, Y.-K. Seo, J.-S. Chang, J.-M. Greneche, I. Margiolaki and G. Ferey, *Chem. Commun.*, 2007, doi: 10.1039/B704325B, 2820.

91. C. Volkringer, D. Popov, T. Loiseau, G. Férey, M. Burghammer, C. Riekel, M. Haouas and F. Taulelle, *Chem. Mater.*, 2009, **21**, 5695.
92. M. R. Maurya, A. Kumar and J. Costa Pessoa, *Coord. Chem. Rev.*, 2011, **255**, 2315.
93. D. J. Xuereb, J. Dzierzak and R. Raja, *Catal. Today*, 2012, **198**, 19.
94. J. A. Martens, J. Jammaer, S. Bajpe, A. Aerts, Y. Lorgouilloux and C. E. A. Kirschhock, *Microporous Mesoporous Mater.*, 2011, **140**, 2.
95. Y. Wan and Zhao, *Chem. Rev.*, 2007, **107**, 2821.
96. F.-J. Ma, S.-X. Liu, C.-Y. Sun, D.-D. Liang, G.-J. Ren, F. Wei, Y.-G. Chen and Z.-M. Su, *J. Am. Chem. Soc.*, 2011, **133**, 4178.
97. Z. Zhang, L. Zhang, L. Wojtas, M. Eddaoudi and M. J. Zaworotko, *J. Am. Chem. Soc.*, 2012, **134**, 928.
98. Z. Zhang, L. Zhang, L. Wojtas, P. Nugent, M. Eddaoudi and M. J. Zaworotko, *J. Am. Chem. Soc.*, 2012, **134**, 924.
99. C.-Y. Sun, S.-X. Liu, D.-D. Liang, K.-Z. Shao, Y.-H. Ren and Z.-M. Su, *J. Am. Chem. Soc.*, 2009, **131**, 1883.
100. J.-S. Qin, D.-Y. Du, W. Guan, X.-J. Bo, Y. Li, L. Guo, Z.-M. Su, Y.-Y. Wang, Y.-Q. Lan and H.-C. Zhou, *J. Am. Chem. Soc.*, 2015, doi: 10.1021/jacs.5b02688.
101. S. S.-Y. Chui, S. M.-F. Lo, J. P. H. Charmant, A. G. Orpen and I. D. Williams, *Science*, 1999, **283**, 1148.
102. S. S. Wang and G. Y. Yang, *Chem. Rev.*, 2015, **115**, 4893.
103. K. Inumaru, T. Ishihara, Y. Kamiya, T. Okuhara and S. Yamanaka, *Angew. Chem. Int. Ed.*, 2007, **46**, 7625.
104. Y. Liu, V. C. Kravtsov, R. Larsen and M. Eddaoudi, *Chem. Commun.*, 2006, 1488.
105. Y. Liu, V. C. Kravtsov and M. Eddaoudi, *Angew. Chem. Int. Ed.*, 2008, **47**, 8446.
106. S. Yang, X. Lin, A. J. Blake, K. M. Thomas, P. Hubberstey, N. R. Champness and M. Schroder, *Chem. Commun.*, 2008, 6108.
107. S. Yang, X. Lin, A. J. Blake, G. S. Walker, P. Hubberstey, N. R. Champness and M. Schröder, *Nat. Chem.*, 2009, **1**, 487.
108. S. Chen, J. Zhang, T. Wu, P. Feng and X. Bu, *J. Am. Chem. Soc.*, 2009, **131**, 16027.
109. J. An and N. L. Rosi, *J. Am. Chem. Soc.*, 2010, **132**, 5578.
110. T.-F. Liu, J. Lü, C. Tian, M. Cao, Z. Lin and R. Cao, *Inorg. Chem.*, 2011, **50**, 2264.
111. S. Yang, X. Lin, W. Lewis, M. Suyetin, E. Bichoutskaia, J. E. Parker, C. C. Tang, D. R. Allan, P. J. Rizkallah, P. Hubberstey, N. R. Champness, K. Mark Thomas, A. J. Blake and M. Schröder, *Nat. Mat.*, 2012, **11**, 710.
112. T. Li and N. L. Rosi, *Chem. Commun.*, 2013, **49**, 11385.
113. J. Yu, Y. Cui, H. Xu, Y. Yang, Z. Wang, B. Chen and G. Qian, *Nat. Commun.*, 2013, **4**.

114. L. Liu, X.-N. Zhang, Z.-B. Han, M.-L. Gao, X.-M. Cao and S.-M. Wang, *J. Math. Chem. A*, 2015, **3**, 14157.

115. D. T. Genna, A. G. Wong-Foy, A. J. Matzger and M. S. Sanford, *J. Am. Chem. Soc.*, 2013, **135**, 10586.

116. J. Yu, Y. Cui, C. Wu, Y. Yang, Z. Wang, M. O'Keeffe, B. Chen and G. Qian, *Angew. Chem. Int. Ed.*, 2012, **51**, 10542.

117. X.-Z. Shu, S. C. Nguyen, Y. He, F. Oba, Q. Zhang, C. Canlas, G. A. Somorjai, A. P. Alivisatos and F. D. Toste, *J. Am. Chem. Soc.*, 2015, **137**, 7083.

118. A. Grigoropoulos, G. F. S. Whitehead, N. Perret, A. P. Katsoulidis, F. M. Chadwick, R. P. Davies, A. Haynes, L. Brammer, A. S. Weller, J. Xiao and M. J. Rosseinsky, *Chem. Sci.*, 2016, **7**, 2037.

119. D. T. Genna, L. Y. Pfund, D. C. Samblanet, A. G. Wong-Foy, A. J. Matzger and M. S. Sanford, *ACS Catal*, 2016, **6**, 3569.

120. J. López-Cabrelles, G. Mínguez Espallargas and E. Coronado, *Polymers*, 2016, **8**, 171.

Chapter 5

Polymerisation Catalysis Using CO_2: Dinuclear Homogeneous Catalysts

Prabhjot K. Saini and Charlotte K. Williams*

Department of Chemistry, Imperial College London, London SW7 2AZ

*c.k.williams@imperial.ac.uk

1 Using CO_2 as a Feedstock

Carbon dioxide is a ubiquitous renewable feedstock and would be a highly abundant, low cost C_1 source. Considering these attributes and its low toxicity, CO_2 is one of the most desirable feedstocks for the production of fine chemicals.[1] From an environmental perspective, replacing petrochemicals with CO_2 is attractive to reduce fossil fuel depletion and, in some cases, greenhouse gas emissions. Currently, the scale of using CO_2 for chemical synthesis remains rather small ($\approx$100 Mt per year), this is not least due to challenges in activating and catalysing reactions using this stable molecule. It should always be made clear that using carbon dioxide to make chemicals is very unlikely to be able to influence net CO_2 levels in the atmosphere, but it can contribute to reducing greenhouse gas emissions associated with chemical manufacture and, if coupled with carbon capture and sequestration (CCS), can provide an economic stimulus for large-scale carbon dioxide sequestration.[1a,1b,2]

The thermodynamic stability of CO_2 makes its use as a feedstock difficult, but not impossible.[1b,3] Before CO_2 can be used in the synthesis of important chemicals, it needs to be reduced, which generally requires high temperatures and pressures; the addition of highly reactive reagents or the use of catalysts that can lower the energy barriers for incorporating CO_2 in chemical reactions. Such catalysts will allow accessing CO_2 as a feedstock at lower physical energy inputs, which are more manageable and less costly.[1b,3a]

2 Polymer Production

Currently 288 Mt of polymers are produced worldwide annually.[1b,4] Approximately, 8% of the world's oil and gas supply annually is used in polymer synthesis.[1b,5] More specifically, 5 Mt of polycarbonates are synthesised by industry per year.[4a–4c] They are used in many applications, such as, clothing, adhesives, packaging and construction.[1b] The conventional route to producing polycarbonates involves copolymerising bisphenol A and phosgene, using a base catalyst (NaOH).[1b]

A desirable alternative process involves the ring opening copolymerisation (ROCOP) of epoxides and CO_2 (Figure 1).[6] This is because approximately 20–40% of the polycarbonate produced is derived from a renewable resource (CO_2), depending on the epoxide monomer used.[7] By using renewable epoxides, such as, limonene oxide provides the potential for synthesising 100% renewably derived polycarbonates.[1b,8] Although, it must be emphasised that the properties of such polycarbonates are quite different to those of the material produced from bis(phenol) A.

Conventionally, low number average molecular weights (M_n) polycarbonates are produced and these products may be used in the synthesis of higher polymers, such as, polyurethanes.[7a,9] Polyurethanes are also a highly desirable commodity (roughly 20 Mt are produced globally each year).

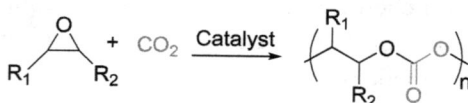

Figure 1:   Polycarbonate formation from the ROCOP of epoxides and CO_2.

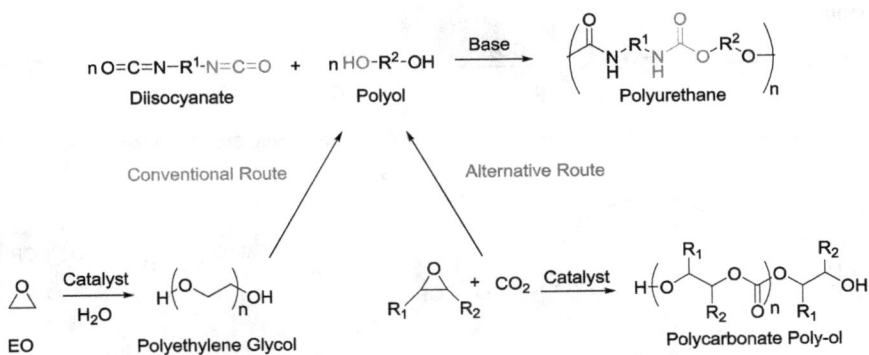

Figure 2: Two synthesis routes for polyurethane production.

Polyurethanes are used to make furniture, coatings, elastomers, adhesives and insulation foams.[4a–4c] Currently, a significant proportion of polyurethanes are synthesised using polyether polyols, which are reacted with diisocyanates. The polyols are usually synthesised by the homopolymerisation of epoxides.[10] However, substituting these polyols with polycarbonate polyols, derived from the copolymerisation of epoxide and CO$_2$, has been demonstrated to significantly improve the sustainability of polyurethane synthesis, including reducing fossil fuel depletion and greenhouse gas emissions by ~20% (Figure 2).[7a] Furthermore, in some cases, it has been shown that the properties of the polycarbonate containing polyols are similar, or even better, than conventional polyether polyols.[11]

3 Catalyst Development for Epoxide/CO$_2$ ROCOP

Catalysts for the ROCOP of epoxides and CO$_2$ have been explored since the late 1960s.[6] Frequently, the catalysts incorporate a Lewis acidic metal halide, carboxylate, alkoxide or aryloxide functionality with an ancillary ligand (L).[1b]

The proposed mechanism involves epoxide monomer coordination to the metal centre and ring opening by the nucleophilic attack of the co-ligand (X — in Figure 3).[12] Then, CO$_2$ insertion occurs to form a metal carbonate species. The metal carbonate species produced then nucleophilically attacks another metal bound epoxide monomer, to form another

Initiation:

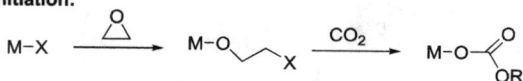

Propagation:

Potential Side Reactions:

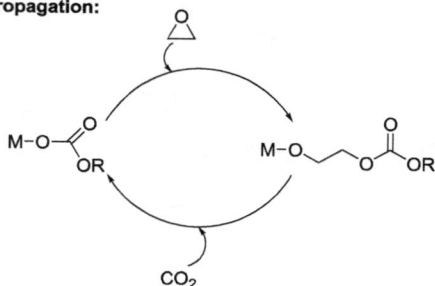

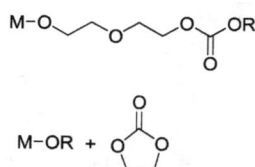

Chain Transfer:

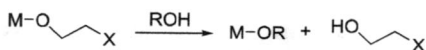

M = Lewis acidic metal centre
X = Halide, carboxylate
ROH = Alcohol

Figure 3:   Mechanism for the ROCOP of epoxide and CO_2. Adapted from Ref. [1b] with the permission from the Royal Society of Chemistry.

metal alkoxide species. This cycling between metal alkoxide and metal carbonate species produces a polycarbonate chain.[1b]

Two side reactions can occur during these copolymerisation reactions. Either the sequential enchainment of epoxide monomers to form ether linkages within the polycarbonate chain, or backbiting within the propagating polycarbonate chain to form a five-membered cyclic carbonate by-product may occur. This latter by-product is thermodynamically more stable than the copolymer and therefore forms more readily at higher temperatures.[1b]

Furthermore, the ROCOP of epoxides and CO_2 also undergo chain transfer reactions with alcohols or water. Chain transfer reactions produce hydroxyl terminated copolymer chains and metal alkoxide or hydroxide species, which can further initiate copolymerisation.[1b] Under such conditions, the M_n of the copolymer chains produced depend on the quantity of alcohol/chain transfer agent added and the dispersity indices (PDIs) remain narrow. Such polymerisations are termed "immortal polymerisations" and are characterised by the rate of chain transfer being more rapid than chain propagation.[13]

The productivity and activity of catalysts are described by the turnover number (TON), which is the number of mols of monomer consumed per mol of catalyst used, and the turnover frequency (TOF), which is the TON/h, respectively. The productivity of the catalyst can also be reported as a g/g value, which may be of more use in an industrial context.[1b]

4 Catalyst Development — Early Discoveries

Many homogeneous and heterogeneous catalysts have been synthesised over the past 60 years and the area has been extensively reviewed.[1b,8d,14] One leading hypothesis, and the topic for this chapter, is the mechanistic proposal that bimetallic or dinuclear catalysts are necessary for high activity and selectivity. Such concepts were articulated since the inception of the field but were not followed for a number of decades[6,8b,15]; here, a selection of relevant studies are presented which demonstrate the potential for bimetallic catalysts. The reader is also referred to other reviews for an extensive coverage and discussion of the development of this hypothesis.[1b,8d,14]

A significant and important class of homogeneous catalyst are the zinc β-diiminate (BDI) complexes, which were first reported by Coates and co-workers (Figure 4). In a series of elegant mechanistic and kinetic studies, Coates and co-workers established that the most active catalysts adopted

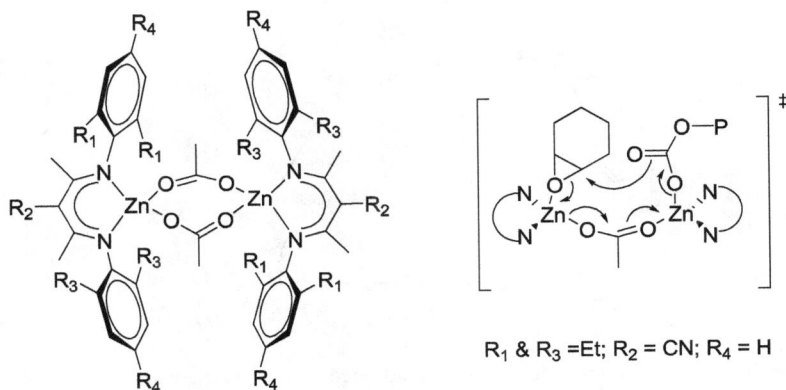

R₁ & R₃ = Et; R₂ = CN; R₄ = H

Figure 4: Zinc β-diiminate complex and proposed bimetallic transition state.[12,16]

dimeric structures under the conditions of the catalysis. Thus, it was proposed that the catalysis was enabled by a pathway involving two metal centres during copolymerisation reactions. It was suggested that one metal coordinates the epoxide and the other metal provides the co-ligand or growing copolymer chain to attack the epoxide for ring opening and thus alkoxide formation (Figure 4).[12,16]

Some years later, Rieger and co-workers re-visited a very well-known class of heterogeneous catalysts — the zinc dicarboxylate surfaces.[17] They proposed that catalysts which featured a high proportion of Zn–Zn surfaces with separations of 4.6–4.8 Å had the highest activity. An ideal Zn–Zn distance of 4.3–5.0 Å was calculated in an accompanying DFT study. It was suggested that these separations may reduce the activation energy barrier to epoxide ring opening. Additionally, it was hypothesised that such heterogeneous zinc dicarboxylate catalysts operate by a bimetallic mechanism (Figure 5).[17]

Recently, a surge into the development of discrete bimetallic catalysts for epoxide/CO_2 copolymerisation reactions has occured.[28] Three main classes of bimetallic complexes have been developed: bimetallic complexes containing a single binucleating ligand, monometallic complexes that form dimeric and multimeric structures and monometallic complexes covalently tethered together via the ligand motif. In the remainder of the

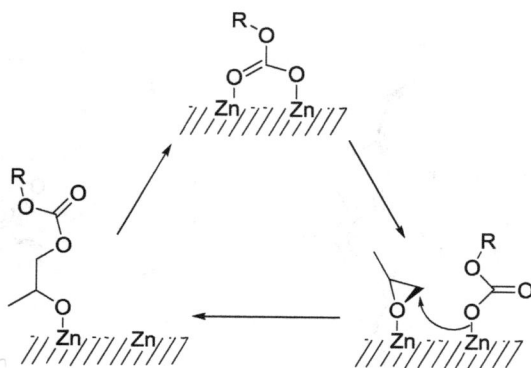

Figure 5:   Proposed bimetallic mechanism adopted by heterogeneous zinc dicarboxylate catalysts. Reprinted with permission from Ref. [17]. Copyright (2011) American Chemical Society.

chapter, we will focuss only on the bimetallic complexes containing a single binucleating ligand.

5 Dinucleating Ligand Motifs for Bimetallic Catalysis

Lee and co-workers reported one of the first bimetallic catalysts based on an anilido-aldimine ligand coordinated to two zinc metal centres (Figure 6).[18] These catalysts are very active (TON = 2980) in cyclohexene oxide (CHO) and CO$_2$ copolymerisation reactions and produced high CO$_2$ content (85–96%) copolymers. The high M_n values (90–280 kg/mol) reported showed broader dispersities (1.30–1.70). Structure-activity studies revealed that bulky groups on the R′ position and electron withdrawing groups (fluorine) on the anildo-amidinate moieties (R″) increased the activity.[18a] Adding bulky substituents on the R position reduced the activity (Figure 6).[18]

Ding *et al.* also reported di-zinc and di-magnesium complexes coordinated by a chiral phenolate ligand (Figure 7).[19] The zinc analogue showed only a low activity in CHO/CO$_2$ copolymerisation at 80°C (TOF = 142 h^{-1}) and was used at high loadings (5 mol%) compared to the zinc anilido-aldimine catalysts (0.006 mol%).[18–20]

Very recently, Lin and co-workers have synthesised bis(benzotriazole iminophenol) bimetallic zinc(II), nickel(II) and cobalt(II) catalysts for

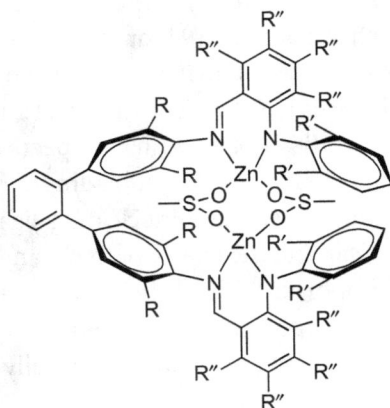

R = Me; R′ = ^{i}Pr; R″ = H

Figure 6:   Zinc anildo-amidinate complexes by Lee and co-workers.[18]

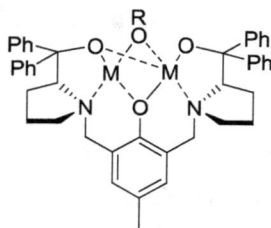

M = Zn, Mg; R = Et, Bu

Figure 7:   Trost phenolate complexes by Refs. [19] and [20].

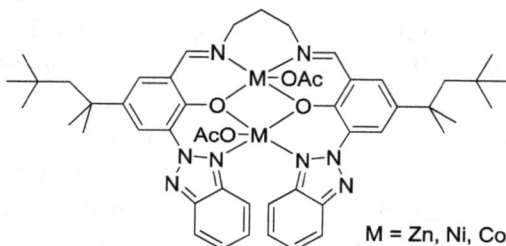

M = Zn, Ni, Co

Figure 8:   Complexes produced by Refs. [21] and [22].

CHO/CO$_2$ copolymerisation reactions (Figure 8).[21] The di-zinc catalyst has poor copolymer selectivity (66% cyclic carbonate) compared to the di-nickel and di-cobalt analogues (94 or >99% copolymer selectivity, respectively). However, the copolymer produced did not contain a high proportion of ether linkages (>99%). The di-nickel and di-cobalt catalysts have similar activities (TOF = 40 and 53 h^{-1}, respectively at 120°C, 21 bar of CO$_2$ pressure and 0.0625 mol% catalyst loading).[21,22]

So far, the range of deliberately dinucleating ligands remains rather narrow; however, it is clear that the choice of metal and substituents on the ligand backbones are important. This is because the electronic properties of the metal centre affect epoxide binding and the lability of the metal co-ligand or metal carbonate bonds to nucleophilically ring open epoxides.

6 Macrocyclic Dinucleating Catalysts

Our research team have also synthesised several homodinuclear catalysts based on a single binucleating macrocyclic ligand.[23] A range of

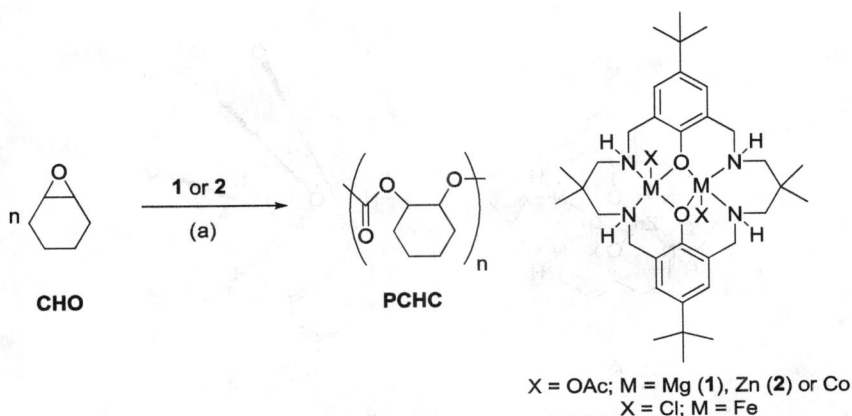

X = OAc; M = Mg (**1**), Zn (**2**) or Co
X = Cl; M = Fe

Scheme 1: Catalyst **1** or **2** in CHO/CO$_2$ copolymerisation reactions using; (a) 80°C, 1 bar CO$_2$ pressure.

metals have been investigated with an order of activity of Co(III/II) > Mg(II) > Zn(II) > Fe(III), at 80°C, 1 bar CO$_2$ pressure and 1:1000 catalyst:epoxide loading. The catalysts are particularly remarkable in being able to operate at 1 bar pressure and show a high tolerance to impurities. The activity and productivity obtained for these catalysts in cyclohexene oxide and CO$_2$ ROCOP reactions can be controlled and maximum values of 27,000 (TON) and 9000 h^{-1} (TOF) have recently been reported (Scheme 1).[23a,23e–i,24]

In a series of unoptimised reactions in glassware, the di-magnesium catalyst (**1**) showed a rate which was double that of the di-zinc derivative.[23g,23h] Various other di-zinc complexes, coordinated by derivatives of the macrocyclic ligand showed slightly reduced activity, for example the electron donating capabilities of the methoxy phenolate substituent was investigated.[23i] Additionally, trimetallic zinc and cobalt(III) catalysts (Figure 9) showed lower activities compared to their dinuclear counterparts. The third Zn or Co(III) metal centre is proposed to hinder the ability of CHO and CO$_2$ to bind to the catalyst and hence result in a lower activity.[23i,24]

Experimental, spectroscopic and computational investigations using theses dinuclear catalysts, specifically, the di-zinc and di-cobalt derivatives, have been carried out in order to gain insight into the mechanism adopted by these catalysts in CHO/CO$_2$ copolymerisation reactions.[23f,25]

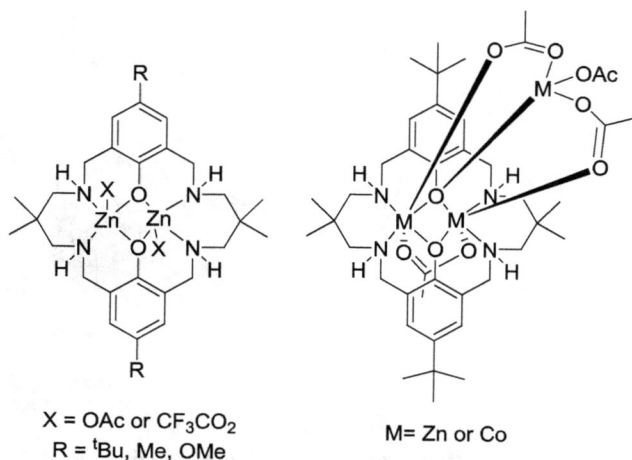

X = OAc or CF$_3$CO$_2$
R = ^{t}Bu, Me, OMe

M= Zn or Co

Figure 9:   Analogues of the produced by Williams and co-workers.[23f,g,i,24]

7 Rate Law for the Di-zinc Catalyst (2)

The rate law for CHO/CO$_2$ copolymerisation reactions using catalyst **2** (di-zinc catalyst) is:

$$\text{Rate} = k_p[\text{CHO}][\text{catalyst}]. \qquad (1)$$

The rate law suggests that the rate determining step of the ROCOP reactions is the ring opening of the epoxide and not CO$_2$ insertion.[25b] The reaction is first order in the concentration of catalyst **2** which indicates that both metal centres of the homodinuclear catalyst are involved in the rate limiting step and that catalyst aggregation does not occur. The first-order dependence on the epoxide monomer concentration (cyclohexene oxide — CHO) implies that only one copolymer chain grows per catalyst molecule and therefore suggests that only one of the acetate co-ligand groups initiates copolymerisation. The other acetate ligand is hypothesised to maintain an octahedral coordination sphere at the zinc centres and to ensure that the overall complex neutral charge state is balanced. It is proposed to maintain a bridging coordination mode between the two metal centres throughout the polymerisation.[25b] The rate of copolymerisation is zero order on the pressure of CO$_2$ (1–40 bar), which indicates that CO$_2$ insertion is non-rate determining.[25b]

8 Support for Dinuclear Mechanism — Di-cobalt (II) Catalysts

Various propagating structures have been reported in literature: (A) — dinuclear mechanism[12,18,20,26]; (B) intermolecular binuclear mechanism;[26–27] or (C) and (D) — mononuclear co-catalyst mechanisms[28] (Figure 10) and were investigated using a series of di-cobalt(II) catalysts.[23f,29]

Di-cobalt(II) derivatives of the Williams' group catalysts have been reported and consist of a macrocyclic ligand, two chloride co-ligands and a neutral nucleophile (Figure 11).[29a,30] These analogues crystallised in a fashion where the ligand adopts a conformation that consists of all the N–H substituents on the same face (concave) of the molecule (Figure 11). The ligand adopts a "bowl-shape" and thus has two different faces.[23f] Moreover, a chloride co-ligand bridges between the two metal centres on the concave face of the catalyst structure and due to its small size, the two axial coordination sites on the convex side of the complex are pushed apart. Therefore, the other chloride co-ligand on the convex face does not bridge between the two metal centres (Figure 11) and hence one axial position is occupied by a chloride co-ligand and the other vacant axial site is occupied by a nucleophile (methyl imidazole — MeIm or dimethyl amino pyridine — DMAP).[23f]

These di-cobalt(II) catalysts are active in CHO/CO_2 copolymerisation reactions under unoptimised conditions in laboratory glassware. Amongst the series of compounds, the TOF values obtained (20 h^{-1} and 104 h^{-1}, DMAP and MeIm, respectively), are lower than the TOF value recorded

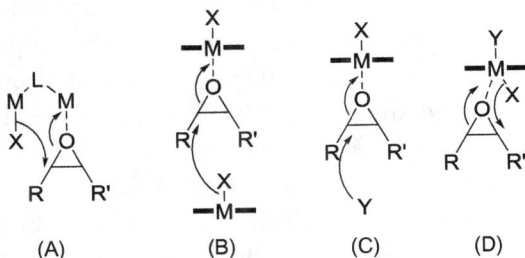

(A)	(B)	(C)	(D)

Figure 10: Proposed mechanisms for epoxide/CO_2 copolymerisation reactions. M = metal, L = ligand, X = co-ligand or initiator, Y = co-catalyst.

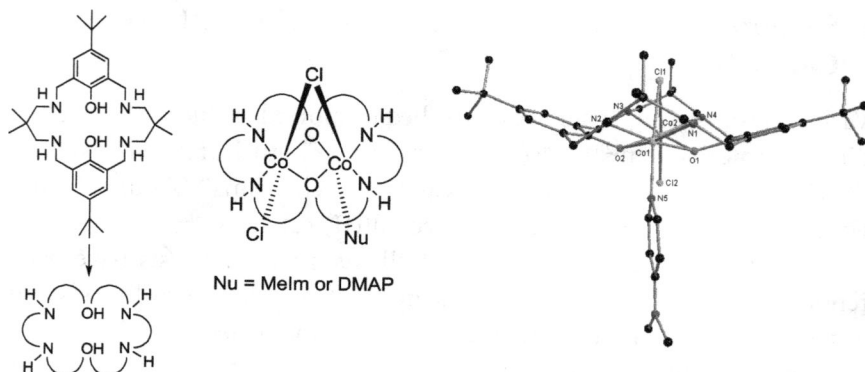

Figure 11: Di-cobalt catalyst with crystal structure showing the "bowl-shape" conformation of the macrocyclic ligand.[23f]

for the di-cobalt catalyst with acetate as the co-ligands (Scheme 1 — 159 h^{-1}).[23f,24] Additionally, the poly(cyclohexene) carbonate (PCHC) samples have a bimodal M_n distribution and MALDI-ToF mass spectrometry showed that one copolymer chain series is Cl end-capped and the other series is hydroxyl end-capped. No chains were end-capped with either of the nucleophilic co-catalysts used (DMAP and MeIm) and hence it can be inferred that the co-catalysts do not initiate the copolymerisation.[2]

From these findings, it was hypothesised that the nucleophile dissociates from the cobalt metal centres, once a vacant coordination site is formed, the epoxide can bind to the metal centre. The DMAP molecule binds strongly to the cobalt metal centres and thus this complex has a low activity (TOF = 20 h^{-1}).[23f] The reduction in activity also implies that both chloride co-ligands do not initiate copolymerisation. Moreover, this observation implicates both metals in the copolymerisation reaction.[23f] Overall, these results suggest an intramolecular dinuclear mechanism is adopted during CHO/CO$_2$ copolymerisation reactions (A — Figure 10).[23f]

Additionally, the rate of propagation is identical for both the di-cobalt analogues (DMAP and MeIm), which is expected as they both have the same chemical structure once the nucleophile dissociates from the metal centre. The difference in the TOF values recorded (20 h^{-1} vs. 104 h^{-1}) is explained by different initiation periods. This is because TOF is a point kinetic measurement and combines initiation and propagation periods.[23f] The DMAP derivative has a longer initiation time (takes longer to dissociate) compared to the MeIm analogue.

9 Computational Studies

Computational studies using the dinuclear catalysts in CHO/CO$_2$ copolymerisation reactions (using DFT and the basis set 6-31G(d)) showed that the "bowl-shape" confirmation was favoured, as was the growth of one copolymer chain per catalyst structure and therefore, one acetate co-ligand remains bridging between the two zinc metal centres.[25a]

Moreover, the potential energy surface (Figure 12) for CHO/CO$_2$ copolymerisation reactions using catalyst **2** was determined. The "bowl-shape" catalyst conformation was used as the initial point with one site for

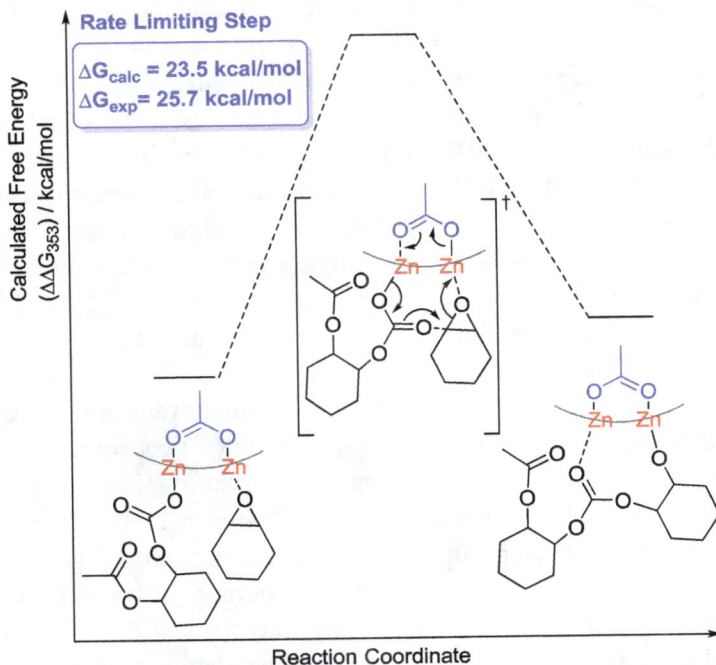

Figure 12: Illustrates the calculated rate limiting step in the copolymerisation cycle. Reprinted with permission from Ref. [25a]. Copyright (2012) American Chemical Society.

copolymer chain growth. The results revealed that the alternating incorporation of epoxide and CO_2 molecules within the growing copolymer occurs more favourably compared to side reactions. Sequential epoxide addition to produce ether linkages requires more energy and sequential CO_2 insertion is reversible and thermodynamically disfavoured. The calculations also suggest that CO_2 insertion is not rate limiting, the insertion process has a low activation energy barrier. In contrast, the rate determining step (process with highest activation energy barrier) is the ring opening of CHO by a zinc-carbonate species, a theoretical finding which is in line with the experimentally derived rate law.[25a]

10 Solid State and *In Situ* IR Spectroscopy Studies

In order to investigate the catalyst structure, in particular the coordination modes of the carboxylate co-ligands, during polymerisation, a series of spectroscopic studies using the di-zinc catalyst (**2**) were carried out.[25a] The solid state IR spectrum and two *in situ* IR spectra of catalyst **2** were recorded. The latter *in situ* IR spectra showed catalyst **2** dissolved in TCE, which was heated for 15 min at 80°C and catalyst **2** dissolved in neat CHO, heated for 2 h at 80°C and stirred for 14 h under 1 bar of CO_2 pressure.[25a] The solid state IR spectrum of catalyst **2** showed resonances at 1,591 and 1,615 cm^{-1} assigned to the two carboxylate groups. The two values are attributed to the two bridging acetate co-ligands being in different chemical environments, which can be explained by the "bowl-shape" conformation of the ligand motif (structure (E) — Figure 13). The *in situ* IR spectrum of catalyst **2** in TCE has a shifted carboxylate resonance (1,601 cm^{-1}), which could be due to solvent effects. A small signal at 1,737 cm^{-1} indicates a terminal acetate (κ^1) coordination mode, assigned to the ligand on the convex face (structure (F) — Figure 13). The second *in situ* IR spectrum produced a signal at 1,741 cm^{-1} assigned to a κ^2 bound acetate group and two resonances at 1,804 cm^{-1} and 1,825 cm^{-1} assigned to a zinc-bound κ^1 carbonate group and a cyclohexene bound acetate group, respectively (structure (G) — Figure 13). These observations support the notion that one acetate group may remain bridging between the two metal centres during the copolymerisation reaction.

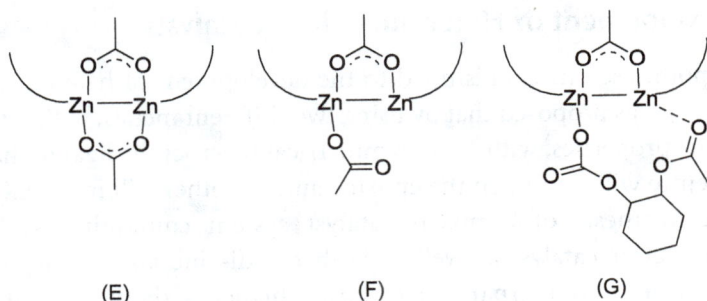

Figure 13:   Catalyst structures observed during solid state and *in situ* IR spectroscopy.

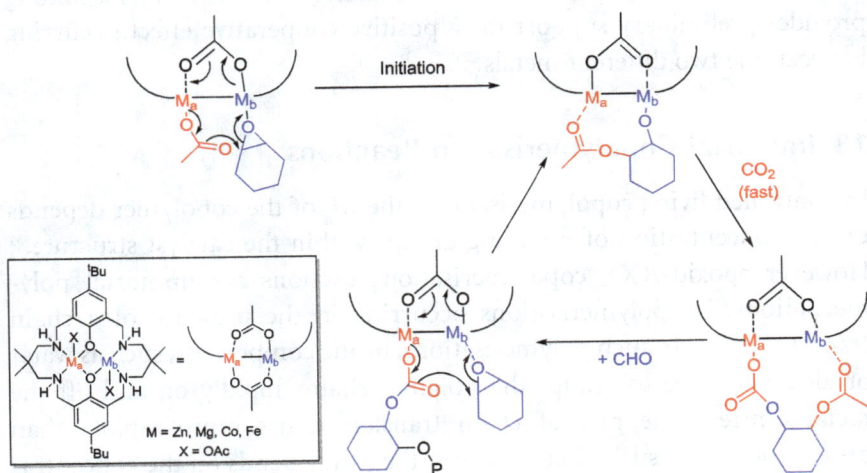

Figure 14:   Hypothesised mechanism for homodinuclear catalysts.[25a]

11 Hypothesised Mechanism

The hypothesised mechanism for catalyst **2** in the ROCOP of CHO and CO$_2$ indicates that both zinc metal centres are involved in the copolymerisation reaction and that one acetate group remains bridging on the concave face of the catalyst. The copolymer chain is proposed to propagate on the convex face of the catalyst and the copolymer chain "shuttles" between the two metal centres of the catalyst twice per catalytic cycle (Figure 14).[25a]

12 Development of Heterodinuclear Catalysts

The hypothesised mechanism led to the development of heterodinuclear catalysts. It was proposed that by using two different metals, with differing electronic properties, within the symmetrical macrocyclic ligand, that one metal centre will ring-open the epoxide and the other will insert CO_2.

The synthesis of a mixed catalyst system containing a Zn/Mg heterodinuclear catalyst as well as both the di-zinc and di-magnesium homodinuclear counterparts has shown higher activity in CHO/CO_2 copolymerisation reactions at 1 bar CO_2 pressure and 80 °C (TOF = 79 h^{-1}) compared to the homodinuclear catalysts alone (TOF = 17 h^{-1} for di-Zn and 52 h^{-1} for di-Mg) or in a 50:50 combination (TOF = 40 h^{-1}). This finding provides preliminary support for a positive cooperative effect occurring between the two different metals.[23j]

13 Immortal Copolymerisation Reactions

In controlled living copolymerisations, the M_n of the copolymer depends on the concentration of initiating groups within the catalyst structure.[31] However epoxide/CO_2 copolymerisation reactions are immortal polymerisations, i.e. polymerisations occurring in the presence of a chain transfer agent.[32] In such polymerisations, protic compounds, such as water or alcohols, serve to "swap" the polymer chains rapidly on and off the metal centres. The rate of chain transfer occurs more rapidly than propagation.[32a] In such circumstances, the Mn depends on the concentration of protic compounds present, both provided by the catalyst and the chain transfer agent.[23-25,25]

Additionally, it is commonly observed that the PCHC polymers also have bimodal M_n distributions even when chain transfer agents are not deliberately added. In the case of the dinuclear catalysts, MALDI-ToF reveals two series of chains: one is acetate end-capped and the other is di-hydroxyl end-capped. The acetate end groups result from initiation by the homodinuclear catalyst. However, the other copolymer series, which was usually double the M_n of the acetate end-capped chains, is hydroxyl end-capped. The latter series forms from chain transfer reactions with cyclohexane-1,2-diol (CHD) and the growing copolymer chain. It is proposed that CHD forms as a result of the epoxide, CHO, reacting with trace

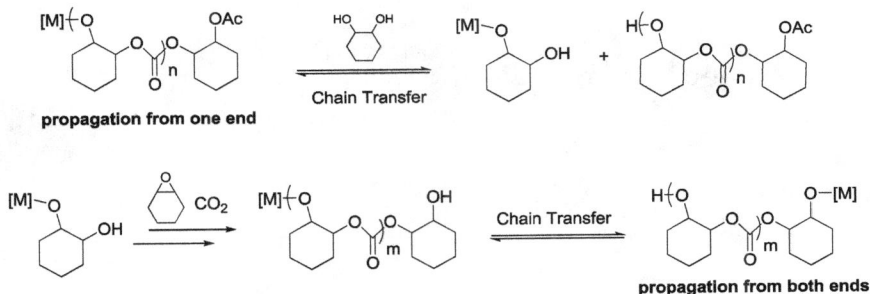

Figure 15:   Cyclohexane-1,2-diol (CHD) mediated chain transfer reactions in CHO/CO$_2$ ROCOP.

amounts of water present in the copolymerisation reaction or in the carbon dioxide. The hydroxyl end-capped series is double the M_n of the acetate end-capped copolymers because the CHD initiated copolymer propagates from both hydroxyl groups to form a telechelic copolymer (Figure 15).[25]

14 Side Reactions During Copolymerisation Reactions

Side reactions within ROCOP form either ether linkages or cyclic carbonates as by-products.[1b] Ether linkages form from the sequential ring opening of epoxide monomers. This presence of ether linkages in the PCHC chains formed can affect the thermal, chemical and mechanical properties of the copolymer, for example, these linkages reduce the T_g (glass transition temperature).[33]

Cyclic carbonate by-products form from backbiting reactions within the growing copolymer chain,[1b] which can be metal bound or dissociated from the metal centre (Figure 16).[34] *Trans*-cyclic carbonate is usually formed and occurs because of backbiting from a metal alkoxide species.[1b,23e,g,29d] *Cis*-cyclic carbonate is rarely observed, but occurs from backbiting from metal carbonate species (Figure 16).[23e,34]

Elevated temperatures are required to form cyclic carbonate by-products because they are the thermodynamic product (Figure 17).[25b] The energy required to convert copolymer to cyclic carbonate with the di-zinc catalyst (**2**) is 137.5 kJmol^{-1}. This is similar to the energy requirements needed to

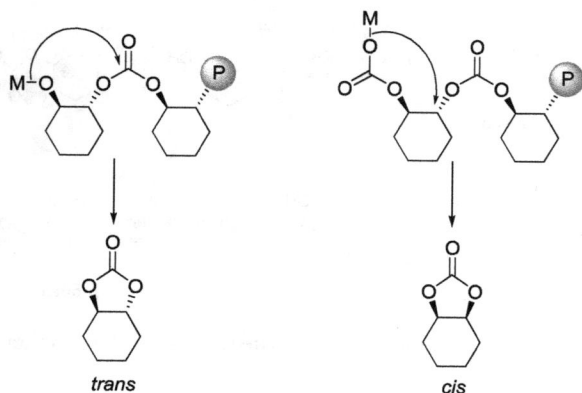

Figure 16:   Formation of *trans*- and *cis*-cyclic carbonate by-product.

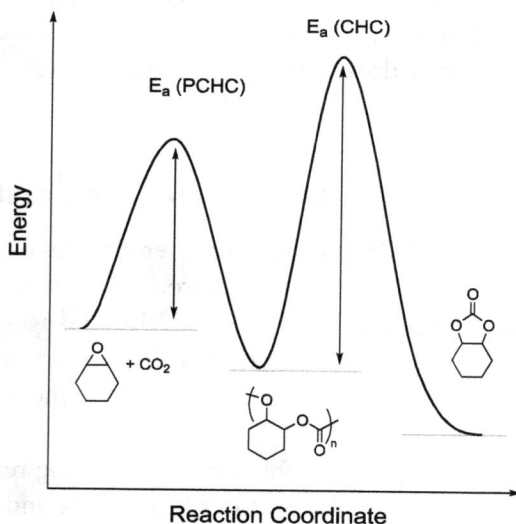

Figure 17:   Reaction co-ordinate vs. energy profile for PCHC and CHC formation.

form cyclohexene carbonate (CHC) using chromium salen catalysts (133.0 kJmol^{-1}).[25b,29c] Using catalyst **2**, the activation energy barrier for PCHC formation is 96.8 kJmol^{-1} and closely matches the computationally calculated value.[25] The activation energy barrier for PCHC synthesis using chromium salen catalysts is 46.9 kJmol^{-1},[19] which is double the value for catalyst **2** and thus higher temperatures are required with the di-zinc catalyst (**2**) (>60°C) compared to the chromium salen complexes (25–50°C) in CHO/CO$_2$ ROCOP.[25b,29c]

When using the di-zinc catalyst (**2**) and increasing the temperature of the copolymerisation reaction from 80°C to 100°C, cyclic carbonate production increases to 6% compared to <1%, as expected. However, the di-magnesium catalyst (**1**) does not (>99% copolymer at 100°C), which implies that the activation energy barrier for CHC production with the di-magnesium catalyst (**1**) is higher than for the di-zinc catalyst (**2**).[23g,23h]

References

1. (a) M. Aresta, *Carbon Dioxide as Chemical Feedstck.* Wiley-VCH: Germany, 2010; (b) M. R. Kember, A. Buchard and C. K. Williams, *Chem. Comm.*, 2011, **47**(1), 141–163; (c) D. J. Darensbourg, *Inorganic Chem.*, 2010, **49**(23), 10765–10780.

2. P. Markewitz, W. Kuckshinrichs, W. Leitner, J. Linssen, P. Zapp, R. Bongartz, A. Schreiber and T. E. Muller, *Energy & Environ. Sci.*, 2012, **5**(6), 7281–7305.

3. (a) H. Arakawa, M. Aresta, J. N. Armor, M. A. Barteau, E. J. Beckman, A. T. Bell, J. E. Bercaw, C. Creutz, E. Dinjus, D. A. Dixon, K. Domen, D. L. DuBois, J. Eckert, E. Fujita, D. H. Gibson, W. A. Goddard, D. W. Goodman, J. Keller, G. J. Kubas, H. H. Kung, J. E. Lyons, L. E. Manzer, T. J. Marks, K. Morokuma, K. M. Nicholas, R. Periana, L. Que, J. Rostrup-Nielson, W. M. H. Sachtler, L. D. Schmidt, A. Sen, G. A. Somorjai, P. C. Stair, B. R. Stults and W. Tumas, *Chem. Rev.*, 2001, **101**(4), 953–996; (b) M. Aresta and A. Dibenedetto, *Dalton Transact* 2007, **28**, 2975–2992; (c) T. Sakakura, J.-C. Choi and H. Yasuda, *Chem. Rev.*, 2007, **107**(6), 2365–2387.

4. (a) https://www.polymersolutions.com/psi-newsletter-archive/january-2014/; (b) http://www.plasticseurope.org/documents/document/2013101 4095824-final_plastics_the_facts_2013_published_october2013.pdf; (c) http://www.basf.com/group/corporate/en_GB/function/conversions:/ publish/content/investor-relations/calendar/images/120905/BASF_Investor_Day_Automotive_Polyurethanes.pdf; (d) M., Okada *Prog. Poly. Sci.*, 2002, **27**(1), 87–133; (e) http://www.petrochemconclave.com/presentation/2013/ Mr.SMoolji.pdf.

5. http://www.bpf.co.uk/Press/Oil_Consumption.aspx.

6. S. Inoue, H. Koinuma and T. Tsuruta, *J. Poly. Sci., Part B: Polym. Lett.*, 1969, **7**, 287–292.

7. (a) N. von der Assen and A. Bardow, *Green Chem.*, 2014, **16**(6), 3272–3280; (b) N. von der Assen, P. Voll, M. Peters, A. and Bardow, *Chem. Soc. Rev.*, 2014, **43**, 7982–7994.

8. (a) C. M. Byrne, S. D. Allen, E. B. Lobkovsky and G. W. Coates, *J. Am. Chem. Soc.*, 2004, **126**(37), 11404–11405; (b) W. Kuran, *Prog. Polym. Sci.*, 1998, **23**(6), 919–992; (c) G. W. Coates and D. R. Moore, *Angew. Chem. Int. Ed.*, 2004, **43**(48), 6618–6639; (d) D. J. Darensbourg, *Chem. Rev.*, 2007, **107**(6), 2388–2410; (e) D. J. Darensbourg and M. W. Holtcamp, *Coord. Chem. Rev.*, 1996, **153**, 155–174; (f) H. Sugimoto and S. Inoue, *J. Poly. Sci. Part A: Poly. Chem.*, 2004, **42**(22), 5561–5573.

9. (a) K. Nakano, M. Nakamura, and K. Nozaki, *Macromolecules*, 2009, **42**(18), 6972–6980; (b) S. J. S. S. Na, A. Cyriac, B. E. Kim, J. Yoo, Y. K. Kang, S. J. Han, C. Lee, and B. Y. Lee, *Inorg. Chem.* 2009, **48**(21), 10455–10465; (c) E. K. Noh, S. J. S. S. Na, S.-W. Kim, and B. Y. Lee, *J. Am. Chem. Soc.*, 2007, **129**(26), 8082–8083; (d) C. T. Cohen, T. Chu, and G. W. Coates, *J. Am. Chem. Soc.*, 2005, **127**(31), 10869–10878.

10. M. Ionescu, *Chem. Technol. Poly. Polyurethanes.* Smithers Rapra Technology, 2005.

11. (a) J. Langanke, A. Wolf, J. Hofmann, K. Bohm, M. A. Subhani, T. E. Muller, W. Leitner, and C. Gurtler, *Green Chem.*, 2014, **16**(4), 1865–1870; (b) S. H. Lee, A. Cyriac, J. Y. Jeon, and B. Y. Lee, *Poly. Chem.*, 2012, **3**(5), 1215–1220.

12. D. R. Moore, M. Cheng, E. B. Lobkovsky, and G. W. Coates, *J. Am. Chem. Soc.*, 2003, **125**(39), 11911–11924.

13. S. Inoue, *J. Poly. Sci. Part A: Poly. Chem.*, 2000, **38**(16), 2861–2871.

14. (a) G. W. Coates and D. R. Moore, *Angew. Chem. Int. Ed.*, 2004, **43**(48), 6618–6639; (b) D. J. Darensbourg and S. J. Wilson, *Green Chem.*, 2012, **14**(10), 2665–2671; (c) K. Nozaki, *Pure and Appl. Chem.* 2004, **76**(3), 541–546; (d) S. Klaus, M. W. Lehenmeier, C. E. Anderson, and B. Rieger, *Coord. Chem. Rev.*, 2011, **255**(13–14), 1460–1479; (e) C. Romain, A. Thevenon, P. K. Saini, and C. K. Williams, *in Topics in Organometallic Chemistry: Carbon Dioxide and Organometallics*, ed. X. B., Lu, Springer DE., 2015.

15. (a) M. Kobayashi, S. Inoue and T. Tsuruta, *J. Poly. Sci. Part A: Poly. Chem.*, 1973, **11**(9), 2383–2385; (b) W. Kuran, S. Pasynkiewicz, J. Skupińska, and A. Rokicki, *Die Makromol. Chem.*, 1976, **177**(1), 11–20.

16. (a) M. Cheng, E. B. Lobkovsky, and G. W. Coates, *J. Am. Chem. Soc.*, 1998, **120**(42), 11018–11019; (b) M. Cheng, D. R. Moore, J. J. Reczek, B. M. Chamberlain, E. B. Lobkovsky, and G. W. Coates, *J. Am. Chem. Soc.*, 2001, **123**(36), 8738–8749.

17. S. Klaus, M. W. Lehenmeier, E. Herdtweck, P. Deglmann, A. K. Ott and B. Rieger, *J. Am. Chem. Soc.*, 2011, **133**(33), 13151–13161.

18. (a) T. Bok, H. Yun, and B. Y. Lee, *Inorg. Chem.*, 2006, **45**(10), 4228–4237; (b) B. Y. Lee, H. Y. Kwon, S. Y. Lee, S. J. Na, S.-I. Han, H. Yun, H. Lee, and Y.-W. Park, *J. Am. Chem. Soc.* 2005, **127**(9), 3031–3037.

19. Y. Xiao, Z. Wang, and K. Ding, *Macromolecules*, 2005, **39**(1), 128–137.

20. Y. L. Xiao, Z. Wang, and K. L. Ding, *Chemistry — Eur. J.*, 2005, **11**(12), 3668–3678.

21. C.-H. Li, H.-J. Chuang, C.-Y. Li, B.-T. Ko, and C.-H. Lin, *Poly. Chem.*, 2014, **5**(17), 4875–4878.

22. C.-Y. Tsai, B.-H. Huang, M.-W. Hsiao, C.-C. Lin, and B.-T. Ko, *Inorg. Chem.*, 2014, **53**(10), 5109–5116.

23. (a) A. M. Chapman, C. Keyworth, M. R. Kember, A. J. J. Lennox, and C. K. Williams, *ACS Catalysis* 2015, **5**(3), 1581–1588; (b) C. Romain and C. K. Williams, *Angew. Chem. Int. Ed.*, 2014, **53**(6), 1607–1610; (c) Y. Zhu, C. Romain, V. Poirier, and C. K. Williams, *Macromol.*, 2015, **48**(8), 2407–2416; (d) Y. Zhu, C. Romain, and C. K. Williams, *J. Am. Chem. Soc.* 2015, 137, 12179–12182; (e) A. Buchard, M. R. Kember, K. G. Sandeman, and C. K. Williams, *Chem. Communicat.*, 2011, **47**(1), 212–214; (f) M. R. Kember, F. Jutz, A. Buchard, A. J. P. White, and C. K. Williams, *Chem. Sci.* 2012, 3, 1245-1255 (g) M. R. Kember, P. D. Knight, P. T. R. Reung, and C. K. Williams, *Angew. Chem. Int. Ed.*, 2009, **48**(5), 931–933; (h) M. R. Kember and C. K. Williams, *J. Am. Chem. Soc.*, 2012, **134**(38), 15676–15679; (i) M. R. Kember, A. J. P. White, and C. K. Williams, *Inorg Chem.*, 2009, **48**(19), 9535–9542; (j) P. K. Saini, C. Romain, and C. K. Williams, *Chem. Comm.*, 2014, **50**(32), 4164–4167.

24. M. R. Kember, A. J. P. White, and C. K. Williams, *Macromolecules*, 2010, **43**(5), 2291–2298.

25. (a) A. Buchard, F. Jutz, M. R. Kember, A. J. P. White, H. S. Rzepa, and C. K. Williams, *Macromolecules*, 2012, **45**(17), 6781–6795; (b) F. Jutz, A. Buchard, M. R. Kember, S. B. Fredrickson, and C. K. Williams, *J. Am. Chem. Soc.*, 2011, **133**, 17395–17405.

26. C. T. Cohen, T. Chu, and G. W. Coates, *J. Am. Chem. Soc.*, 2005, **127**(31), 10869–10878.

27. (a) K. Nakano, S. Hashimoto, and K. Nozaki, *Chem. Sci.*, 2010, **1**(3), 369–373; (b) S. I. Vagin, R. Reichardt, S. Klaus, and B. Rieger, *J. Am. Chem. Soc.*, 2010, **132**(41), 14367–14369.

28. (a) X.-B. Lu, L. Shi, Y.-M. Wang, R. Zhang, Y.-J. Zhang, X.-J. Peng, Z.-C. Zhang, and B. Li, *J. Am. Chem. Soc.* 2006, **128**(5), 1664–1674; (b) D.-Y. Rao, B. Li, R. Zhang, H. Wang, and X.-B. Lu, *Inorg. Chem.*, 2009, **48**(7), 2830–2836; (c) C. Chatterjee and M. H. Chisholm, *Inorg. Chem.*, 2011, **50**(10), 4481–4492; (d) T. Aida and S. Inoue, *J. Am. Chem. Soc.* 1983, **105**(5), 1304–1309.

29. (a) D. J. Darensbourg, R. M. Mackiewicz, J. L. Rodgers, and A. L. Phelps, *Inorg. Chem.*, 2004, **43**(6), 1831–1833; (b) D. J. Darensbourg and J. C. Yarbrough, *J. Am. Chem. Soc.*, 2002, **124**(22), 6335–6342; (c) D. J. Darensbourg, J. C. Yarbrough, C. Ortiz, and C. C. Fang, *J. Am. Chem. Soc.*, 2003, **125**(25), 7586–7591; (d) D. J. Darensbourg, *Chem. Rev.* 2007, **107**(6), 2388–2410.
30. (a) D. J. Darensbourg, Mackiewicz and J. L. Rodgers, *J. Am. Chem. Soc.*, 2005, **127**(49), 17565–17565; (b) D. J. Darensbourg, R. M. Mackiewicz, J. L. Rodgers, C. C. Fang, D. R. Billodeaux, and J. H. Reibenspies, *Inorg. Chem.*, 2004, **43**(19), 6024–6034.
31. J. O'Donnell, *J. Am. Chem. Soc.*, 2010, **132**(29), 10206–10207.
32. (a) S. Inoue, *J. Polym. Sci. A Polym. Chem.* 2000, **38**(16), 2861–2871; (b) B. Iván, *Macromol. Symp.*, 1994, **88**(1), 201–215.
33. J. K. Varghese, A. Cyriac and B. Y. Lee, *Polyhedron* 2012, **32**(1), 90–95.
34. X.-B. Lu and D. J. Darensbourg, *Chem. Soc. Rev.*, 2012, **41**(4), 1462–1484.

Chapter 6

Precious Metals for Environmental Catalysis: Gold

Catherine Davies, Simon Kondrat and Peter Miedziak

School of Chemistry, Cardiff University, Main building,
Park Place, Cardiff, CF10 3AT, UK.

Numerous environmental problems exist which may be solved using catalysis, in particular by supported precious metal nanoparticles. We focus here on the use of gold as a catalyst, which since the 1980s has become an important field of research. The use of gold-based catalysts for tackling environmental issues is discussed, within the context of projects currently being undertaken within the environment theme of the UK Catalysis Hub; the applications of gold catalysis for utilisation and valorisation of biomass derivatives, treatment of waste water using photocatalysis and vehicle emissions control are considered.

While supported metal catalysts such as platinum and palladium have been used in industrial processes for a number of decades, the potential use of gold in catalyst systems is a more recent development. Two key discoveries were made that led to the current levels of research being carried out in this field; Hutchings predicted and subsequently confirmed that based on electrode potentials, gold should be the most active metal for the hydrochlorination of acetylene,[1,2] and Haruta *et al.* demonstrated that gold nanoparticles supported on transition metal (TM) oxides, e.g. Fe_2O_3, are active for the catalytic oxidation of CO at low temperatures.[3] Previously, gold had largely been considered unreactive, although the use of gold catalysts had been reported for oxidations of ethylene and propylene[4] and reductions of olefins.[5] However, these new findings illustrated the potential use of gold as a catalyst when present in nanoparticle form. In particular, Haruta showed that the activity of the nanoparticles is determined by their size.

There are many different methods that have been reported for the preparation of nanoparticles of gold, but the most commonly used amongst these are impregnation, deposition precipitation (DP), co-precipitation and sol-immobilisation, with new techniques such as nanoalloying being developed recently. Variations and modifications to these methods are frequently reported, since the variables within each method have been shown to significantly affect the activity of the resulting nanoparticles; the main advantages and disadvantages are further described briefly.

Impregnation is by far the simplest method of preparing supported metal catalysts, and can be used with any porous support. The metal precursor is made into a solution (in the case of gold catalysts $HAuCl_4$ is commonly used), which is then mixed with the support to be used. Metal particles form on the external surface and within the pores of the support, when the sample is dried (in order to remove the solvent) and calcined (to decompose the precursor). Gold catalysts prepared by impregnation methods tend to have a range of nanoparticle sizes, including larger nanoparticles, which are likely to be in part due to the high chloride content which may cause sintering of the gold. There are numerous variations of the impregnation process, for example in incipient wetness impregnation the minimum amount of solvent to fill the pores of the support is used; this method is favoured industrially due to the minimal waste and the accurate dosing achieved.

DP is a method of producing smaller supported metal particles than afforded by impregnation; it involves the modification of the pH of a stirred mixture of the support in a solution of the metal precursor in order to form the desired species (the metal hydroxide), which precipitates onto the support. This technique has been successfully used to produce gold catalysts on a range of supports. Catalysts prepared *via* this method are often found to be more active than those prepared *via* impregnation for various reactions; for example, this has been attributed to smaller metal particle size in the case of Au/titania catalysts for CO oxidation.[6] There are a number of variables within the DP preparation procedure which may affect a catalyst's activity, for example the pH of the mixture, the base used and the temperature at which the preparation is carried out. Moreau *et al.* investigated the effect of pH during preparation of Au/TiO_2 catalysts by the deposition–precipitation method.[7] Catalysts were prepared with a range of initial pH values of a $HAuCl_4$ solution, varied by addition of NaOH; gold particles with a smaller particle size were attained at higher pH, with pH 9 the optimum in terms of catalytic activity.

In co-precipitation, the metal particles and the support are formed simultaneously. This is a method commonly used for metal oxide supported catalysts, such as Au/Fe_2O_3, and was the method used by Haruta in the early work on gold catalysts for low temperature CO oxidation.[3] For gold catalysts, an aqueous solution of $HAuCl_4$ is typically mixed with a solution of the corresponding nitrate to the desired support and then the pH is adjusted by addition of a precipitating agent, such as sodium carbonate, so that precipitation of the desired product occurs. The mixture is then filtered, washed, dried and may be calcined. A benefit of using this method is that good dispersion of small metal particles (<10 nm) within the support can be achieved.[8] Important variables in this method that can affect the activity of catalysts are the precipitating agent used,[9] the speed at which precipitation is carried out and the aging time,[10] all of which can affect the gold loading, particle size and residual chloride content.

The sol immobilisation (SI) method was first described by Prati *et al.* as a novel way of forming sufficiently small particles of gold on a carbon support.[11] Whilst SI is one of the most complex and sensitive methods used for preparation of catalysts, it has the unique advantage that the size of the metal particles that are being prepared may be strictly controlled. The method involves first making a sol of metallic gold particles, which is done by adding a stabilising ligand, such as poly vinyl alcohol (PVA), and then a reducing agent, usually $NaBH_4$, to a dilute solution of $HAuCl_4$ in water. By adjusting variables such as the ligand that is used and the ratio of ligand to metal, different sizes of metal particles are produced. The support is then added to the sol, to immobilise the metal particles on it. It is important that the metal particles bind to the support, so the pH of the mixture may be adjusted in order to encourage this, by addition of small amounts of acid or base. The pH used is dependent on the iso-electric point of the support being used, in order to encourage interactions between the support and the metal particles. For example, when using a carbon support a small amount of concentrated H_2SO_4 is added to reduce the pH to approximately 2. In a number of cases, it has been demonstrated that catalysts prepared using SI techniques are more active than those prepared by other methods, such as impregnation and DP. For example, when Miedziak *et al.* prepared Au–Pd/TiO$_2$ catalysts by impregnation, deposition–precipitation and SI methods, the order of activity for benzyl alcohol oxidation was SI > DP > impregnation.[12]

An alternative strategy, which has gained popularity in recent years, for producing active unsupported gold catalysts is the preparation of nanoporous gold.[12,13] This type of catalyst has been found to be active for a range of oxidation reactions. Formed from the dealloying of AuAg, the presence of residual Ag within the nanoporous gold network is inevitable. This has resulted in much discussion as to whether Ag nanoparticles are in fact responsible for much of the activity observed with these catalysts.[14] Despite this, the area of catalysis by nanoporous gold is significant and requires attention within the context of this review.

1 Biomass Derivatives

In addition to the remediation of environmentally harmful waste products, produced by industry and society, the principles of green chemistry propose the mitigation of waste products.[15] Through atom efficiency, lowering reaction temperatures and pressures, and using more renewable feedstocks, catalysis fundamentally fulfils the requirements of green chemistry. The use of heterogeneous supported gold catalysts for the utilisation and valorisation of biomass derivatives will be discussed within this section. For a broader view of the use of catalysts in biomass conversion, the reader is directed to the excellent review articles by Corma and co-workers and Dumesic and co-workers.[12,16] Gold nanoparticles supported on a range of TM oxide or activated carbons are known to be active for a range of selective oxidation reactions.[17–20] With respect to biomass conversion, this chapter will consider research into gold catalysis with three principle bio-mass substrates: glycerol, glucose and 5-hydroxymethylfurfural (HMF). In each case, the nature of

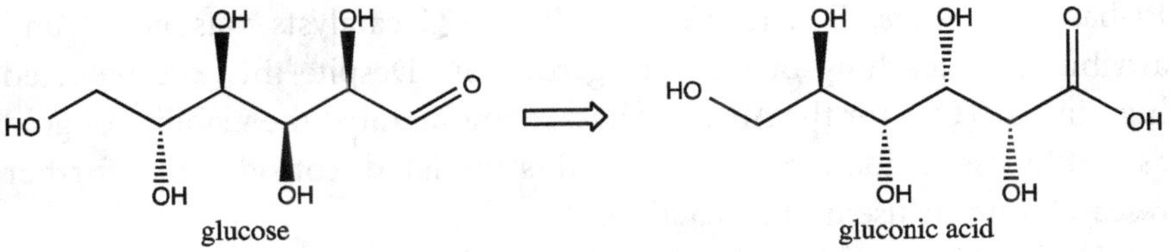

Figure 1: Scheme of glucose oxidation.

the supported gold catalyst and the variation of reaction conditions have been found to have a significant effect on the activity and selectivity of the catalysts.

1.1 Glucose

Glucose, derived from starch, cellulose, lactose and sucrose by a hydrolysis process, is a widely available bio-derived platform molecule. The selective oxidation of glucose can produce gluconic acid (Figure 1), an important compound for the food, pharmaceutical and hygiene industries.[21] Initial studies into the use of a heterogeneous catalyst for glucose oxidation focused on platinum supported by carbon at high reaction pH.[22] However, this catalyst was found to be prone to significant deactivation, which could be slowed but not eliminated by maintaining pH 9 throughout the reaction.

The first reported use of gold catalysts for glucose oxidation was by Biella *et al.*, where a 1 wt.% Au/C catalyst prepared by SI was used.[23] The selective oxidation of glucose to gluconic acid, with no isomerisation side-reaction products, was successfully carried out with O_2 as the oxidant at mild pressures and temperatures (1–3 bar and 50–100°C). Similar to previously reported monometallic (Pt or Pd) and bimetallic (Pt–Bi and Pd–Bi) catalysts, alkaline conditions of pH 9.5 were found to be most favourable for high conversion. As the reader will observe, alkaline conditions are favourable for all of the discussed biomass oxidation reactions. However, unlike Pd or Pt-based systems reported for glucose oxidation, the Au/C catalyst was found to have some activity in less alkaline media, with some activity being observed even under neutral conditions. Initial TOFs of the Au/C catalyst were significantly higher than comparable Pd or

Pt-based catalysts. Deactivation of the Au/C catalysts was noted and attributed to leaching of Au during reaction. Despite this, the reported high initial TOFs for the Au/C catalysts demonstrated the viability of gold as a glucose oxidation catalyst and stimulated considerable further research into its use in this reaction.

Design of the gold catalyst forms a significant part of research that has stemmed from the initial work. Önal *et al.* studied the effect of changing the sol-immobilisation technique and the carbon support type on the activity of the catalysts.[24] The activity of the catalysts studied was correlated to the surface area of gold, calculated from TEM analysis, with a strong dependence of initial activity on high gold surface area being found. Prüße and co-workers investigated alteration of support type, and its effect on the gold catalysts' stability under reaction conditions.[25-28] Initial studies suggested that gold supported on alumina, by DP or wet impregnation, produced catalysts stable over 20 reaction cycles.[25,26] However, the reaction was oxygen mass transfer limited, which could have masked potential deactivation phenomena. Studies with gold supported on titania have further demonstrated that stable glucose oxidation catalysts could be prepared on metal oxide supports.[29] While batch reactions provide ideal conditions for catalyst screening, stability can be more reliably determined by testing under continuous flow conditions. Glucose oxidation testing has been performed on Au/Al_2O_3 catalysts under flow conditions, and confirms that good stability can be maintained for 110 days on line.[28]

The dependence of the support-metal interaction in catalysed reactions is an often discussed and researched area.[30] With respect to glucose oxidation by gold, several interesting reports have been published which suggest that gold itself is intrinsically active for this reaction. Comotti *et al.* demonstrated that short lived "naked gold" colloids (i.e. in the absence of stabilising agents such as PVA) are highly active for glucose oxidation, when compared to similarly prepared Cu, Pt, Pd and Ag colloids.[31] Initial rates were found to be constant with those of an Au/C catalyst. Interestingly, the initial TOF of the gold catalysts competed with an enzymatic catalyst under comparable conditions (30°C, 1 bar and pH 7). However, "naked gold" was found to agglomerate and deactivate rapidly in the absence of the support. Hence, it could be concluded that gold is the active component in the reaction with metal-support interactions providing stability of Au particle size only. Yin *et al.* have reported on the selective oxidation of glucose to gluconic acid using nanoporous gold. This nanoporous gold catalyst was found to behave in a similar way to the Au/C catalyst reported by Biella *et al.*[23] The catalyst showed the same dependence on reaction pH, with the catalyst being most active at pH 9, but still having notable activity at pH 7. In addition, notable deactivation occurred on catalyst re-use. Direct comparison between this nanoporous gold and a conventional supported gold catalyst has not-to-date been provided.

A growing body of evidence, such as that described before, suggests that for glucose oxidation, the gold itself is the active component and that the support is integral to the long-term stability of the catalyst. However, the support can play additional roles in the reaction if designed correctly. For a range of oxidation reactions (also see sections on HMF and glycerol oxidation), basic supports can be used to enhance activity, while removing the requirement for sacrificial base (such as NaOH). Specifically for glucose oxidation, Kondrat *et al.* showed that gold supported on MgO ($Mg(OH)_2$ under reaction conditions) gave high conversion and selectivity to gluconic acid under mild conditions, without the addition of sacrificial base.[32] However, it was found that the catalyst deactivated due to a combination of product inhibition and the leaching of the support by the gluconic acid formed during reaction.

The paper by Kondrat *et al.* also reported that partial alloy formation with Pd enhanced the activity of the gold catalysts. Similarly, Comotti *et al.* showed that alloying with Pt could modestly enhance the activity of Au catalysts for glucose oxidation.[33] It is interesting to note that for many other alcohol oxidations, Au alloying has a far more substantial synergistic effect.[18,20,34]

1.2 5-Hydroxymethylfurfural (HMF)

HMF is a platform molecule obtained from the dehydration of fructose or glucose. 2,5-Furandicarboxylic acid (FDCA), formed from the selective oxidation (see Figure 2) of HMF, has become an alternative to terephathalic acid in the manufacture of polymers.[35]

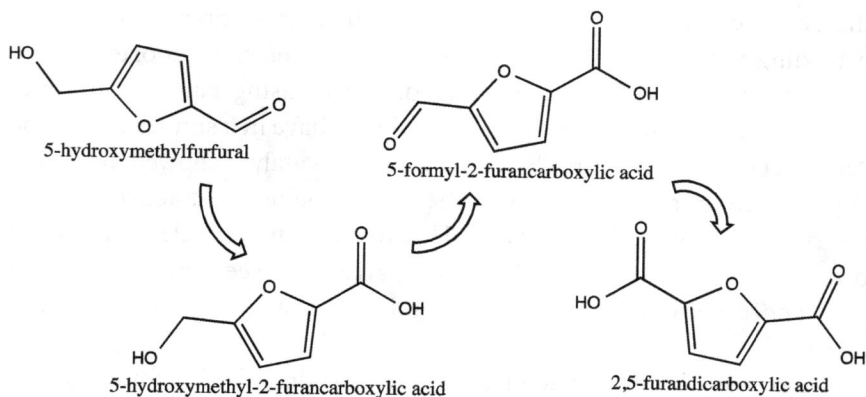

Figure 2: Scheme of HMF oxidation pathway.

Several heterogeneous catalysts, including supported gold, have been investigated for this reaction. In general terms, it has been observed that oxidation of HMF is relatively easy, though only to the single carboxylic acid product 5-hydroxymethyl-2-furancarboxylic acid (HFCA). Further oxidation to the dicarboxylate (FDCA) is much more difficult, with more vigorous conditions required.

Gorbanev *et al.* reported the first direct oxidation of HMF to FDCA using a gold catalyst, supported on titania.[36] At a mild temperature of 30°C, with O_2 as oxidant and water as solvent, it was noted that selectivity towards FDCA or the single carboxylic acid intermediate was highly dependent on oxygen pressure and high base to substrate ratio. Under optimal conditions, which required significant excess of sacrificial base, 71% yield of FDCA was produced. Davis and co-workers investigated HMF oxidation with monometallic Pd, Pt and Au catalysts and found that, under specific conditions (NaOH:HMF ratio of 2:1), while the Au catalyst had a significantly higher turnover frequency than the other metals the reaction predominantly stopped at HFCA, while proceeding to FDCA with Pd and Pt catalysts.[37] Increasing the concentration of base within the reaction then facilitated the alcohol group oxidation to convert HFCA to FDCA for the Au catalysed reaction.

Davis and co-workers further developed the work on supported monometallic Au and Pt with a mechanistic study. These authors asserted that molecular oxygen was not directly involved in the mechanism. It was shown that the aldehyde oxidation (HMF to HFCA) could proceed in the absence of oxygen, with only NaOH. While the oxidation of the alcohol arm, to proceed from HFCA to FDCA, was shown to require oxygen, it was still asserted that O_2 did not play a direct role in the mechanism. De-protonation of the alcohol arm to form an alkoxy intermediate can be facilitated by base. Then hydroxide ions on the metal surface were thought to activate the C–H bond required to produce the aldehyde, with further oxidation to carboxylic acid and FDCA proceeding in the same manner as the aldehyde group in HMF. Oxygen was thought to participate by removing the electrons from the metal during the adsorption and reaction of hydroxide species. The higher selectivity to FDCA with the Pt catalysts compared to Au catalysts, under comparable base:substrate ratios, can be attributed to the ability of Pt to activate geminal hydrogen associated with alcohols.

Given the undesirably high base substrate ratio required to obtain good FDCA yield with Au based catalysts, several strategies have been employed to limit or eliminate the requirement of sacrificial base. As discussed previously when reviewing glucose oxidation, a heterogeneous base as the support for Au can be employed. Gupta *et al.* showed that both hydrotalcite and magnesium hydroxide could successfully be used in the base-free oxidation of HMF to FDCA, though MgO was not as selective as the hydrotalcite.[38] While this work initially appears as promising, it must be considered that the good FDCA selectivities were observed at substrate to metal ratios of only 40:1. An alternative approach could be alloying of Au, with AuCu alloys showing a clear synergistic effect with respect to activity.[39,40]

1.3 Glycerol

Glucose and HMF oxidation, while receiving significant scientific attention, have not been developed to the same extent as polyol oxidation. Polyols, the main example of which is glycerol, are produced by the transesterification or saponification of tryglycerides in processes including biodiesel production. Crude glycerol from this process is traditionally considered a waste product and consequently its valorisation could be

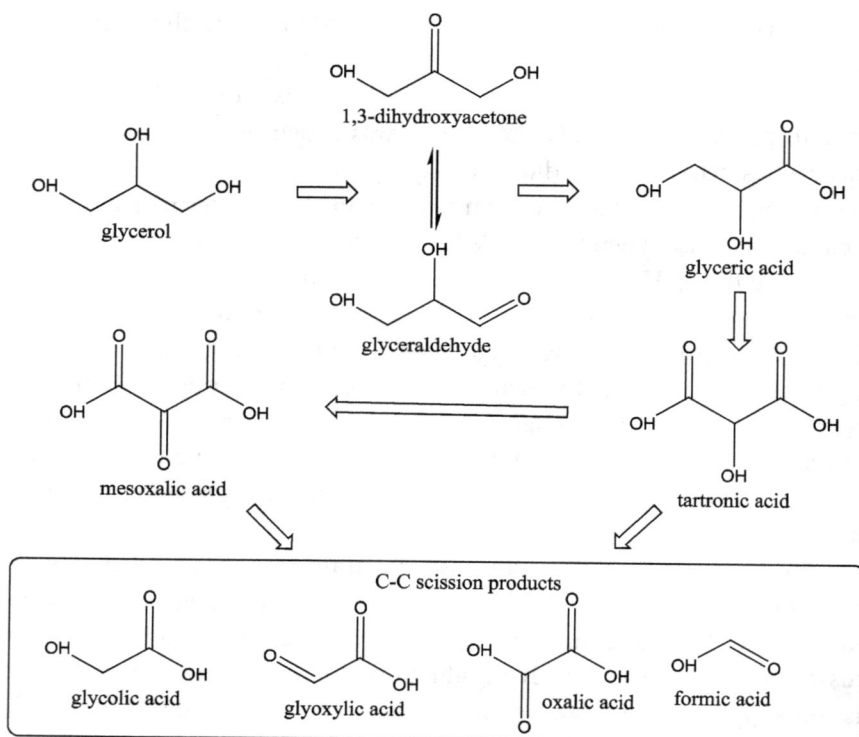

Figure 3:   Scheme of glycerol oxidation pathway.

considered to make biodiesel production more economical. As with many bio-derived molecules there are many transformation reactions, aside from selective oxidation, that can be performed with glycerol. For a broader view of glycerol valorisation and transformations the reader is directed to the review by Corma *et al.*[41] In the context of this chapter, the discussion will be focused on the selective oxidation of glycerol with heterogeneous gold catalysts.

Potential oxidation products of glycerol and the relevant reaction pathways are shown in Figure 3. Initial oxidation of glycerol produces glyceraldehyde or dihydroxyacetone (which exist in equilibrium with each other); sequential oxidation results in glyceric acid, then tartronic acid and finally mesoxalic acid. Further, oxidation then results in C–C scission and the formation of C_2 glycolic or oxalic acid and C_1 products such as formic acid or carbon dioxide. An interesting side reaction, which occurs at elevated temperature and high reaction pH, is the dehydration of glyceraldehyde or dihydroxyacetone followed by rearrangement to lactic acid.

Early work with monometallic Pd or Pt catalysts showed promise for the selective oxidation of glycerol with O_2 or H_2O_2 as oxidant. However, the catalysts were prone to deactivation from product inhibition and over oxidation.[42,43] Carrettin *et al.* investigated glycerol oxidation using Au, Pd and Pt catalysts all supported on graphite and showed that Au provided the highest selectivity to glycerate. They also noted that base was essential for Au catalyst activity, with no conversion of glycerol being detected in its absence. It was hypothesised that initial hydrogen abstraction from glycerol required base.[44,45] Davis and co-workers, in a similar study to that published on the HMF oxidation mechanism, showed that surface hydroxide ions were the active oxygen species and not molecular oxygen.[46]

Initial catalyst development, as with reactions with other substrates discussed here, focused on the gold deposition method and correlations of activity with particle size. Ketchie *et al.* observed that high turnover frequency was associated with small (*ca.* 5 nm) Au particles but selectivity to glycerate was better with larger (*ca.* 45 nm) particles.[47] What was not considered within the work was that selectivity-conversion relationships could result in misleading conclusions and that iso-conversion selectivities would have provided clearer evidence for any structure–selectivity effect.

Bimetallic AuPd and AuPt catalysts have been shown to have a strong synergistic effect on both activity and selectivity for glycerol oxidation.[20] While such a strong effect was not observed for glucose oxidation, it has been frequently noted in other alcohol oxidation reactions.[18] Given the dramatic improvement in catalyst performance on introducing other metals to Au, a substantial body of work has been carried out on alloy formation and its effect on the reaction mechanism. This body of work is far too substantial to be included within this book and the reader is recommended a recent review dedicated to the topic of bimetallic catalysts for bio-mass transformations.[48]

One exception, that helps with the continuity of the chapter, is the work by Brett *et al.* on the use of MgO as a heterogeneous basic support for AuPt catalysts.[20] As with the work on glucose oxidation, it was shown that good conversion of glycerol to glyceric acid occurred. However, unlike with glucose oxidation, Mg^{2+} leaching was found to be negligible in the glycerol oxidation reaction.

Several overarching conclusions can be made, with respect to Au cata-lysed reactions with the substrates discussed. First, gold is active with good selectivity to desired products, compared to other noble metal catalysts. However, alloy formation with other noble metals often, but not exclusively, improves activity and selectivity towards desired products. The presence of hydroxide, in the form of a sacrificial or catalytic species, is paramount for activation of HMF and glycerol (glucose reactions can proceed, all be it slowly, without base).

2 Waste Water Treatment

Photocatalysis is the use of a catalyst for the conversion of light into chemical energy; generally a semiconductor is illuminated with photons of energy that is greater than its band gap energy. This creates electron–hole pairs, which can either recombine or migrate to the surface where they can interact with adsorbate species to give reduction and oxidation products. Typical examples of photocatalysts are metal oxides, particularly TiO_2 and ZnO; these have the advantage of a suitable band gap (3.2 eV for both TiO_2 (anatase) and ZnO) and are stable under irradiation. TiO_2 is the most widely used oxide owing to its aforementioned stability, but also its avail-ability and lack of toxicity. The most frequently used form of titania is the commercially available P25 (Degussa), which is 80% anatase and 20% rutile; the advantage of this is that is a commercially available standard and should have good reproducibility. The main disadvantage of TiO_2 is that it is most active when irradiated by UV light. Ideally a photocatalyst would be able to utilise solar light. Most of the light in the solar spectrum is at longer wavelengths than the UV, in the visible region; in fact only about 4% of solar irradiation is in the UV range.

Strategies to improve the photocatalytic response in the visible region include doping of the titania with either metallic or non-metallic ele-ments. Doping with metallic elements can lead to metal leaching or corro-sion problems, while doping with non-metallic elements adds considerable complexity to the preparation procedures, with specific doping levels required for optimum performance making them hard to reproduce.

Characterisation can also be difficult with these doped titanias. The second approach, the deposition of metal nanoparticles onto the surface of the titania, is where the metal is in a separate phase but in interfacial contact with the titania. The deposited metals can act as electron sinks, preventing recombination with the electron holes and increasing the probability that they can be used for the desired chemical reactions. To form a stable catalyst, this metal should be chemically inert with respect to photocatalysis, making noble metals excellent candidates. For these reasons, gold is a good candidate for such reactions; as outlined earlier in the chapter, the use of gold in catalysis has received increased attention in recent years and this trend is just beginning to extend to the field of photocatalysis.

2.1 Photodecomposition of organic compounds

The addition of gold to titania for photocatalysis has been reported for various organic compounds and the effect of the addition of gold varies from compound to compound. Hidalgo *et al.* investigated the effect of the addition of gold to titania for the photodecomposition of phenol.[49] The catalysts were prepared by a photo deposition method with variation of several parameters, most notably light intensity and deposition time. While the authors reported better initial reaction rates for some of the catalysts, others were worse than bare P25 titania; this shows the importance of the parameters in the preparation method and indicates that the

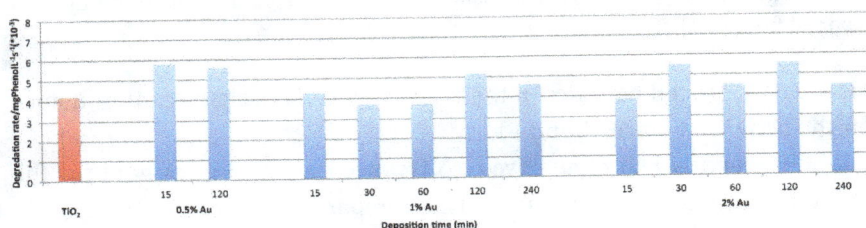

Figure 4: Initial reaction rates for phenol degradation (mg of phenol per litre per second) over the indicated catalysts prepared by photodeposition with high light intensity. Reaction conditions: [phenol]$_0$ = 50 mg/L, V = 0.2 L, [catalyst] = 1 g/L and I = 140 W/m^2. Reaction conditions: [phenol]0 = 50 mg/L, V = 0.2 L, [catalyst] = 1 g/L and I = 140 W/m^2. Reproduced based on the data reported from Ref. [49].

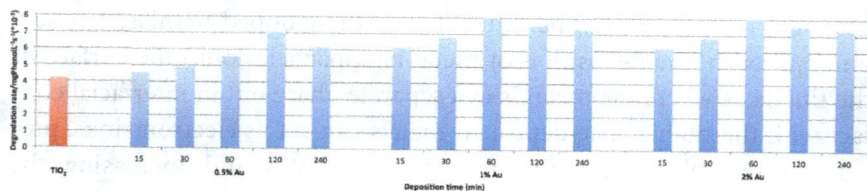

Figure 5: Initial reaction rates for phenol degradation (mg of phenol per litre per second) over the indicated catalysts prepared by photodeposition with low light intensity. Reaction conditions: [phenol]^0 = 50 mg/L, V = 0.2 L, [catalyst] = 1 g/L and I = 140 W/m^2. Reproduced based on the data reported from Ref. [49].

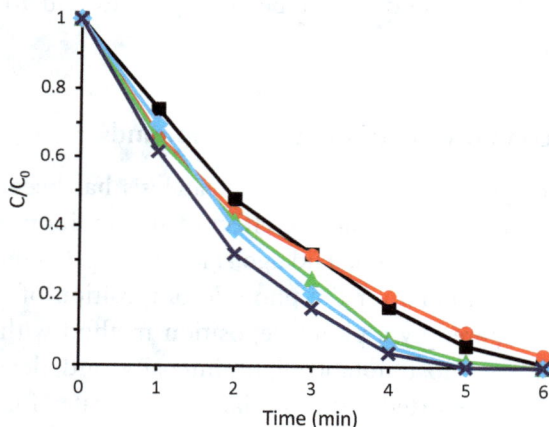

Figure 6: Photodegradation of phenol under UV illumination by pure and modified P25 photocatalysts. Black squares, P25; red circles, Au/P25; green triangles, AuCu1:1/P25; dark blue crosses, AuCu1:3/P25; light blue diamonds, Cu/P25. Reproduced based on the data reported from Ref. [50].

nature of the metal particles in the final catalyst has a significant influence on the activity (Figures 4 and 5).

Subsequently Hai *et al.* have also reported on the use of gold for the photodecomposition of phenol.[50] They prepared gold, copper and gold–copper bimetallic catalysts by a sol-immobilisation methodology. Figure 6 shows that while the gold catalysts displayed a very similar activity to titania for the photodecomposition, the combination of both metals, particularly with a 1:3 Au:Cu ratio, resulted in a more active catalyst.

Gold supported on TiO$_2$ has also been reported for the photodecomposition of methyl *tert*-butyl ether and 4-chlorophenol, using a fixed bed flow-through reactor by Orlov *et al.* In this case, the authors coated the walls of the reactor with the catalyst and reported a significant improvement in the reaction rate with only a low loading of gold.

2.2 Nitrate removal

The contamination of ground water with nitrates from highly soluble nitrogen based fertilisers is a major problem. Nitrate and nitrite, which is formed by reduction of nitrate, are toxic to the human body, particularly infants. For a comprehensive review on the photocatalytic destruction of nitrates using titania based materials, the readers are directed to the review by Anderson and Shand.[51] The World Health Organisation (WHO) has introduced a recommended maximum nitrate concentration of 10 ppm[52] and the European Union recommends levels of 50 ppm and 0.1 ppm for nitrate and nitrite respectively.[53] Current methods to reduce the nitrate levels in drinking water include ion-exchange,[54] biodegradation[55] and reverse osmosis,[56] however, all these methods have the disadvantage of producing waste brines, which subsequently need to be disposed of. Photocatalysis provides an alternative to these methods, its main advantage being its simplicity which would enable the transfer of the technology to the required point of use where more complicated technology may not be available.

There have been several examples of photocatalytic nitrate removal using metal supported on titania. Cu, Pt, Pd and Ag photocatalysts have been reported, with the most active and selective of these being Ag which has a selectivity towards nitrogen of almost 100%.[57] The first example of the use of gold supported on titania was reported by Kominami *et al.*[57] who reported the use of P25 titania as the catalyst along with various metal salts with the pH adjusted with sodium hydroxide. Under the reported conditions, gold was found to be less effective as a co-catalyst than either silver or copper, however, the greatest selectivity towards nitrogen was achieved when using a palladium copper alloy (Table 1).

Anderson subsequently reported 1% Au/TiO_2 prepared by a DP method using two different types of titania support, P25 and Hombikat; these were also compared to a commercially available Au/TiO_2 catalyst in project AuTEK.[58] These were used for the simultaneous removal of nitrate and oxalic acid, a pollutant that can also be found in waste water. With

Table 1: Photocatalytic reduction of nitrate in an aqueous suspension of 0.5 wt.% metal-loaded P-25 TiO_2 in the presence of oxalic acid for 12 h reproduced from Ref. [57].

Metal loaded	pH	NO_3^- (μmol) [a]	NO_2^- (μmol)	NH_3 (μmol)	N_2/ (μmol)	H_2/ (μmol)	Nitrogen balance (%)	NO_2^- selectivity[b] (%)	N_2 selectivity[c] (%)	Hydrogen overvoutage (V[d])
Cu	1	28	0	20	0	3	96	0	0	0.43
Cu	5	24	5	8	0	1	75	13	0	0.43
CU	11	36	13	0	0	0	98	99	0	0.43
Pt	11	48	0	0	0	26	96	0	0	0.01–0.09
Pd	11	47	0	0	0	14	94	0	0	0.04
Au	11	42	6	0	0	6	96	0	0	0.18
Ag	11	38	13	0	0	0	102	50	0	0.30
Pd–Cu[e]	11	24	2.4	0	10	0	93	4.6	95	
Pd–Cu[f]	11	22	1	0	13	0	98	1.5	98	

Note: [a]Nitrate remaining after the reaction for 12 h (initial amount: 50 mmol). [b]Based on consumption of photogenerated electron. Calculated from the equation $100 \times 2NO_2^-/(10N_2 + 2NO_2^- + 8NH_3 + 2H_2)$ [c]Based on consumption of photogenerated electron. Calculated from the equation $100 \times 10N_2/(10N_2 + 2NO_2^- + 8NH_3 + 2H_2)$ [d]literature values, refer to source paper for details. [e]0.5 wt. % Pd-0.5 wt. % Cu. [f]1.0 wt. % Pd-0.5 wt. % Cu.

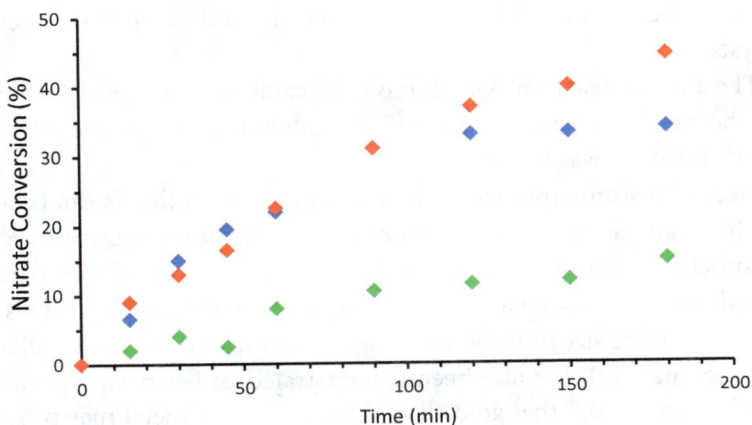

Figure 7: Conversion of nitrate at 15°C over Au/P25, Au/Autek and Au/Hombikat using 0.008 m oxalic acid as hole scavenger. Red, Au/TiO$_2$ Autek P25; blue, Au/TiO$_2$ P25; green, Au/TiO$_2$ Homb. Reproduced based on the data reported from Ref. [58].

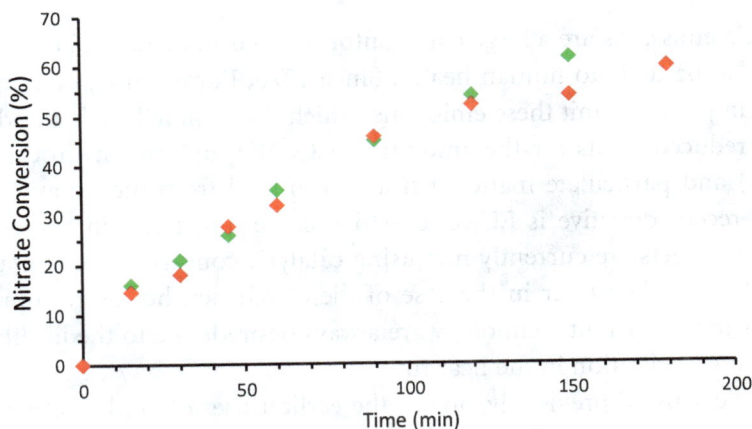

Figure 8: Nitrate conversion at 30°C over Au/Autek and Au/Hombikat using 0.008 m oxalic acid as hole scavenger. Red, Au/TiO$_2$ Autek P25; green, Au/TiO$_2$ Homb. Reproduced based on the data reported from Ref. [58].

oxalic acid as the hole scavenger, different nitrate reduction rates were reported at different temperatures (Figures 7 and 8). While photocatalysis should be independent of temperature, the author ascribed this affect to the rate of desorption of the final product at low temperatures (15°C); at higher temperatures (30°C) there was very little difference between the catalysts.

The rate of reaction with these gold catalysts compared favourably with the previously mentioned Ag/TiO$_2$ catalyst, although the selectivity towards nitrogen was lower.

Overall, it seems apparent that gold supported on titania can be active as a photocatalyst, however, it is more effective for some substrate systems than others. In almost all of the currently reported papers, the preparation method seems to have a large effect of the activity of the catalyst. This suggests that particle size of the gold is a significant factor on the overall activity of the catalyst. It has also been demonstrated, as has been reported for oxidation reactions,[18] that gold alloyed with a second metal may provide a significant improvement in the activity and this may be an avenue of future research in this subject area that could lead to significant breakthroughs.

3 Vehicle Emissions Control

Vehicle emissions are a large contributor to environmental pollution and can also be toxic to human health. Since 1970, European legislation has been in place to limit these emissions, which is continually being updated with reduced limits for the amount of CO, NO$_x$ unburnt hydrocarbons (HCs) and particulate matter that are permitted from the exhaust. The most recent directive is EURO 6, which came into effect in September 2014.[53] Targets are currently met using catalytic converters, and an additional particulate filter in the case of diesel vehicles, however, improvements to the current technology are always desired, due to the likelihood of stricter legislation in the near future.

As discussed previously, one of the earliest uses of gold catalysis was for low temperature CO oxidation,[8] which is a key process in emissions catalysis. Gold has therefore been investigated as a potential catalyst for emissions treatment particularly during cold starts, not only for CO oxidation[59–61] but also for hydrocarbon oxidation[62–64] and NO$_x$ reduction.[65–68] However, other precious metals were found to be preferable because they combine low light-off temperatures with durability at high temperatures, and so current "three-way catalysts" used in petrol vehicles for simultaneous oxidation of CO and HCs and reduction of NO$_x$ utilise Pt, Pd and Rh.

Other potential uses of gold in automotive applications include as a preferential CO oxidation catalyst to avoid poisoning of fuel cells,[69] and as electrodes in electrochemical cells which act as ammonia sensors in heavy duty diesel systems,[70] which use ammonia in the reduction of NO_x.

Diesel emissions treatment systems are typically more complex than those used in petrol vehicles, with separate catalysts for oxidation (diesel oxidation catalysts, DOCs) and reduction (selective catalytic reduction, SCR), and a particulate filter (diesel particulate filter, DPF). More recent publications regarding the use of gold as an emissions treatment catalyst are about its potential use in the DPF. After a period of use, DPFs become blocked by the soot particles they have trapped and need to be regenerated — this can be either passive, such as is the case for heavy duty vehicles, whereby the NO_2 in the exhaust gas is used to oxidise the trapped soot, or active, which is more common in light duty vehicles where the exhaust gas is cooler and contains less NO_x, in which case some fuel is injected into a catalyst upstream of the filter where it combusts and generates the high temperatures needed in order to oxidise the soot. However, it is considered that a catalyst coating on the DPF would be beneficial in order to reduce the temperature that the soot can be oxidised at; this would improve fuel efficiency by eliminating the need for fuel injections for regeneration, and enable the oxidation in passive systems to use the oxygen in the gas stream rather than NO_2 — this would reduce the amount of NO emitted. A catalyst-coated filter also has potential use in a gasoline particulate filter (GPF) which may be required in the near future due to legislation changes — particulate matter will be regulated in terms of particle number rather than particle mass, as it currently is, which means that the large amount of very small particles typically emitted from petrol exhausts will need to be mitigated in order to meet targets.

Research concerning soot oxidation is not always directly comparable, due to significant variations in the methods adopted for catalyst testing. For example, diesel soot can be difficult or time consuming to accumulate, so researchers use a variety of carbon-based model soot materials, most commonly types of carbon black including a commercial material used for printing, Printex-U. In some cases, bench-top soot generators are used and on occasion test engine set-ups are used. It should also be noted whether the catalyst is in powder or coated monolith form. Where powders are used, important details to note are the catalyst:soot ratio used (commonly around 10:1 by mass) and the type of mixing used — this is described as either tight contact, where the materials are ground together in a pestle and mortar, or loose contact, where they are simply mixed with a spatula. Loose contact is generally accepted to be the more realistic of the two mixing modes so is often used for this reason, however tight contact generally gives much lower temperatures for soot oxidation so a catalyst tested this way may not be as active as it first appears.

The use of Au/perovskite catalysts for DPF regeneration was investigated by Russo *et al.*,[71] who investigated the activity of catalysts for both soot and CO oxidation. Four $LaBO_3$ perovskites were used (B=Cr, Mn, Fe and Ni) both with and without Au (2% weight loading), and it was found that the best of these was the $LaNiO_3$ perovskite, which had a peak soot combustion temperature of 431°C. For all cases, the addition of Au did not affect the soot combustion temperature, but did significantly increase the activity for CO oxidation. The $Au/LaNiO_3$ catalyst, being the most active, was also prepared on a SiC monolith which was then tested on a diesel engine bench, enabling the DPF to be loaded with soot and regenerated under real-life conditions; it was found that the filter was able to be regenerated in a third of the time of a non-catalytic filter, and performed better than a Pt/Al_2O_3 catalytic filter tested previously by the same group.[72] This is illustrated in Figure 9, which shows the pressure drop and temperature in non-catalysed and catalysed filters with respect to time during loading and regeneration.

The high activity of the $LaNiO_3$ catalysts for soot oxidation was considered to be due to its ability to desorb suprafacial oxygen at low temperatures (approximately 250°C), as identified using oxygen temperature-programmed desorption (TPD) experiments and indicative of an increased amount of weakly chemisorbed oxygen species which was thought to be key for soot oxidation. The "spongy" morphology of the catalyst, as obtained using the solution combustion synthesis (SCS) preparation method which involves a sudden release of gases, was also considered to contribute to the activity by favouring contact between the catalyst and soot particles. Microscopy images of a catalyst prepared by this method are shown in Figure 10, which illustrate this structure; the final image is from TEM and the gold nanoparticles can be clearly seen, with

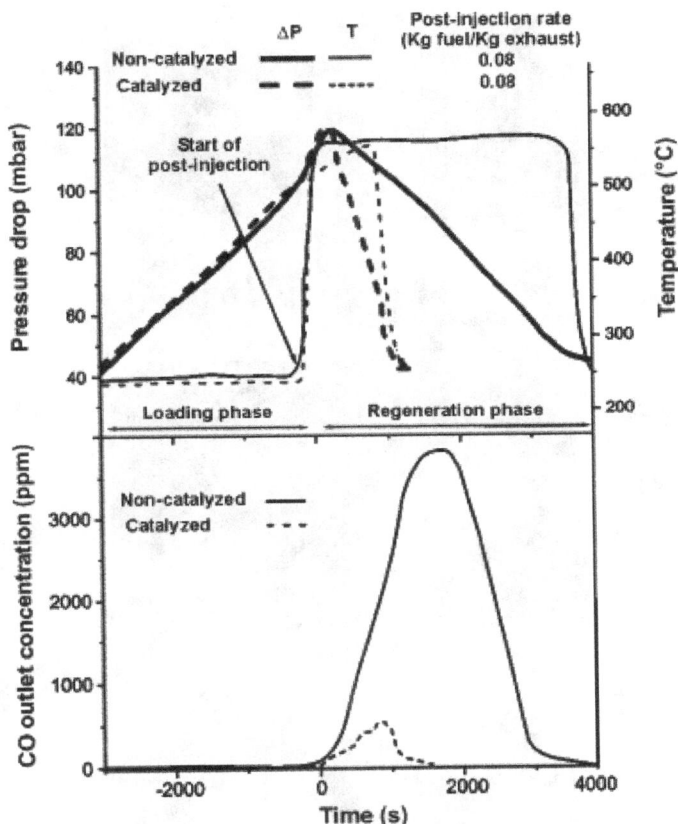

Figure 9: Results of the loading and regeneration runs for the 2 wt.% Au–LaNiO$_3$-catalyzed and the non-catalyzed wall-flow traps, reproduced with permission from Elsevier (Ref. [71]).

an estimated size of 5–25 nm. A similar structure was obtained when the catalyst was prepared on a SiC monolith support.

A number of papers regarding the use of gold in catalysts for soot oxidation have recently been published by Wei *et al.*, a common theme of which is the use of three-dimensionally ordered macroporous (3DOM) structures, which are used in order to improve the contact efficiency between soot particle and catalyst. A range of support materials have been used, including LaFeO$_3$[73,74] and CeZrO$_2$.[75] The 3DOM materials have uniform, large pore sizes of >50 nm which enable the soot to come into

Figure 10: FESEM views of the 2 wt.% Au–LaNiO$_3$ catalyst: (A) 5000× magnification; (B) 150,000× magnification; (C) TEM micrograph, reproduced with permission from Elsevier Ref. [71].

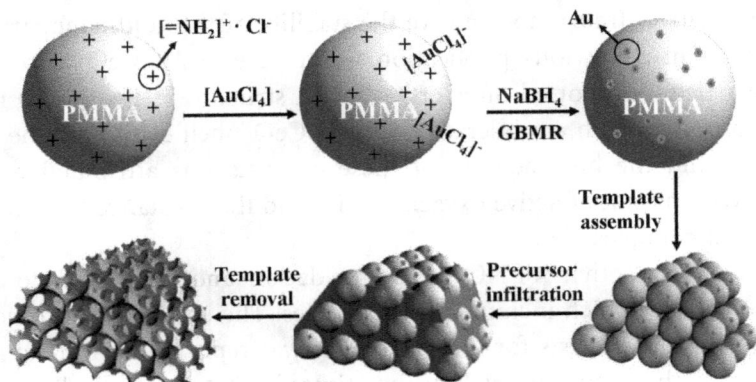

Figure 11:   CCT catalyst preparation method, reproduced with permission from Elsevier (Ref. [73]).

contact with more active sites. The materials exhibit good activity; however, they have poor stability, which is attributed to the sintering of Au particles during the reaction. Therefore, the same group further developed the method of catalyst synthesis in order to produce more stable catalysts, which is reported for $Au/LaFeO_3$.[73] The novel method is described as *in situ* colloidal crystal template (CCT) and involves the assembly of polymethyl methacrylate (PMMA) spheres with Au deposited on the surface into a face-centred cubic template, the voids of which are then filled with an inorganic precursor solution containing La and Fe nitrates. Finally, calcination steps remove the PMMA template, resulting in the $Au/LaFeO_3$ catalyst structure. This process is illustrated in Figure 11.

A 1.25 wt.% loading $Au/LaFeO_3$ catalyst was found to be the most active of the materials tested; the catalyst prepared by the CCT method had higher activity than $LaFeO_3$ particles, and the activity increased with an increase in Au content. In addition, re-usability of the catalyst was tested and for three temperature-programmed oxidation (TPO) testing cycles, the soot oxidation temperature did not change significantly, indicating that the catalyst has good stability. The high activity of the catalysts is rationalised in terms of the macroporous network allowing highly efficient contact between the catalyst and soot particles, and the supported Au nanoparticles improving the redox ability of the support. It is also demonstrated that the presence of Au increases the selectivity to CO_2, to >99%.

An alternative strategy to improve the stability of the gold nanoparticles described in a previous publication was to create Au–CeO$_2$ core-shell nanoparticles, supported on a ZrO$_2$ 3DOM structure.[76] The Au/Ce ratio was optimised, with a moderate thickness CeO$_2$ shell exhibiting the best activity, and the high activity of these catalysts was attributed to the improved amount of active oxygen species and the resistance to sintering of the Au nanoparticles.

Due to the activity of NO$_x$ in soot oxidation, another route is to consider a catalyst which includes nitrate species. The use of gold-promoted cesium nitrate catalysts for soot oxidation was reported by Ruiz *et al.*,[77] with gold/lithium nitrate catalysts investigated subsequently.[78] Alkali metals are known to have good activity for soot oxidation and are frequently investigated as components of catalysts for this purpose. The authors report previous work which demonstrates the high activity of catalysts prepared using alkali metal nitrate precursors (with cesium nitrate found to be the most active), which was explained by the ability of the nitrates to reduce to nitrite species upon oxidation of the soot.[79] However, such materials typically have low selectivity to CO$_2$ and so the addition of gold was investigated due to its known activity for CO oxidation and its lower cost and higher availability compared to other precious metals. Cesium nitrate (23.6 wt.%) and gold (2 wt.%) were impregnated both individually and together onto either a zirconia or silica support and catalyst activity was tested using TGA and TPO. The catalyst that was found to be the most active by TGA, with tight contact conditions in an air atmosphere, was the cesium nitrate/zirconia. However, by TPO testing, which is considered to be more lifelike due to using loose contact mixing of catalyst and soot and including NO in the gas stream, it was found that the addition of gold to this catalyst enhanced the soot oxidation activity and increased the CO$_2$ selectivity to 100%. This effect did appear to be somewhat support-dependent however, since the same trend was not observed for all cases — for a different (hydrated) zirconia support the addition of gold led to an increase in soot oxidation temperature by TPO. Lithium nitrate catalysts were then considered, due to the differences in behaviour between cesium and lithium nitrate precursors upon calcination; while caesium nitrate remains abundant after calcination, lithium nitrate decomposes into lithium oxide, nitrogen dioxide and oxygen. The presence of NO$_x$ in the exhaust can then generate nitrate ions *in situ*, which have good catalytic activity. The decomposition of the nitrate was only observed where gold was present in the catalyst as well, and was therefore considered to be due to reaction between the lithium nitrate and chlorine generated from thermal decomposition of the $HAuCl_4$ precursor. As for the cesium catalysts, the $Au–LiNO_3/ZrO_2$ was found to be most active under the same testing conditions (TPO, NO in the feed, loose contact), with 100% selectivity to CO_2 obtained. There is little difference in the peak soot oxidation temperatures observed for these catalysts however, with the lithium and cesium based catalysts at 357°C and 360°C, respectively.

Overall, it can be seen that in cases where gold-containing catalysts are investigated for soot oxidation, the main role of the gold is as a CO oxidation catalyst, to increase the CO_2 selectivity of the reaction to values of or close to 100%. This is an important feature of the catalysts, since CO emissions are targeted by the legislation, and is a good example of a new use for one of the longest-known reactions that can be catalysed by gold. It is also evident that the morphology of the whole catalyst is important, and structures that enable efficient contact between the soot and the catalyst are preferable.

4 Summary/Conclusions

Compared to photocatalytic water treatment and automotive soot remediation reactions, selective oxidation of bio-mass derived platform molecules has been far more extensively researched; however, through all these subject areas, the overall theme seems to emerge that the key factors affecting the performance of the catalysts are the particle size, morphology, oxidation state and the metal/support interaction of the gold, which are strongly affected by the preparation method. All the preparation methods described in this chapter have disadvantages, the main one amongst these being that even what we describe as good control of particle size contains an inherent massive variation, with SI resulting in particles of a range 2–10 nm; a 10 nm particle will contain 25 times the number of gold atoms compared to a 2 nm particle. To fully utilise the potential of gold, preparation methods with better control of the particle size will need to be developed. Despite the large increase in the volume of papers published in the area of gold catalysis since the 1980s, the majority are still focused on the traditional areas such as alcohol oxidation. It is the opinion of the authors of this chapter that there are still considerable discoveries that can be made by applying gold catalysts to different reaction systems which have until now received little attention.

References

1. G. J. Hutchings, *J. Catal.*, 1985, **96**, 292–295.
2. B. Nkosi, N. J. Coville and G. J. Hutchings, *Appl. Catal.*, 1988, **43**, 33–39.
3. M. Haruta, T. Kobayashi, H. Sano and N. Yamada, *Chem. Lett.*, 1987, 405–408.
4. N. W. Cant and W. K. Hall, *J. Phys. Chem.*, 1971, **75**, 2914–2921.
5. G. C. Bond, P. A. Sermon, G. Webb, D. A. Buchanan and P. B. Wells, *J. Chem. Soc., Chem. Commun.*, 1973, 444–445.
6. S. Shimada, T. Takei, T. Akita, S. Takeda and M. Haruta, *Stud. Surf. Sci. Catal.*, 2010, **175**, 843–847.
7. F. Moreau, G. C. Bond and A. O. Taylor, *J. Catal.*, 2005, **231**, 105–114.
8. M. Haruta, N. Yamada, T. Kobayashi and S. Iijima, *J. Catal.*, 1989, **115**, 301–309.
9. J. Hua, Q. Zheng, K. Wei and X. Lin, *Chin. J. Catal.*, 2006, **27**, 1012–1018.
10. A. Jain, X. Zhao, S. Kjergaard and S. M. Stagg-Williams, *Catal. Lett.*, 2005, **104**, 191–197.
11. L. Prati and G. Martra, *Gold Bulletin*, 1999, **32**, 96–101.
12. D. M. Alonso, J. Q. Bond and J. A. Dumesic, *Green Chem.*, 2010, **12**, 1493–1513.
13. A. Wittstock, V. Zielasek, J. Biener, C. M. Friend and M. Bäumer, *Science*, 2010, **327**, 319–322.
14. M. Haruta, *Chem. Phys. Chem.*, 2007, **8**, 1911–1913.
15. P. Anastas and J. Warner, *Green Chemistry: Theory and Practice*, Oxford Univ. Press, Oxford, 1998.
16. G. W. Huber, S. Iborra and A. Corma, *Chem. Rev.*, 2006, **106**, 4044–4098.
17. M. D. Hughes, Y.-J. Xu, P. Jenkins, P. McMorn, P. Landon, D. I. Enache, A. F. Carley, G. A. Attard, G. J. Hutchings, F. King, E. H. Stitt, P. Johnston, K. Griffin and C. J. Kiely, *Nature*, 2005, **437**, 1132–1135.
18. D. I. Enache, J. K. Edwards, P. Landon, B. Solsona-Espriu, A. F. Carley, A. A. Herzing, M. Watanabe, C. J. Kiely, D. W. Knight and G. J. Hutchings, *Science*, 2006, **311**, 362–365.
19. L. Kesavan, R. Tiruvalam, M. H. Ab Rahim, M. I. bin Saiman, D. I. Enache, R. L. Jenkins, N. Dimitratos, J. A. Lopez-Sanchez, S. H. Taylor, D. W. Knight, C. J. Kiely and G. J. Hutchings, *Science*, 2011, **331**, 195–199.
20. G. L. Brett, Q. He, C. Hammond, P. J. Miedziak, N. Dimitratos, M. Sankar, A. A. Herzing, M. Conte, J. A. Lopez-Sanchez, C. J. Kiely, D. W. Knight, S. H.

Taylor and G. J. Hutchings, *Angew. Chem., Int. Ed.*, 2011, **50**, 10136–10139, S10136/10131-S10136/10111.

21. S. Ramachandran, P. Fontanille, A. Pandey and C. Larroche, *Food Technol. Biotechnol.*, 2006, **44**, 185–195.

22. A. Abbadi and H. van Bekkum, *J. Mol. Catal. A: Chem.*, 1995, **97**, 111–118.

23. S. Biella, L. Prati and M. Rossi, *J. Catal.*, 2002, **206**, 242–247.

24. Y. Onal, S. Schimpf and P. Claus, *J. Catal.*, 2004, **223**, 122–133.

25. C. Baatz and U. Pruesse, *J. Catal.*, 2007, **249**, 34–40.

26. C. Baatz, N. Thielecke and U. Pruesse, *Appl. Catal., B*, 2007, **70**, 653–660.

27. A. Mirescu, H. Berndt, A. Martin and U. Prüße, *Appl. Catal., A*, 2007, **317**, 204–209.

28. N. Thielecke, K.-D. Vorlop and U. Prüße, *Catal. Today*, 2007, **122**, 266–269.

29. A. Mirescu, H. Berndt, A. Martin and U. Prüße, *Applied Catalysis A: General*, 2007, **317**, 204–209.

30. M. Haruta, *Catal. Today*, 1997, **36**, 153–166.

31. M. Comotti, C. Della Pina, R. Matarrese and M. Rossi, *Angew. Chem., Int. Ed.*, 2004, **43**, 5812–5815.

32. P. J. Miedziak, H. Alshammari, S. A. Kondrat, T. J. Clarke, T. E. Davies, M. Morad, D. J. Morgan, D. J. Willock, D. W. Knight, S. H. Taylor and G. J. Hutchings, *Green Chem.*, 2014, **16**, 3132–3141.

33. M. Comotti, C. Della Pina and M. Rossi, *J. Mol. Catal. A: Chem.*, 2006, **251**, 89–92.

34. P. Miedziak, M. Sankar, N. Dimitratos, J. A. Lopez-Sanchez, A. F. Carley, D. W. Knight, S. H. Taylor, C. J. Kiely and G. J. Hutchings, *Catal. Today*, 2011, **164**, 315–319.

35. A. Gandini and M. N. Belgacem, *Prog. Polym. Sci.*, 1997, **22**, 1203–1379.

36. Y. Y. Gorbanev, S. K. Klitgaard, J. M. Woodley, C. H. Christensen and A. Riisager, *Chem. Sus. Chem.*, 2009, **2**, 672–675.

37. S. E. Davis, L. R. Houk, E. C. Tamargo, A. K. Datye and R. J. Davis, *Catal. Today*, 2011, **160**, 55–60.

38. N. K. Gupta, S. Nishimura, A. Takagaki and K. Ebitani, *Green Chem.*, 2011, **13**, 824–827.

39. S. Albonetti, T. Pasini, A. Lolli, M. Blosi, M. Piccinini, N. Dimitratos, J. A. Lopez-Sanchez, D. J. Morgan, A. F. Carley, G. J. Hutchings and F. Cavani, *Catal. Today*, 2012, **195**, 120–126.

40. T. Pasini, M. Piccinini, M. Blosi, R. Bonelli, S. Albonetti, N. Dimitratos, J. A. Lopez-Sanchez, M. Sankar, Q. He, C. J. Kiely, G. J. Hutchings and F. Cavani, *Green Chem.*, 2011, **13**, 2091–2099.

41. A. Corma, S. Iborra and A. Velty, *Chem. Rev. (Washington, DC, U. S.)*, 2007, **107**, 2411–2502.

42. B. Katryniok, H. Kimura, E. Skrzynska, J.-S. Girardon, P. Fongarland, M. Capron, R. Ducoulombier, N. Mimura, S. Paul and F. Dumeignil, *Green Chem.*, 2011, **13**, 1960–1979.

43. S. E. Davis, M. S. Ide and R. J. Davis, *Green Chem.*, 2013, **15**, 17–45.

44. S. Carrettin, P. McMorn, P. Johnston, K. Griffin and G. J. Hutchings, *Chem. Commun.*, 2002, 696–697.

45. S. Carrettin, P. McMorn, P. Johnston, K. Griffin, C. J. Kiely and G. J. Hutchings, *Phys. Chem. Chem. Phys.*, 2003, **5**, 1329–1336.

46. B. N. Zope, D. D. Hibbitts, M. Neurock and R. J. Davis, *Science*, 2010, **330**, 74–78.

47. W. C. Ketchie, M. Murayama and R. J. Davis, *Top. Catal.*, 2007, **44**, 307–317.

48. M. Sankar, N. Dimitratos, P. J. Miedziak, P. P. Wells, C. J. Kiely and G. J. Hutchings, *Chem. Soc. Rev.*, 2012, **41**, 8099–8139.

49. M. C. Hidalgo, J. J. Murcia, J. A. Navío and G. Colón, *Appl. Catal., A*, 2011, **397**, 112–120.

50. Z. Hai, N. El Kolli, D. B. Uribe, P. Beaunier, M. Jose-Yacaman, J. Vigneron, A. Etcheberry, S. Sorgues, C. Colbeau-Justin, J. Chen and H. Remita, *J. Mater. Chem. A*, 2013, **1**, 10829–10835.

51. M. Shand and J. A. Anderson, *Catal. Sci. Technol.*, 2013, **3**, 879–899.

52. WHO, *Chemical Fact Sheets — Nitrate and Nitrite*, 2003.

53. *Report from the Commission to the Council and the European Parliament. On implentation of Council Directive 91/676/EEC concerning the protection of waters against pollution caused by nitrates from agricultural sources based on Member State Reports for period 2004–2007*, 2011.

54. A. A. Hekmatzadeh, A. Karimi-Jashani, N. Talebbeydokhti and B. Kløve, *Desalination*, 2012, **284**, 22–31.

55. J. Dou, A. Ding, X. Liu, Y. Du, D. Deng and J. Wang, *J. Environ. Sciences*, 2010, **22**, 709–715.

56. L. A. Richards, M. Vuachère and A. I. Schäfer, *Desalination*, 2010, **261**, 331–337.

57. H. Kominami, T. Nakaseko, Y. Shimada, A. Furusho, H. Inoue, S. Murakami, Y. Kera and B. Ohtani, *Chem. Commun.*, 2005, 2933–2935.

58. J. A. Anderson, *Catal. Today*, 2012, **181**, 171–176.

59. J. C. Bauer, T. J. Toops, Y. Oyola, J. E. Parks, II, S. Dai and S. H. Overbury, *Catal. Today*, 2014, **231**, 15–21.

60. B. L. Moroz, P. A. Pyrjaev, V. I. Zaikovskii and V. I. Bukhtiyarov, *Catal. Today*, 2009, **144**, 292–305.

61. A. Wichmann, A. Wittstock, K. Frank, M. M. Biener, B. Neumann, L. Maedler, J. Biener, A. Rosenauer and M. Baeumer, *Chem. Cat. Chem.*, 2013, **5**, 2037–2043.

62. Y. Azizi, C. Petit and V. Pitchon, *J. Catal.*, 2010, **269**, 26–32.
63. V. R. Choudhary, V. P. Patil, P. Jana and B. S. Uphade, *Appl. Catal., A*, 2008, **350**, 186–190.
64. F. Ying, S. Wang, C.-T. Au and S.-Y. Lai, *Microporous Mesoporous Mater.*, 2011, **142**, 308–315.
65. P. Miguel, P. Granger, N. Jagtap, S. Umbarkar, M. Dongare and C. Dujardin, *J. Mol. Catal., A: Chem.*, 2010, **322**, 90–97.
66. D. L. Nguyen, S. Umbarkar, M. K. Dongare, C. Lancelot, J. S. Girardon, C. Dujardin and P. Granger, *Catal. Commun.*, 2012, **26**, 225–230.
67. D. L. Nguyen, S. Umbarkar, M. K. Dongare, C. Lancelot, J. S. Girardon, C. Dujardin and P. Granger, *Top. Catal.*, 2013, **56**, 157–164.
68. M. Sridhar, J. A. van Bokhoven and O. Kroecher, *Appl. Catal., A*, 2014, **486**, 219–229.
69. P. Landon, J. Ferguson, B. E. Solsona, T. Garcia, A. F. Carley, A. A. Herzing, C. J. Kiely, S. E. Golunski and G. J. Hutchings, *Chem. Commun.*, 2005, 3385–3387.
70. D. Schoenauer-Kamin, M. Fleischer and R. Moos, *Sensors*, 2013, **13**, 4760–4780.
71. N. Russo, D. Fino, G. Saracco and V. Specchia, *Catal. Today*, 2008, **137**, 306–311.
72. E. Cauda, S. Hernandez, D. Fino, G. Saracco and V. Specchia, *Environ. Sci. Technol.*, 2006, **40**, 5532–5537.
73. Y. Wei, Z. Zhao, J. Jiao, J. Liu, A. Duan and G. Jiang, *Catal. Today*, 2015, **245**, 37–45.
74. Y. Wei, J. Liu, Z. Zhao, Y. Chen, C. Xu, A. Duan, G. Jiang and H. He, *Angew. Chem., Int. Ed.*, 2011, **50**, 2326–2329.
75. Y. Wei, J. Liu, Z. Zhao, A. Duan and G. Jiang, *J. Catal.*, 2012, **287**, 13–29.
76. Y. Wei, Z. Zhao, X. Yu, B. Jin, J. Liu, C. Xu, A. Duan, G. Jiang and S. Ma, *Catal. Sci. Technol.*, 2013, **3**, 2958–2970.
77. M. L. Ruiz, I. D. Lick, M. I. Ponzi, E. Rodriguez-Castellon, A. Jimenez-Lopez and E. N. Ponzi, *Appl. Catal., A*, 2011, **392**, 45–56.
78. M. L. Ruiz, I. D. Lick, M. S. Leguizamon Aparicio, M. I. Ponzi, E. Rodriguez-Castellon and E. N. Ponzi, *Ind. Eng. Chem. Res.*, 2012, **51**, 1150–1157.
79. I. D. Lick, A. L. Carrascull, M. I. Ponzi and E. N. Ponzi, *Ind. Eng. Chem. Res.*, 2008, **47**, 3834–3839.

Chapter 7

Precious Metal Catalysts for Sustainable Energy and Environmental Remediation

Hasliza Bahruji*, Shaoliang Guan*
and Vinod Kumar Puthiyapura†

*School of Chemistry, Cardiff University, Main building,
Park Place, Cardiff, CF10 3AT, UK
†School of Chemistry and Chemical Engineering,
Queens University Belfast, Belfast, BT9 5AG, UK

Precious metals have been attracting significant interest as active catalysts for sustainable fuel production and environmental remediation. This chapter is divided into two parts encompassing the use of precious metals, mainly Pt, Pd and Rh in the utilisation of CO_2 for fuels production and for sustainable H_2 production. For CO_2 utilisation, we focus on CO_2 hydrogenation using heterogeneous catalysts, photocatalytic CO_2 reduction and electrochemical reduction of CO_2. The influence of metal oxide support, alkaline promoter and bimetallic catalysts is discussed in detail. Basic principles and application of precious metal catalysts for electrochemical energy application (fuel cells, FCs) are discussed briefly.

1 CO_2 Utilisation to Value-added Molecules on Precious Metal Catalysts

1.1 Introduction

Direct fixation of CO_2 has received tremendous attention, largely to tackle the rising level of CO_2 in the atmosphere. Recognised as a major greenhouse gas, the utilisation of CO_2 as a carbon source for sustainable fuels production is seen as an ideal solution to address this issue. Conversion of CO_2 to value-added products requires very active catalysts for the activation of the thermodynamically stable C=O bond. There is a growing interest in utilising CO_2 as a sole carbon source for methanol synthesis assuming a renewable or sustainable source of H_2 is available. Methanol is currently produced from synthesis gas, which consists of a mixture of CO_2, CO and H_2 gases derived from natural gas. The first commercial methanol production using CO_2 as a feedstock was carried out by ICI using Cu-based catalysts on an Al_2O_3 support. In industry, the reaction takes place over Cu/ ZnO/Al_2O_3 catalyst at ~50 bar pressure and 250°C.[1,2] In this case, CH_3OH is produced by hydrogenation on CO_2 with the CO acting as a promoter to enhance the rate of methanol production.[3] The kinetics of methanol production in the presence of CO is more advantageous in comparison to the process without CO.[4] Methanol formation enhances with an increase in the concentration of CO_2 in the proportion of CO in the feed gas. However, deactivation over time is observed, mainly caused by the sintering of the Cu metal.[5] The reaction scheme for methanol production from CO_2, CO and H_2 mixture are given in Eqs. (7.1)–(7.3).

$$CO + 2H_2 \leftrightarrow CH_3OH, \quad \Delta H^{\circ}_{298K} = -90.77 \text{ kJmol}^{-1}, \quad (7.1)$$

$$CO_2 + 3H_2 \leftrightarrow CH_3OH + H_2, \quad \Delta H^{\circ}_{298K} = -49.58 \text{ kJmol}^{-1}, \quad (7.2)$$

$$CO_2 + H_2 \leftrightarrow CO + H_2O, \quad \Delta H^{\circ}_{298K} = -41.19 \text{ kJmol}^{-1}. \quad (7.3)$$

CO_2 as the carbon source for methanol production demands a catalyst to deoxygenate and hydrogenate CO_2 molecules and preserve methanol from reverse water–gas shift reaction (RWGS). CO_2, being a thermodynamically stable molecule, requires a substantial amount of energy to activate the C=O bond. Temperatures of more than 240°C are needed to activate CO_2 molecule for methanol production. Also low temperature and high-pressure conditions are required as the hydrogenation of CO_2 is an exothermic reaction. However, as the temperature increases, the formation of CO as a result of RWGS is favoured. The catalysts must also have high resistance towards water. Methanol synthesis may also compete with further C–O bond dissociation and hydrogenation reactions that could lead to methanation process.

$$CO_2 + 4H_2 \rightarrow CH_4 + 2H_2O. \qquad (7.4)$$

Cu on ZnO support is a popular choice of catalysts that gives higher conversion and selectivity towards methanol.[6,7] The strong interaction between Cu–Zn on the catalyst surface is believed to influence the formation of methanol.[8–10] CuZn species may possess a bifunctional catalytic property which was created by the interaction between Cu and ZnO in a closed vicinity.[12] The hydrogenation of CO_2 to formate species occur on the Zn-deposited Cu surface.[12] Cu^0 provides active sites for H_2 dissociation for hydrogenation steps, meanwhile Cu+ stabilises the intermediates such as formate or methoxy species.[11,12] The presence of both species is important and to maintain the oxidation charge throughout the reaction is rather difficult as the Cu species tends to oxidise to CuO. Metal oxide supports have an important role during CO_2 conversion, normally accepting the hydride H-spillover for second hydrogenation steps into formaldehyde species to form the methoxy intermediate. Hydroxyl groups on the support have been shown to be advantageous as they facilitate the hydrogenation of the methoxy intermediate to form methanol. CO_2 reduction may also be activated with reducible supports such as CeO_2 which can accept oxygen from the CO_2.[13] The concentration of oxygen vacancies created on ceria after reduction at high temperatures enhance catalytic activity by the annihilation of these oxygen vacancies by oxygen from CO_2.[14]

Due to the strongly connected reaction network, it is important to consider the thermodynamic properties of CO_2, methanol and related molecules that may be produced by CO_2 reduction. These are summarised in Table 1. CO_2 being a stable molecule requires a substantial amount of energy to activate the C=O bond. Any chemical reaction is driven by

Table 1: The Gibbs free energy of CO_2, methanol and related molecules from CO_2 hydrogenation.

Molecules	Gibbs free energy ΔG°_f (kJ mol^{-1})	Molecules	Gibbs free energy ΔG°_f (kJ mol^{-1})
$C_2O_4(l)$	−671	CH_3OH	−166
$HCO_3^-(l)$	−586	$CH_3OCH_3(g)$	−113
$CO_3^=$	−528	$HCHO(l)$	−102
$(CH_3)_2CO_3(l)$	−492	$CH_4(g)$	−51
$CO_2(l)$	−386	$C_6H_{14}(l)$	−4
$CO_2(g)$	−394	$H_2(g)$	0
$HCOOH(l)$	−361	$H_2(l)$	+18
$H_2O(g)$	−237	$C_2H_4(g)$	+68
$H_2O(l)$	−228	$C_6H_6(l)$	+124
$CO(g)$	−137	$C_6H_6(g)$	+130

differences in the Gibbs free energy of the reactant and products under reaction conditions. This can be calculated by the Gibbs–Helmholtz relationship as shown in Eq. (7.5):

$$\Delta G^\circ = \Delta H^\circ - T\Delta S^\circ. \qquad (7.5)$$

In the formation of methanol, the conditions must avoid CO formation and for the reaction to follow the formate intermediates, the C=O bond cleavage and H insertion should be equally efficient. ΔG° provides the information on the yield of product at equilibrium through the relationship as shown in Eq. (7.6).

$$\Delta G^\circ = -RT\ln K. \qquad (7.6)$$

The calculated thermodynamic equilibrium conversions of CO_2 to methanol and CO within the range of 250–650 K and 10–90 bar are shown in Figure 1. As shown, the reaction favours a low temperature condition to force the reaction equilibrium towards methanol. However, this must obviously be balanced by the lower rates of reaction at low temperature and thus consideration of the thermodynamic limitations and the kinetics must be considered.

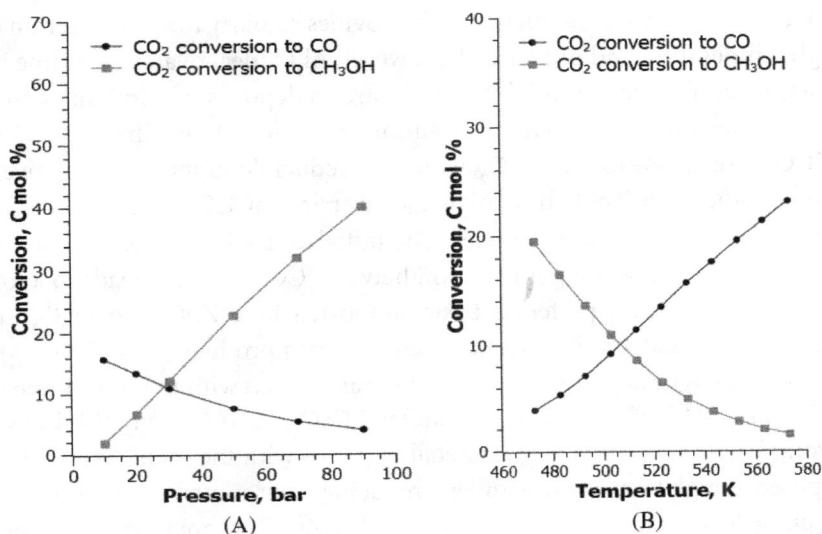

Figure 1: Calculated CO_2 conversion to CO and CH_3OH at equilibrium for the feed gas of H_2/CO_2 of 3.0.[15] (a) Influence of pressure on equilibrium yield at 250°C; (b) Influence of reaction temperature on equilibrium yield at 20 bar.

1.2 CO_2 Conversion on precious metal-based heterogeneous catalysts

1.2.1 Pd, Pt and Rh catalysts

CO_2 is selectively reduced to methanol, methane and carbon monoxide depending on the active metal and type of support used to catalyse the reaction. Precious metals, Pt,[16,17] Pd,[18,19] and Rh[20] provide facile hydrogen dissociation and strongly promote the hydrogenation of CO_2. Pd is the most common precious metal studied for CO_2 hydrogenation[21–24] with the resulting products dependant on the type of support and the promoter used.[25]

1.2.2 Role of support

CO_2 is converted via two kinetic pathways; deoxygenation and hydrogenation. As noted before, reducible supports such as Fe_2O_3, TiO_2, CeO_2, MnO_x[26–28] provide an alternative pathway for the deoxygenation of CO_2 due to facile surface oxygen vacancy formation compared with irreducible oxides such as ZnO, ZrO, Ga_2O_3, Al_2O_3.[2,9,29,30] Al_2O_3 is a common metal oxide support for CO_2 reduction as it provides stability towards reaction at high temperature and pressure.[2,31] However, Al_2O_3 deactivates over time by reacting with water or is inhibited with carbon deposits.[32,33] ZrO_2 has better resistant towards water than Al_2O_3 although the activity for the conversion of CO_2 is relatively lower.[34] TiO_2, which is a reducible support and relatively more acidic than ZnO show higher conversion of CO_2 than ZnO when impregnated on Au nanoparticles. The initial activation of CO_2 occurs on the support which strong interaction between CO_2 and ZnO leads to a low reactivity. Methanol prefers a basic support, where ZnO and Fe_2O_3 are much better catalysts than TiO_2, which the main product was CO.

The catalytic activity of Pd is strongly associated with the type of metal oxide support[13,21,23,24,36] including ZnO,[21,23] CeO_2[13] and Ga_2O_3.[24,29] Pd has a strong inclination to form intermetallic species with the metal oxide when exposed at high temperature under a reducing environment. PdGa bimetallic phase formed in Pd/Ga_2O_3 catalysts and PdZn alloys are formed on ZnO. These are formed via hydrogen from the Pd surface to the support and reduction of the oxide at the interface of the Pd and oxide support initiating intermetallic phase formation. This formation subsequently reduces both CH_3OH decomposition and CO production.[24,29,39] In addition, high CO_2 conversion to methanol in Pd/Ga_2O_3 has been ascribed an optimal amount of Pd^{n+} ($0 < n < 2$) species stabilised by a Ga_xO_y suboxide layer on the surface of Pd.[29] The Pd–Zn alloy formed is strongly dependent on the reduction temperature with PdZn formed at 200°C and at higher temperatures of 300°C leading to a close bulk value of a 1:1 PdZn alloy.[40] At higher temperatures of 550°C, the alloy shows a loss of Zn at forming a Pd_2Zn binary phase.[40] There are different proposals concerning the role of PdZn alloys towards CO_2 hydrogenation. Kim and co-workers suggested that the PdZn bimetallic species led to the deactivation of catalysts for methanol synthesis.[21] In contrast, Liang and co-workers indicated that the active site for methanol formation was the PdZn alloy rather than metallic Pd.[23]

1.2.3 Influence of promoter

As well as changed in the support used, promoters have been explored including using alkaline promoters.[41,42] Alkali adatoms increase the binding energy of adsorbed CO_2 and subsequently promote dissociation of CO_2.[44] Surface studies of CO_2 adsorption on Pd metal surface have revealed that CO_2 does not adsorb strongly on Pd; however, this can be enhanced by the presence of alkali metals, for instance, Na.[43] In this case, CO_2 activation proceeds via a bent anionic $CO_2^{\delta-}$ at intermediate Na coverages[43] and the adsorbed CO_2 is further dissociated into CO and chemisorbed oxygen.[43] At high Na coverages, surface carbonate is formed whereas at low Na coverages spontaneous dissociation into CO and O occurs.[43] Studies on CO_2 adsorption on Pd, Pt and Rh show weak physical absorption of CO_2 on the clean metal surface.[44] Lithium has also been shown to promote Rh supported on zeolites wherein CO selectivity is enhanced by suppressing the methanation pathways.[45] At Li/Rh atomic ratios of 10%, the CO selectivity was found to be 86% and leading to CH_3OH and C_2H_5OH formation. *In situ* FTIR studies indicated that the presence of Li on the Rh surface provided the active site for CO_2 adsorption and stabilisation of adsorbed CO species.[45] Adding Li, however, results in lower catalyst activity for CO_2 conversion.

Ca has also been used as a promoter on Pt/C catalysts leading to a high activity and selectivity for CO in contrast to the absence of Ca.[50] This effect has also been observed on supporting Pt on CaO. Therein, the CO_3^{2-} formed on the basic support dissociates into CO but further hydrogenation was not observed.[50]

Koizumi and co-workers have investigated the effect of Ca, Mg and/or K promoters on the activity of Pd catalysts deposited onto mesoporous silica supports.[22] While the small mesoporous structure of the support ensures the formation of small Pd^0 nanoparticles, the presence of Ca, Mg and/or K promoters were essential to increase the methanol yield.[22]

The surface studies of CO_2 adsorption on model precious metal catalysts is in agreement with the catalytic activity that suggest the addition of alkaline metal promoter enhances the activity by providing a site for CO stabilisation. In general, the addition of alkaline metal promoters on precious metal catalysts have been found to improve the selectivity to a range of products (methane, carbon monoxide, methanol) but the catalyst activity is decreased.

1.3 Bimetallic catalysts

In a pursuit of developing improved catalyst for CO_2 conversion to methanol, catalysts with both deoxygenation and dehydrogenations sites have been designed. In particular, bimetallic catalysts have been examined. For monometallic and bimetallic surfaces, the surface d-band centre value, with respect to the Fermi level, is a useful parameter to describe the electronic properties of a metal catalyst. The d-band centre parameter is a measure of the average energy of the unoccupied and occupied states in the d-level of a metallic surface. The position of the d-band centre relative to the Fermi level determines the interaction with the adsorbate. The adsorbate binds more strongly on metal surface when the d-band centre is above the Fermi level. Bimetallic catalyst allows this to be tuned by the interaction of the two metals and thus allows changes to the binding energy of reactants and reactive intermediates which, in turn, influence the surface chemistry.[51] It has been shown that the binding energy of an adsorbate to the metal surface is linearly correlated to the d-band centre of the metal.[52]

Studies on selective hydrogenation of C=O bond in acrolein on Pt–Ni–Pt and Ni–Pt–Pt shows a correlation between the experimental studies and the theoretical calculation of d-band centre for the bimetallic substrate.[53] The surface d-band centre of the Pt–3d–Pt(111) subsurface structure determined from DFT calculations shows the value shifts away from the Fermi level as the subsurface 3d metal moves toward the left-hand side of the periodic table.[54] Hydrogenation of the C=O only occurs on the surface Pt–Ni–Pt(111) but negligible activity is found on Pt(111) or Ni–Pt–Pt(111) surfaces.[53] HREELS has been used to differentiate between the bonding configurations of acrolein on Pt–Ni–Pt (111). It reveals that the C=O bond in acrolein strongly interacts on Pt–Ni–Pt(111) surface at 200 K via the formation of di-σ-bonded acrolein through the C=O moiety, whereas the Ni–Pt–Pt(111) shows weak interaction with the C=O.[53]

The analysis of bimetallic catalyst is interesting because it opens a new aspect of the design of catalysts for CO_2 hydrogenation. Pt–Ni, Pt–Co, Pd–Ni bimetallic catalysts for CO_2 reduction are active to activate C=O bond for CO and CH_4 formations.[55] For these bimetallic systems, the subsurface structures, Pt–Ni–Pt, Pt–Co–Pt and Pd–Ni–Pd were identified as the thermodynamically stable structures in the presence of hydrogen. Using the combinations of these metal, catalysts based on a reducible CeO_2 support showed higher activity than on a non-reducible Al_2O_3 support. The ability of CeO_2 to uptake oxygen facilitates a greater extent of CO_2 hydrogenation.

Well-dispersed Pd–Cu bimetallic species increased selectivity to methanol compared with Pd or Cu catalysts.[56] The alloy is thought to increase the chemisorption of CO_2 and enhancing the hydrogenation process by providing a weakly-bound hydrogen atom. The Pd–Cu alloy formation is dependent on Pd(Pd+Cu) ratios which reflect the chemisorption of CO_2 and hydrogen on the intermetallic surface. In addition, the activity is found to be strongly influenced by the preparation method with co-impregnation found to lead to a homogeneous alloy which is thought to be important to high methanol selectivity.[56]

1.4 Photocatalytic reduction of CO_2

Photocatalysis offers an alternative route for CO_2 conversion and thus a green sustainable approach compared with thermal activation. Photocatalytic reactions require photon energy from sunlight that excite electrons in the valence band (VB) to the conduction band of a semiconductor. This photogenerated energy carrier is responsible for the photocatalytic reaction. Potential reducing agents for CO_2 are H_2, H_2O, NH_3 or CH_3OH and TiO_2, WO_3, CdS, GaP, SiC and ZrO_2 are among the metal oxide semiconductors that have been shown to reduce CO_2.[7,57–62] Photocatalytic reduction of CO_2 is a thermodynamically unfavourable reaction involving multi electron redox processes. The products vary from CO, CH_4 to higher hydrocarbons in the gas phase, and various oxygenates in the liquid phase such as alcohols, aldehydes and carboxylic acids. CO_2 reduction does not originate from photocatalytic water splitting as in this case the H_2O acts as a reducing agent by providing H^+ as opposed to H_2. The CO_2 reduction process is considerably more difficult that water splitting due to the high endoergonicity of the process.

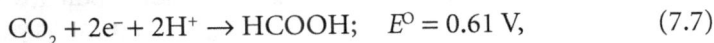

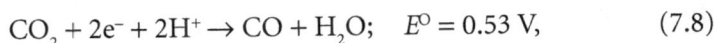

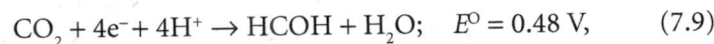

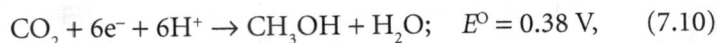

$$CO_2 + 2e^- + 2H^+ \rightarrow HCOOH; \quad E^\circ = 0.61 \text{ V}, \quad (7.7)$$

$$CO_2 + 2e^- + 2H^+ \rightarrow CO + H_2O; \quad E^\circ = 0.53 \text{ V}, \quad (7.8)$$

$$CO_2 + 4e^- + 4H^+ \rightarrow HCOH + H_2O; \quad E^\circ = 0.48 \text{ V}, \quad (7.9)$$

$$CO_2 + 6e^- + 6H^+ \rightarrow CH_3OH + H_2O; \quad E^\circ = 0.38 \text{ V}, \quad (7.10)$$

$$CO_2 + 8e^- + 8H^+ \rightarrow CH_4 + 2H_2O; \quad E^\circ = 0.24 \text{ V}. \quad (7.11)$$

Although photocatalysis for CO_2 reduction appears to be an ideal approach to take full advantage of sunlight energy for green sustainable fuel production, the conversion remains relatively low. The main challenge for CO_2 photoreduction is that it is competing with water reduction with the standard reduction potential of H_2O to form H_2 being considerably lower ($E^o_{red} = 0V$) than the standard reduction of CO_2 to form CO_2^- ($E^o_{red} = -1.9V$).

The rates are also dependent on the solubility of the CO_2 in water as this determines the surface concentration on the photocatalyst surface. Conducting the reaction in the alkaline NaOH solution helps to improve solubility of CO_2 in water.[63] However, the resulting species from CO_2 dissolution, CO_3^{2-} or HCO_3^-, are thermodynamically more stable than CO_2.

TiO_2 has received significant attention as a photocatalyst for CO_2 hydrogenation. The conduction band is above the redox potential for CO_2 reduction to methanol and the VB is well below the redox potential of water oxidation. Precious metals including Pd,[64,65] Rh[66] and Pt[67,68] are often added to TiO_2 to prolong the lifetime of the photogenerated energy carrier and these systems have been examined for CO_2 reduction with the photo generated products dependent on the metal used. There is an uncertainty in the literature on the actual source of carbon that contributes to the product formation.[69,70] Since the level of conversion is relatively low <1% conversion, in general, the presence of carbon contamination on the TiO_2 surface may contribute to the product formation. For example, Yui et al.[71] highlighted the discrepancy of products formed from untreated and air annealed TiO_2 due to the presence of carbon species mostly in the form of acetate on the surface of fresh TiO_2 catalyst. CH_4 was formed as main product from untreated fresh TiO_2 photocatalyst via the photo-Kolbe decomposition of acetate.[71] CO becomes a major product when TiO_2 calcined in air at 350 °C was used. Deposition of Pd on TiO_2 initiates the formation of CH_4.[71]

Pd and Rh metal supported on TiO_2 improved the photoactivity of TiO_2 for CO_2 photoreduction under UVA irradiation compared with the pure oxide.[72] In addition, improved activity was found, TiO_2 ceramic honeycomb monolithic structures threaded with optical fibres were employed rather than the powder suspended in water. Therein, the quantum efficiency of the reaction was improved significantly which indicates the importance of the need to increase the surface area of the catalyst and thus enhance the rate of CO_2 adsorption.[72] CO_2 photo reduction on Pt-doped TiO_2 nanotube catalysts have also been shown to result in a higher yield of CH_4 compared with a TiO_2 nanoparticle support.[73]

1.5 Electrochemical reduction of CO_2

Electrochemical reduction of carbon dioxide provides an excellent method for new energy storage by using renewable energy sources.[74–76] The process involves CO_2 gas and uses H_2 gas or aqueous electrolytes as the source of H^+. There are several advantages that this process may provide. First, it gives better chemical efficiency than the traditional Sabatier process (>80%), which has been considered heavily for producing alternative fuels, involves the conversion of CO_2 and hydrogen gas into methane and water catalysed by nickel at high temperatures and high pressure.[77] Second, depending on the reduction method, high Faradaic efficiency can be achieved. Faradaic efficiency is the energy efficiency with which charge (electrons) is transferred in a system facilitating an electrochemical reaction. Hence, high Faradaic efficiency suggests that the process requires lower energy to complete the reaction so that it makes the chemical process more feasible. Third, compared to the Sabatier process, which involves both high temperature and pressure, the electrochemical process of CO_2 reduction can be achieved at low temperature.[78,79]

It has been found that the products being formed in the electrochemical reduction of CO_2 include a wide variety of hydrocarbons and alcohols, as well as environmentally friendly water.[80–82] Different metal electrodes have been used in this process and these play a significant role of determining the types and yields of the products. It is reported that, at most metal electrodes, the major reaction products are carbon monoxide and formic acid.[79,83–87] For example, CO is the predominant reduction product on Ag, Au, Pd and Zn electrodes, while the main product on Hg, Pb, Sn, In and Cd electrodes is formic acid. Almost no reduction of CO_2 takes place on Ti, Mo, Rh, Fe, Ni and Pt. According to Hori *et al.*[80,88] and confirmed by other studies workers,[89–92] copper electrodes can reduce CO_2 to CH_4,

C_2H_4 and alcohols in aqueous electrolytes at ambient temperature Later, Cook and co-workers improved the efficiency of methane production. They reported that cumulative CO_2 reduction to CH_4 and C_2H_4 had high Faradaic efficiencies on *in situ* electrodeposited Cu layers on glassy carbon electrodes.[89,93] It has been shown that electroplated Ru metal electrode is a good candidate for methanol and CH_4 production.[85] Currently, the molecular mechanism for each electrochemical reduction of CO_2 is not fully understood. Therefore, the properties of metal electrodes which determine the electrocatalytic activity and selectivity remains unclear.

However, previous studies have shown that many factors may determine the product distribution, which include the temperature, the electrolyte, the electrode potential and the electrode surface.[88,91,92,94] For example, the partial current densities for CO_2 reduction at a pressure of 1 atm are not enough for the reaction to occur (4.7 mA cm^{-2} on Cu electrodes[94] and 10 mA cm^{-2} on Au[81]) even though the Faradaic efficiencies are high. Therefore, it is expected that CO_2 reduction at large current densities can take place at high pressure due to the high concentration of CO_2 in the electrolyte.[95,96] Ito and Mayorova investigated the electrochemical reduction of CO_2 at high pressure (<60 atm) on group 8–10 metals (e.g. Fe, Ni, Co, Pd and Pt) in aqueous solutions.[97-99] They have found that formic acid and CO were formed with high Faradaic efficiencies under high pressures whilst only small quantities were produced at a CO_2 pressure of 1 atm. The results have shown that the product selectivities changed dramatically with increasing CO_2 pressure. Sakata and co-workers reported electrocatalytic activities of 32 metal and non-metal electrodes for the electrochemical reduction of CO_2 in aqueous $KHCO_3$ medium at a pressure of 30 atm.[79] Table 2 summarises the current efficiencies for reduction products on some of the metal electrodes at ~0°C.

It was concluded that the formation of methane and ethylene is observed on almost all metal electrodes, although the efficiency is very low (except for Cu). HCOOH is the main reduction product on the Hg electrode. It can be seen from Table 2 that the temperature effect on the CO_2 reduction efficiency has also been studied. The current efficiency of CO_2 reduction on Ni, Ag, Pb and Pd increased substantially with decreasing temperature. Kaneco *et al.* also reported that, on Au, Ag and electrodes, the formation efficiencies for CO increased as the temperature decreased.[100,101] Temperatures of <0°C were also found to effective for the suppression of

Table 2: Typical current efficiencies (%) for CO_2 reduction products at -2.2 V vs. SCE in a CO_2-saturated 0.05 M $KHCO_3$ aqueous solution at about 0°C.[79]

Metal	T/°C	CH_4	CO	C_2H_4	C_2H_6	HCOOH	H_2	Sum
Cu	0	24.7	16.5	6.5	0.015	3.0	49.3	100
	20	17.8	5.4	12.7	0.039	10.2	52.0	98
Ni	0	0.71	21	0.069	0.18	13.7	61.7	97
	20	0.13	0.60	0.010	0.021	0.10	98.8	100
Ag	0	1.4	40.7	0.0052	0.013	20.5	32.6	95
	20	1.1	30.0	0.0090	0.0027	16.0	50.0	98
Pb*	0	0.39	0.12	0.008	0.0014	16.5	82.9	100
	20	0.06	0.10	0.001	0.0003	9.9	93.3	103
Pd*	0	0.083	11.6	0.011	0.014	16.1	73.3	101
	20	0.31	3.2	0.061	0.078	8.6	90.3	103
Hg	0	0.0004	0.20	t	t	90.2	9.5	100
	20	0.0035	0.64	0.0002	0.00006	87.6	7.9	96

Note: *At -2.0 V vs. SCE.

t: trace.

hydrogen formation in the KOH-methanol electrolyte. On Pb electrode, not only C_1 and C_2 reduction products, but also, C_3 products such as propylene and propane were observed.[79]

For the reduction of CO_2 into HCOOH on a Hg electrode in an aqueous medium, there are two possible mechanisms proposed.[79] They are as follows:

Electron transfer mechanism

$$CO_2 + e^- \rightarrow CO_2^-, \tag{7.12}$$

$$CO_2^- + H_2O(\text{or } H^+) \rightarrow HCOO, \tag{7.13}$$

$$HCOO + e^- \rightarrow HCOO^-, \tag{7.14}$$

$$HCOO^- + H_2O(\text{or } H^+) \rightarrow HCOOH. \tag{7.15}$$

Adsorbed hydrogen mechanism,

$$H^+(\text{or } H_2O) + e^- \rightarrow H_{ads}, \tag{7.16}$$

$$H_{ads} + CO_2 \rightarrow HCOO_{ads}, \qquad (7.17)$$

$$HCOO_{ads} + H_{ads} \rightarrow HCOOH. \qquad (7.18)$$

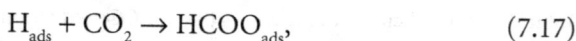

The first mechanism may be favourable in a neutral and weakly alkaline solution while the latter reaction is favourable in acidic conditions.

Since, the low temperature can increase the efficiency of CO_2 reduction, Kaneco and co-workers investigated the electrochemical reaction at the Cu electrode in a methanol-based electrolyte using various supporting alkaline salts at 243 K.[102] Methanol was used as the electrolyte in this case because organic aprotic solvents dissolved much more CO_2.[103–105] It was discovered that methane formation tends to increase in the order Cs^+, Rb^+, K^+, Na^+ and Li^+, i.e. in decreasing order of the cation size. In all sodium supporting electrolytes tested, the methane current efficiency was relatively high (≥43.4%). In contrast, for potassium, rubidium and cesium supporting electrolytes preferential formation of ethylene was observed, with a highest Faradaic efficiency of 35.7% in the methanol-based catholyte with potassium hydroxide supporting salt. In halide and thiocyanate/methanol-based electrolytes, the trends that the current efficiency for CO increased with an increasing cation size was also examined. The main function for different anionic and cationic species added in the electrolyte was to depress hydrogen formation so that the applied energy was used for CO_2 reduction instead of hydrogen evolution.[102]

Finding high selectivity of the reduction product is crucial so that practical and industrial application of CO_2 reduction can be possible in the future. Fundamental studies of single crystal electrodes have been used to reveal the unique electrocatalytic properties of metals. Since Cu electrodes favourably reduce CO_2 to CH_4, C_2H_4, and alcohols,[80,88] the electrochemical reduction at single crystal Cu electrodes, Cu{111}, Cu{110} and Cu{100}, was studied by Frese et al.[106] Therein, it was found that CH_4 formation was favoured in the order of Cu{111}>Cu{110}>Cu{100}. Hori and co-workers also reported that CH_4 is predominantly produced on the {111} surface, and C_2H_4 is formed preferentially on the {100} surface.[107] Interestingly, as the intermediate of the electrochemical reduction of CO_2, CO is also reduced to form similar products at these single crystal Cu electrodes of low index planes.[91,94,108] It was also found that the introduction of {111} steps to Cu{100} basal plane may significantly promote C_2H_4 formation whilst suppressing CH_4 formation. The selectivity ratio C_2H_4/CH_4 on Cu^7 in terms of the current efficiency was found to reach 14 in this case, which is 2 orders of magnitude higher than that on the Cu{111} surface.[107]

2 Electrocatalysts for Direct Fuel Cell Anodes

2.1 Introduction

Fuel cells (FC's) are energy conversion devices (galvanic cells) which continuously and directly convert chemical energy in a fuel e.g. H_2, methanol, ethanol, etc.) to electrical energy bypassing the intermediate, energy intensive, heat to mechanical energy conversion, as in conventional heat engines.[109] Since FCs are not limited by the Carnot limit as in the internal combustion (IC) engine, they are more efficient than ICs with efficiency of about 40–50% in electrical energy and 80–85% in total energy (electric + heat production).[110] Although, the state of the art FCs are limited by their low efficiency, insufficient durability and high cost which makes them unsuitable commercial applications,[111] significant improvements in the performance have been achieved in the last decades with the advancement in engineering and chemistry. Among different FC, proton exchange membrane fuel cells (PEMFC) are considered as the most efficient and promising fuel cell for low-temperature operation of up to 200°C. Schematic of a typical PEMFC is given in Figure 2.

The electrochemical reactions take place on the anode and cathode catalyst layers. The fuel (H_2, methanol, ethanol, etc.) are fed into the anode side, which oxidise to form H^+ and is transported through the proton exchange membrane to the cathode side where it reduces to H_2O in the presence of O_2/air (Figure 2). The net reactions taking place in an alcohol/O_2 fuel cell can be written as in Eqs. (7.19)–(7.21).

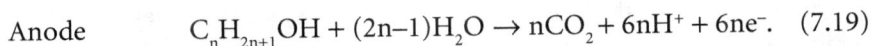

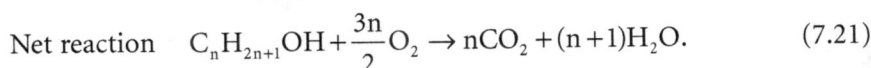

Anode $\qquad C_nH_{2n+1}OH + (2n-1)H_2O \rightarrow nCO_2 + 6nH^+ + 6ne^-.$ (7.19)

Cathode $\qquad O_2 + 4H^+ + 4e^- \rightarrow 2H_2O.$ (7.20)

Net reaction $\quad C_nH_{2n+1}OH + \dfrac{3n}{2}O_2 \rightarrow nCO_2 + (n+1)H_2O.$ (7.21)

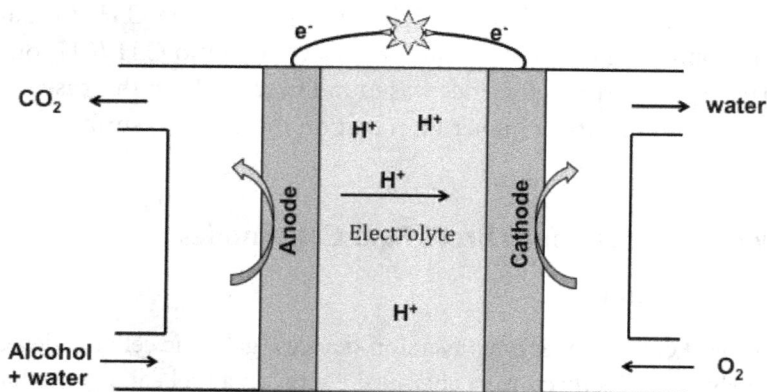

Figure 2: The schematic of a direct alcohol fuel cell with acidic membrane.

The major challenges in Direct Alcohol Fuel cells (DAFC's) are[112] their (i) lower performance compared to H_2/O_2 fuel cell (poor alcohol oxidation kinetics) (ii) mixed potential at cathode due to the alcohol permeation to the cathode (thus lower fuel utilisation) (iii) difficult water and thermal management due to the alcohol cross over (iv) catalyst detachment from the electrode due to the membrane swelling in presence of alcohol.

Pt and Pt alloys are found to be the most active catalyst for alcohol oxidation in acidic media. According to a study by US Department of Energy (DOE), in 2007 based on a projected cost for large scale fuel cell production, about 56% of the cost in a FC stack comes from the Pt-based catalyst layers.[113] Increasing the catalyst activity, stability, durability and reducing the catalyst loading on the electrode layer are the major challenges in the commercialisation of FCs. Also, there are drawbacks associated with the Pt catalyst such as poor CO tolerance and Pt dissolution at high potential.[114] It is significant to reduce the metal loading while maintaining the activity in the catalyst in order the system to be industrially viable and is a challenge in electrocatalysis research. In this section, electrocatalysis for the alcohol oxidation reaction is briefly reviewed with a major focus on methanol and ethanol electro oxidation as they can be considered as model systems for other organic molecules with 2 or more carbon atoms. Since the state of the art of FCs are running on acidic based membrane system (e.g. Nafion), only active catalysts in acidic media are being discussed.

2.2 Electrocatalysts for alcohol oxidation reaction

Though H_2 is considered as the most suitable fuel for PEMFC with higher efficiency, the production, storage and distribution are still limitations. Ultimately, the fuel for PEMFC should be chosen based on the electrochemical activity, availability, production, storage, transportation and distribution of fuel, its safe handling and its effect on the environment, the convenience of refuelling and the suitability under any working conditions. In addition, effects of the emissions of intermediate products as well as leakage problems, etc., should also be taken into account.[112] In this context, hydrogen carriers such as low molecular weight alcohols (e.g. ethanol, methanol, etc.) are proposed as promising alternate fuels to H_2 due to their liquid nature (which simplifies the problem of storage), high energy density (5–8 kWhl^{-1}) compared to liquid hydrogen (2.6 kWhl^{-1}),[109,115] low toxicity (except methanol), bio-availability, ease of handling, storage and transport.[115,116]

To recover the maximum energy in an alcohol, complete oxidation leading to CO_2 formation should be achieved as given in Eq. (7.19). The formation of CO_2 by alcohol oxidation involves the breaking of C–C bonds (except methanol) and goes through various intermediate species depending on the alcohol molecule. Complete oxidation to CO_2 is generally difficult to achieve and, to date, only methanol has been reported to form CO_2 by complete oxidation on Pt in acidic media. In practice, alcohols undergo partial oxidation to give intermediate products without the C–C bond cleavage. Poor reaction rates and the difficulty in breaking C–C bond at low temperature (25–80°C) leads to the need for high over-potentials, which decreases the fuel cell efficiency. The breaking of C–C in the alcohol molecule to achieve complete oxidation is a challenging task in alcohol oxidation catalysis and is the central problem in DAFC.

When an alcohol molecule interacts with the Pt catalyst surface, various destruction processes occur (dehydrogenation, destruction of the molecule at C–C bond and hydrogenation).[117] and various intermediate species are formed as a result. Even small alcohols such as methanol oxidation to CO_2 involves a 6e$^-$ transfer indicating a complex reaction mechanism which involves multiple e$^-$ transfer and consecutive adsorption/surface reaction/desorption of several reaction products or by-products. In addition, poisoning species such as CO are produced during the alcohol oxidation reaction which irreversibly adsorb on Pt surface blocking the catalyst surface.[118] It is clear from Eq. (7.19), that the number of electrons exchanged during the electro oxidation of alcohol increases with increase in the number of carbon atoms in the alkyl chain. However, extending the carbon chain increases the complexity of the reaction pathway as well as the number of possible intermediates which reduce the overall conversion efficiency of alcohol oxidation to CO_2. For example, on a Pt electrode operated in acidic media within the fuel cell operating potential range, the number electrons obtained for a range of fuels were lower than the theoretical e^- yield including CH_3CH_2OH (4e$^-$), HCHO (4e$^-$), HCOOH (2e$^-$), n-PrOH (4e$^-$), i-PrOH (2e$^-$), ethylene glycol (4e$^-$).[119] Only methanol gives the theoretical yield of 6e$^-$ to form CO_2. Ethanol is predominantly oxidised by 2e$^-$ and 4e$^-$ pathways to give acetaldehyde (AAL) and acetic acid (AA), respectively, as the major products (Figure 3).[120,121] Further oxidation of these intermediates to CO_2 are generally difficult. An active electro-catalyst thus should be able to increase the reaction rate and modify the reaction step to reach the final step rapidly (i.e. production of CO_2).[116] It was found that most active catalyst for alcohol oxidation, to date, are Pt and Pt alloys in acidic media.

Various experimental techniques have been adopted to identify the intermediates/products formed during the oxidation reaction to elucidate the mechanism of alcohol oxidation which involve both electrochemical (e.g. cyclic voltammetry, chronoamperometry, Coulometry, etc.) as well as analytical (e.g. HPLC, electrochemical quartz crystal microbalance, chemical radiotracers, etc.) and spectroscopic techniques (e.g. FT-IR, differential electrochemical mass spectrometry, etc.).[122]

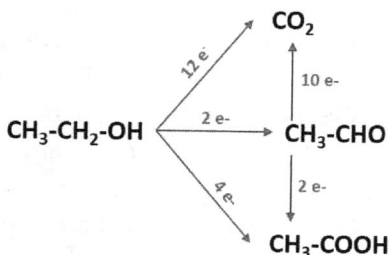

Figure 3:   Number of electrons transferred during the ethanol electro-oxidation in acidic media and the corresponding products formed.[121]

The electro-oxidation mechanism of methanol[116,123] and ethanol have been widely studied and the intermediates and products formed have been clearly identified. From IR studies, it was shown that the main poisoning species during primary alcohol oxidation is CO, either linearly bonded or bridge bonded to the catalyst surface.[124,125] The CO_{ads} formed blocks the Pt active sites deactivating the system and revealing the need for high CO tolerant catalysts for alcohol oxidation.

2.2.1 CO oxidation on platinum

CO oxidation reaction is an important and widely studied reaction in heterogeneous catalysis.[126,127] It is the simplest molecule that can be electrochemically oxidised in a low temperature fuel cell at reasonable (but not practical) potential.[128] CO is also an intermediate formed during CO_2 reduction and direct alcohol oxidation in FCs. In addition, hydrogen produced by reforming processes normally contain a small amount of CO, which affect the Pt catalyst activity in H_2/O_2 FCs as a high anodic potential is required to remove the strongly adsorbed CO from Pt. In contrast, alcohol oxidation in DAFCs produces CO as an intermediate species during fuel oxidation at the anode which blocks the Pt sites and prevents further adsorption of the alcohol molecule. High CO tolerant catalysts are thus required if H_2 generated from reforming or alcohols are to be used as fuel in PEMFC.

CO preferentially adsorbs on Pt compared with H_2, with adsorption probability 15 times higher than that of H_2 and the Gibbs free energy of CO adsorption decreases with temperature.[129] Linear or bridged type CO are formed on Pt catalysts. Such strong bond between CO and metal was explained due to the electron donation from 5σ CO orbital to the metal and subsequent transfer of 2 electrons (back-donation) from the d metal atomic orbital to the $2\pi^*$ antibonding orbital of CO.[130] Two mechanisms have been proposed for the oxidation of CO on Pt, bi-functional mechanism and ligand effect or electronic effect. Gilman[131] first proposed that the oxidation of CO involves the reaction between an adsorbed O-containing species and CO_{ads} molecule to form CO_2 through either a non-competitive or competitive Langmuir–Hinshelwood (L–H) type reaction.[132] Under electrochemical conditions in aqueous media, this O containing species is believed to be OH_{ads}, formed from the dissociation of water molecule on Pt in acid media, or OH^- in alkaline media.[131-133] It has been demonstrated that the addition of oxophilic metals such as Sn, Ru, Mo and W to Pt have higher CO tolerance compared with clean Pt and was attributed to the bi-functional mechanism and electronic effects in bimetallic catalysts.[54,134,135] In these cases, the oxophilic metal oxidises water at lower potentials forming OH_{ads} species which then reacts with the CO_{ads} to form CO_2 (bi-functional mechanism) as in Eqs. (7.22) and (7.23).

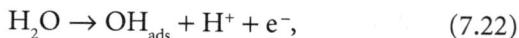

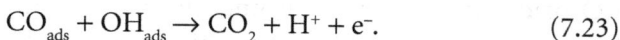

$$H_2O \rightarrow OH_{ads} + H^+ + e^-, \qquad (7.22)$$

$$CO_{ads} + OH_{ads} \rightarrow CO_2 + H^+ + e^-. \qquad (7.23)$$

The electronic effect results from the charge transfer from Pt to oxophilic metal resulting in the weakening of Pt–CO bond which helps in the easy removal of CO_{ads}.[136] The high activity of bimetallic catalyst for CO oxidation compared to pristine Pt is well documented in literature.[137-140] This positive effect of bimetallic catalysts for CO oxidation has been successfully used for the oxidation alcohol molecule on Pt-M binary catalysts.

2.2.2 Methanol oxidation

The mechanism of methanol oxidation reaction (MOR) is generally believed to occur in two parallel pathways, (i) direct oxidation in which the methanol oxidises to CO_2 directly without going through any intermediate species and (ii) indirect pathway in which poisoning species such as CO_{ads} is formed.[141] The CO is formed by the dissociative adsorption of methanol on Pt[142,143] and it can reach up to 90% coverage on a smooth Pt surface. The CO_{ads} on Pt was clearly observed using *in situ* FTIR spectroscopy.[144] Methanol can adsorb through C atom or O atom and successively loses H atoms to give various intermediate species as shown in Figure 4 where $-CHO_{ads}$ is a common intermediate species in both O and C adsorption geometries. The Pt–CHO_{ads} then loses a hydrogen to form Pt–CO_{ads}.[116] The CO strongly binds to Pt and its oxidation to CO_2 requires a high over potential and the methanol oxidation on Pt was observed only at potentials > 0.5 V vs. RHE. Intermediate reaction products such as HCHO and HCOOH are also observed during MOR using HPLC analysis.[115] The methanol adsorption is

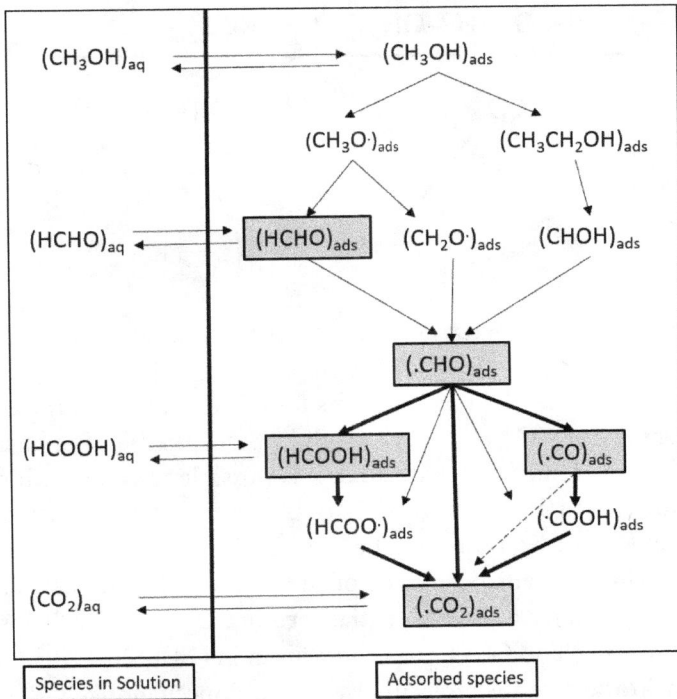

Figure 4: The electro-oxidation mechanism of methanol on Pt in acidic media. Adapted from Ref. [116].

also sensitive to the crystal plane of the Pt electrode. Pt(110) shows a high coverage of linearly bonded CO_L whereas CO_B and multi-bonded CO are observed on other crystal planes.[116]

In order to oxidise the CO on Pt as CO_2 and free the Pt active sites, OH_{ads} is necessary as discussed in Sec. 2.2.1, and this is provided by the dissociation of water on Pt. Though Pt is active towards the decomposition of alcohol molecule (C-bound species), it is less active for water activation (O-bound species),[145] thus, water oxidation on Pt requires high anodic over potential. In practical FCs, however, a high potential is not desirable. Multi-metallic catalysts have been proposed as the solution to this problem where the secondary oxophilic metal can oxidise at lower potentials than on bare Pt.

Watanabe and Motoo[146] first explained the enhancement of methanol oxidation on PtRu based on bifunctional mechanism as well as the ligand

$$\text{Pt-CO}_{ads}+\text{M-OH}_{ads} \rightarrow \text{Pt} + \text{CO}_2 +\text{H}^+ +\text{e}^-$$

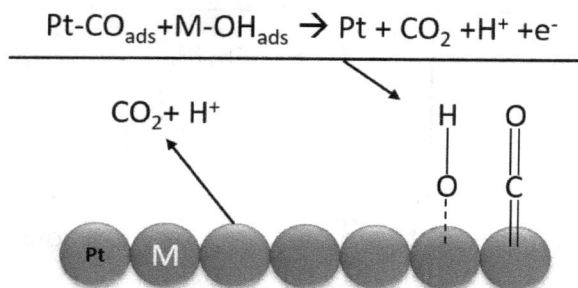

Figure 5:   Bi-functional mechanism on Pt-M catalyst. Schematic representation.[129]

effect. The over potential for MOR was found to depend on the chemical binding energy of CO* and OH* on Pt.[145] Since the binding energies of CO* and OH* are not closely related, it is possible to tune the bimetallic catalyst towards a high MOR activity. According to the bi-functional mechanism (Figure 5), both the Pt and second metal play their role independently with each adsorbing one of the reactant. The Ru atom adsorbs and dissociates water at low potential forming OH_{ads} which helps in the oxidation of CO_{ads} and CHO_{ads} to CO_2 as given in (24)–(26).[144,147] The O–H activation in the alcohol molecule by metal alone is unlikely due to the high activation energy whereas a surface bound OH reduces the activation energy by activating the O–H bond by a proton transfer reaction.[148] Among the Pt-M catalysts, PtRu was found to be the best catalyst for methanol oxidation[137,149–151] and was attributed to both bi-functional mechanism[147,152–158] and electronic effects[134] as described before. PtRu(1:1) and PtRu(3:1) was found to be the best catalyst for methanol oxidation.[145] The activity of PtRu is also dependent on the composition, morphology, particle size and alloying degree.[159] Other bimetallic Pt-M catalysts (M=Sn, Os, W, Mo) are also found to help in the CO oxidation.[145]

However, some studies have reported that, in the case of PtRu catalysts, the CO adsorb and oxidise on both Pt and Ru. Once the CO is oxidised on Ru, the active sites become free and are active for OH adsorption and enhance the CO oxidation.[137] i.e. Ru has dual role for OH and CO adsorption. This is slightly different from the bi-functional mechanism where each metal adsorbs only one of the reactant.

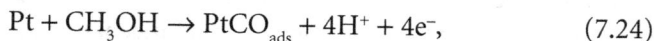

$$\text{Pt} + \text{CH}_3\text{OH} \rightarrow \text{PtCO}_{ads} + 4\text{H}^+ + 4\text{e}^-, \qquad (7.24)$$

$$Ru + H_2O \rightarrow Ru(OH)_{ads} + H^+ + e^-, \qquad (7.25)$$

$$PtCO_{ads} + Ru(OH)_{ads} \rightarrow CO_2 + Pt + Ru + H^+ + e^-. \qquad (7.26)$$

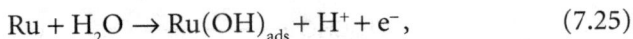

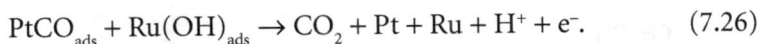

The higher activity was also attributed to a ligand effect. In this case, the second component alters the electronic property of the catalytically active metal. This can be caused by two effects such as broadening and lowering (or narrowing and increasing) the d-band energy of the parent metal or a direct influence of the second metal on the electronic properties of parent metal.[54,135,160] A shift in the d-band centre affects the strength of interaction of the parent metals with adsorbates (e.g. H_2, CO)[160,161] which in turn help in the easy removal of CO by weakening the Pt–CO bond as mentioned in Sec. 2.2.1

Recently, Adzic *et al.*[162] observed a significant improvement in MOR activity in acidic media for Pt monolayer on Au substrate. This was attributed to a substrate induced lattice strain in the Pt monolayer and also a possible ligand effect. The Pt_{ML}/Au catalyst was assumed to help in the dehydrogenation of the intermediates and formed CO_2 without going through the CO_{ads}. Also, a high OH coverage was observed on Pt_{ML}/Au catalysts which helped in the oxidation of –COH_{ads} directly to CO_2. A DFT calculation on Pt_{ML} on various substrates showed that the Pt/Au catalyst effected the d-band energy centre of Pt which increased the bond strength of Pt–CO and Pt–OH.[162]

2.2.3 Ethanol oxidation

A complete oxidation of ethanol to CO_2 yields 12e$^-$ per molecule (Eq. (7.19)). Unlike for methanol, the complete oxidation requires C–C bond cleavage (which requires a higher activation energy than the C–H bond cleavage) and formation of two C–O bonds from the methyl fragment of ethanol, as well as the oxidation of the CO, formed from the alcoholic fragment.[163] The effectiveness of the catalyst in the C–C bond breaking reaction is key in making ethanol oxidation reaction (EOR) useful in FC. However, in practice, partial intermediates such as AA (4e$^-$ oxidation) and AAL (2e$^-$ oxidation) are obtained predominantly with only about 1–10% of CO_2 yield on Pt in acidic media.[120,164] A dual path mechanism with C_1 and C_2 intermediate (which represent fragments with one

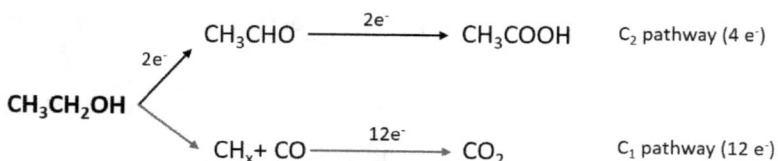

Figure 6: Schematic of dual path mechanism for EOR on Pt in acidic media.[166]

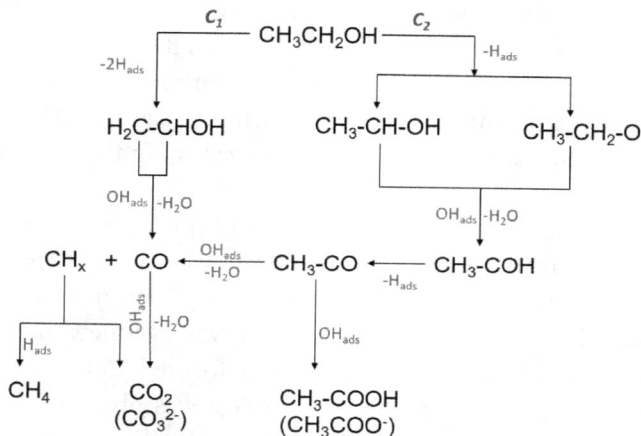

Figure 7: EOR mechanism in aqueous acidic and alkaline media.[117] Reproduced with permission from John Wiley and Sons.

and two carbon atoms respectively)[165] is widely accepted for EOR as given in Figure 6.

In the first pathway, the ethanol oxidises directly without C–C bond breakage to give C2 intermediates such as acetaldehyde (2e⁻ oxidation) and acetic acid (4e⁻ oxidation)[120] and in the second pathway, strongly bound species CO_{ads} (C_1 intermediate) is formed by dissociative adsorption of ethanol which is then further oxidised to give CO_2.[167] The C1 pathway is the desirable pathway for the complete oxidation of ethanol to CO_2 via CO (in acidic media) or carbonates (in alkaline media) (Figure 7). The adsorbates formed during EOR were identified to be $-OCH_2CH_3$, ($=COHCH_3$), $-COCH_3$[168] and CO on Pt. The product distribution also depends on the ethanol concentration. At low EtOH concentrations, AA and CO_2 are formed preferably whereas at high EtOH concentration

(> 0.1 M), AAL is formed predominantly on Pt in acidic media.[164] This was assumed to be due to the poor water availability at high ethanol concentration which leads to the partial oxidation (Langmuir–Hinshelwood mechanism).[163]

As described previously, the removal of poisoning species CO_{ads} requires the formation OH_{ads} which is facilitated by the use of multimetallic catalysts over clean Pt.

Bimetallic catalysts such as PtSn and PtRu[169] have been found to be more active than the pure Pt for EOR. PtRu shows a lower onset potential for ethanol oxidation to form CO_2 and AAL. In this case, the presence of Ru promotes the oxidation of strongly adsorbed intermediate and increases the selectivity towards CO_2 formation.[170] However, a higher Ru content decreases the catalytic activity due to the reduced number of Pt active sites as Ru does not adsorb ethanol strongly. Between 8% and 15% Ru was found to be optimal to promote the EOR by the bi-functional mechanism.[170] A lower onset potential and higher current density were also associated with Pt_3Sn for EOR reaction.[120] The AA production was found to be higher on these bimetallic catalysts indicating the oxidation of adsorbed intermediate species with the help of OH_{ads} formed on Ru or Sn. Souza *et al.*,[171] on the other hand, observed only CO_2 in the FTIR of co-deposited PtRu catalyst during EOR which appears at a potential 0.40 V in comparison to 0.70 V for pure Pt electrode. This again confirms the effect of Ru addition to Pt in improving the selectivity of CO_2 formation during EOR. The absence of C2 products, in this case, could indicate that, in addition to the bi-functional mechanism, other mechanisms may be responsible for the higher activity.[171]

To date, PtSn has been found to be the best catalyst for EOR in acidic media[143,144,172–177] as shown in Figure 8 for a comparison of Pt/C, PtSn/C and PtRu/C catalysts. The high activity of PtSn was explained based on the bi-functional mechanism and ligand effect as explained for methanol oxidation in Sec. 2.2.2. The Sn being an oxophilic metal activates the water molecule at lower potential producing OH_{ads} species which helps in oxidising strongly adsorbed CO to CO_2 and AAL to AA (bi-functional mechanism).[172,175,178] Also, the Pt electronic structure can be modified by the Sn ad-atom which reduces the Pt-CO bond strength and helps in the removal of CO (ligand effect).[172,178,179] Various methods have been adopted in the

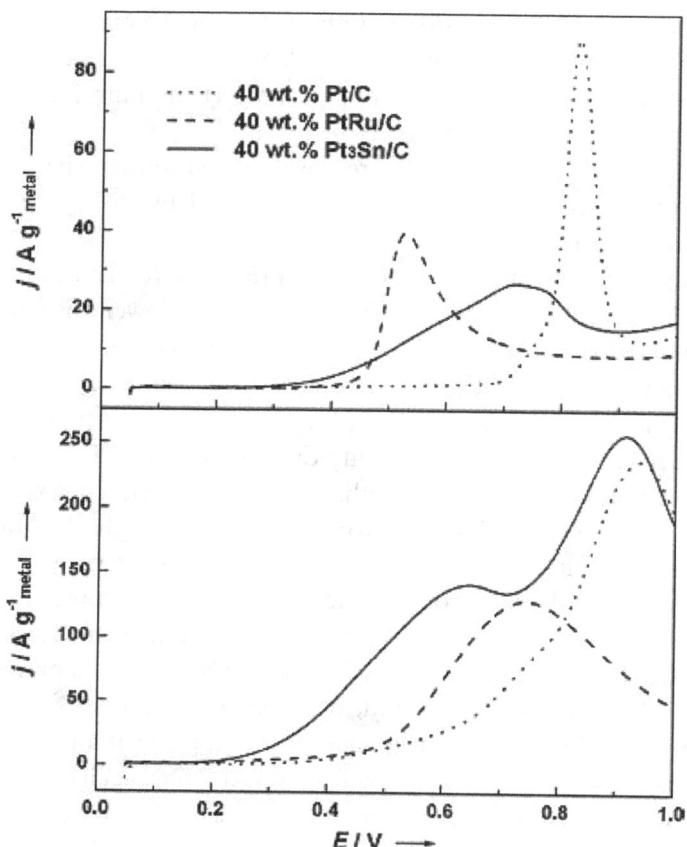

Figure 8: CO stripping (top) and linear sweep voltammogram of EOR (bottom) on Pt and Pt bimetallic catalysts.[120] Reproduced with permission from Wiley and Sons.

literature for the preparation of PtSn catalysts namely (a) electrochemical co-deposition (b) under potential deposition of Sn on Pt and (c) alloy formation. Depending upon the preparation method and the composition, different results may be obtained due to the different states of PtSn.[180] It was observed from X-ray absorption spectroscopy that alloying of Pt with Sn results in the filling of Pt 5d-band vacancy due to the transfer of the electron from Sn to Pt involving Sn 2p states which result in increased Pt–Pt bond distance (The electronegativity of Pt = 2.28 and Sn = 1.96).[180] In contrast, under-potential deposition (UPD) of Sn has little effect on the electronic structure of Pt. The UPD Sn interacts with oxygen species at all potentials and the Pt–O bond strength increases with increases in potential. The increased Pt–Pt distance and lower Pt d-band vacancy result in an unfavourable structure for initial alcohol adsorption and C–H dissociation.[180] The selectivity of alcohol oxidation also depends on the crystallographic orientation of the Pt surface in the catalyst.[117] For C–C bond cleavage, high-index facets are most active. The Pt(110) face exhibits the maximum selectivity with respect to CO_2 formation in acidic media.[117,181,182] However, molecular orbital theory calculations[183] disagree with the bi-functional mechanism and it was concluded that the Sn has a lower activity for OH_{ads} than Pt in a PtSn alloy. However, a reduced activation energy for CO oxidation on PtSn electrode was observed although it was not clear as to the mechanism involved in this case. It was also observed that Sn diluted the Pt active sites which reduced the ethanol dissociative adsorption. This was clear from the product distribution on Pt and PtSn catalyst (Table 3)[110] whereby the acetic acid yield increased and CO_2 yield decreased with Sn addition. A similar effect was also reported for PtRu catalyst.[169]

Other Pt-M catalysts, such as PtRu,[169,170,184–188] PtMo[189] and PtRh[188,190] have also been studied for ethanol oxidation. Ternary alloys were found to be more active than bimetallic for EOR where the third component usually modifies the electronic and structural characteristics of the alloy leading to a better catalyst.[191] These studies include Pt-based alloys such as PtRuSn,[192] PtRhSnO$_2$,[193–196] PtSnIn,[197] PtSnNi[198] PtMoIr,[199] PtRuMo[200–202] and PtSnMo.[203]

Adzic *et al.*[193,194] observed that Pt–Rh–SnO$_2$ can effectively split the C–C bond in ethanol in acidic media and the highest activity was obtained for a composition Pt:Rh:Sn= 3:1:4.[193] A higher oxidation current for

Table 3: Chemical yields in AA, AAL and CO_2 for Pt/C and Pt–Sn (9:1)/C catalysts during the electro-oxidation of ethanol for 4 h at a constant current density (8 mA cm^{-2} for Pt and 32 mA cm^{-2} for Pt–Sn) in a 25 cm^2 surface area DEFC working at 80°C.[110]

Electrocatalysts	Chemical yield in AA (%)	Chemical yield in AAL (%)	Chemical yield in CO$_2$ (%)
60 wt.% Pt/C XC-72	33	47	20
60 wt.% Pt–Sn (9:1)/C XC72	77	15	8

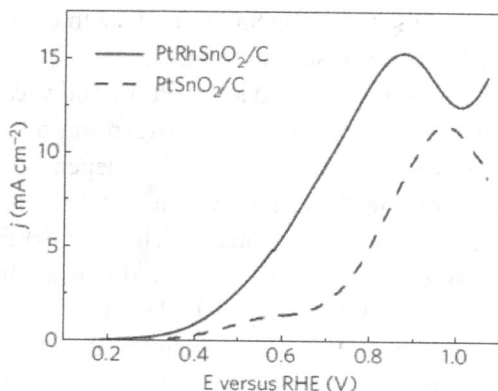

Figure 9: Polarisation curve for the oxidation of ethanol on PtRhSnO$_2$/C and PtSnO$_2$/C in 0.2 m ethanol + 0.1M HClO$_4$, sweep rate 50 mVs^{-1}. Catalysts composition, 30 nmol Pt, 8 nmol Rh and 60 nmol SnO$_2$.[194] Reproduced with permission from Nature Publishing group.

ethanol oxidation was also observed for the ternary catalyst compared to that for bimetallic Pt-SnO$_2$ (Figure 9). *In situ* infrared reflection–absorption spectroscopy (IRRAS) showed the presence of CO$_2$ at potentials as low as 0.30 V on RhSnO$_2$/Pt(111) catalyst whereas on Pt(111) electrode CO$_2$ only appeared at potentials > 0.78 V indicating the effectiveness of SnRh in C–C bond cleavage. The FTIR spectra also showed that on PtRhSnO$_2$/C catalyst, the generation of acetic acid was negligible and acetaldehyde formation was significantly decreased whereas CO$_2$ formation improved confirming the effective C–C bond cleavage in ternary catalysts.[193,194] The higher activity of the ternary catalyst was attributed to the synergy between all the three components. The SnO$_2$ strongly adsorb the water molecule and providing the OH$_{ads}$ and freeing the Pt and Rh sites for ethanol adsorption. The OH$_{ads}$ on Sn helped in the removal of CO formed on Rh whereas Pt facilitated the ethanol dehydrogenation. The PtRh interaction also modified the electronic structure of Rh which helped in the moderate bonding of ethanol and intermediates and products facilitating the C–C bond cleavage.[194]

2.2.4 Oxygen Reduction Reaction (ORR)

The ORR is a multi e-transfer process in which O$_2$ in converted into H$_2$O or OH$^-$ depending on the pH of the solution. In acidic solution, O$_2$

$$O_{2\,(bulk)}$$

$$H_2O_{(bulk)}$$

$$O_{2\,(sur)} \underset{K_5}{\overset{k_5}{\rightleftharpoons}} O_{2(ads)} \underset{k_2}{\overset{k_2}{\rightleftharpoons}} H_2O_{2(ads)} \underset{k_6}{\overset{k_6}{\rightleftharpoons}} H_2O_{2(sur)}$$

(with k_1, k_3 branches to $H_2O_{(bulk)}$, k_4 branch, and $H_2O_{2(bulk)}$)

$$H_2O_{2(bulk)}$$

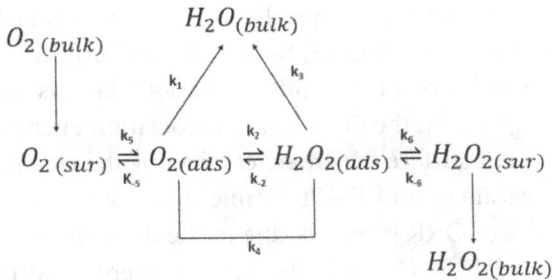

Figure 10: The ORR mechanism on Pt catalyst.[128,205] Reproduced with permission from Electrochemical Society from Ref. [205].

reduced to H_2O by 4 e$^-$ process as given in Figure 10 and Eq. (7.27). O_2 can also undergo partial oxidation with 2e$^-$ transfer to form H_2O_2 followed by another 2e$^-$ transfer to form H_2O as given in Eqs. (7.28) and (7.29).[204,205] The H_2O_2 can also decompose chemically as given in Eq. (7.30) to form water and oxygen.[205] The desired pathway in a fuel cell is the 4e$^-$ pathway to form H_2O and Pt catalysts in acidic media are normally follow this pathway. The partial reduction of oxygen to H_2O_2 not only reduces the energy efficiency but also produces reactive intermediate species that can form harmful free radical species.[136]

$$O_2 + 4H^+ + 4e^- \rightarrow 2H_2O; \quad E_o = 1.229\ \text{V}, \tag{7.27}$$

$$O_2 + 2H^+ + 2e^- \rightarrow 2H_2O_2; \quad E_o = 0.67\ \text{V}, \tag{7.28}$$

$$H_2O_2 + 2H^+ + 2e^- \rightarrow H_2O; \quad E_o = 1.77\ \text{V}, \tag{7.29}$$

$$2H_2O_2 \rightarrow 2H_2O + O_2. \tag{7.30}$$

Pt was found to be the most active catalyst for the (ORR) in PEMFCs. ORR is sluggish compared with hydrogen oxidation reaction (HOR) and leads to significant voltage loss in FC and is one of the major reasons for their lower efficiency. Typically, high metal loadings are used to improve the kinetics and to obtain significant current density. About 28 mg cm^{-2} loading have been used in the initial stage of FC development, however, after the significant developments in high surface area, conductive support materials such as carbon in 1990's, the Pt loading has been significantly reduced (0.2–0.4 mg cm^{-2}).[206] Typical support materials are high surface area carbon such as Vulcan XC-72, Ketjen black, Cabot, etc.

The ORR activity of Pt depends on various factors such as the O_2 adsorption energy on Pt, the O–O bond dissociation energy, and binding energy of OH on Pt surface which are effected by the electronic structure of Pt (d-band vacancy) and Pt–Pt atomic distance (geometric effect).[207] The poor kinetics of ORR is mainly due to the difficulty in O_2 adsorption on the catalyst, high O–O bond dissociation energy and difficult oxide removal.[136] The ORR activity is also strongly depending on the electrolyte used. For example, in $HClO_4$ electrolyte, the activity of low index surface are in the order Pt(110) > Pt(111) > Pt(100)[208] whereas in the presence of bisulphate and halide ions, the order changes to Pt(110) > Pt(100) > Pt(111). In alkaline KOH, the order changes to Pt(111) > Pt(110) > Pt(100).[122]

Figure 11 shows the ORR activity of various metal catalyst as a function of oxygen binding energy.[136,209] A clear volcano type plot was observed and it is clear that the ORR activity follows the order Pt > Pd > Ir >

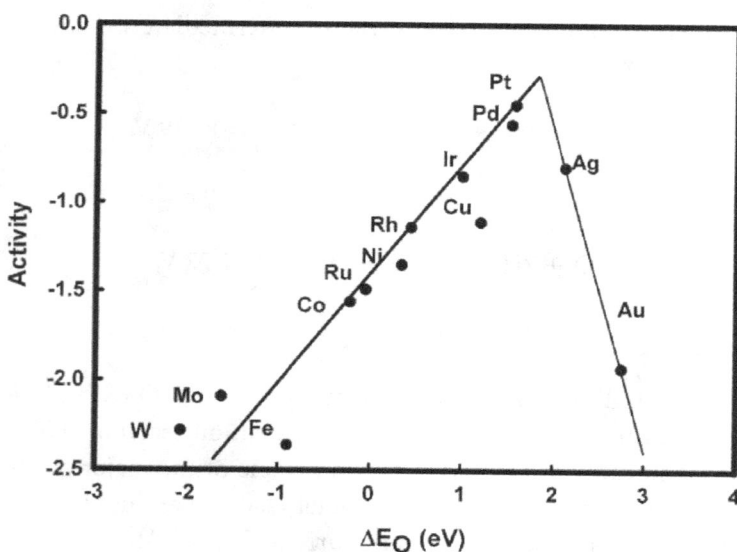

Figure 11:   Trends in oxygen reduction activity plotted as a function of the oxygen binding energy.[209] Reproduced with permission from American Chemical Society.

Rh.[113,136,209] ORR on other metals such as Au, Ir, Rh, etc. has been also studied, however, they show lower activity for ORR and are also electrochemically less stable than Pt. i.e. more easily oxidised than Pt.[136] Alloying Pt with other metals (Fe, Co, Cr, Ni, etc.) has also been found to enhance the ORR activity. Alloying leads to a change in Pt electronic structure and an increased d-band vacancy which can lead to strong Metal-O interaction and a weakening of the O–O bond. Alloying can also lead to a change in the Pt–Pt interatomic distance (geometric effect) and lattice contraction.[136] However, long-term stability of the second metal is generally poor and leads to poor stability of the catalysts. Optimising the composition and making a stable bimetallic catalyst are the current challenges in ORR catalyst research.

Electrocatalysis is the most challenging aspect in the development of direct alcohol fuel cell as both alcohol oxidation reactions on the anode and oxygen reduction on the cathode side are limited by their sluggish reaction and demand of precious metal catalysts. In order to make DAFC viable for the industry, these challenges need to be addressed. In the short run, reducing the Pt loading (multi-metallic and supported catalysts) are the priority and in the long run, the design of new catalyst and non-Pt catalyst has to be developed. Significant progress has been achieved in the recent past in these areas and with the utilisation of more modern characterisation and modelling methods, detailed understanding of the mechanism of alcohol oxidation and ORR is possible and is presumed to help in the development of more active and cheaper catalyst in future.

References

1. S. Lee, J. G. Speight and S. K. Loyalka, *Handbook of Alternative Fuel Technol.*, CRS Press, Taylor & Francis Group, Abingdon 2007.
2. M. Behrens, F. Studt, I. Kasatkin, S. Kühl, M. Hävecker, F. Abild-Pedersen, S. Zander, F. Girgsdies, P. Kurr, B.-L. Kniep, M. Tovar, R. W. Fischer, J. K. Nørskov and R. Schlögl, *Sci.*, 2012, **336**, 893–897.
3. G. Liu, D. Willcox, M. Garland and H. H. Kung, *J. Catal.*, 1985, **96**, 251–260.
4. Y. Zhang, Q. Sun, J. Deng, D. Wu and S. Chen, *Appl. Catal. A: Gen.*, 1997, **158**, 105–120.
5. J. T. Sun, I. S. Metcalfe and M. Sahibzada, *Indus. Eng. Chem. Res.*, 1999, **38**, 3868–3872.

6. Y. Ma, Q. Sun, D. Wu, W.-H. Fan, Y.-L. Zhang and J.-F. Deng, *Appl. Catal. A: Gen.*, 1998, **171**, 45–55.

7. X. Guo, D. Mao, G. Lu, S. Wang and G. Wu, *Catal. Commun.*, 2011, **12**, 1095–1098.

8. T. Fujitani and J. Nakamura, *Catal. Lett.*, 1998, **56**, 119–124.

9. T. Fujitani and J. Nakamura, *Appl. Catal. A: Gen.*, 2000, **191**, 111–129.

10. Y. Choi, K. Futagami, T. Fujitani and J. Nakamura, *Appl. Catal. A: Gen.*, 2001, **208**, 163–167.

11. M. Saito, T. Fujitani, M. Takeuchi and T. Watanabe, *Appl. Catal. A: Gen.*, 1996, **138**, 311–318.

12. J. Nakamura, I. Nakamura, T. Uchijima, Y. Kanai, T. Watanabe, M. Saito and T. Fujitani, *J. Catal.*, 1996, **160**, 65–75.

13. L. Fan and K. Fujimoto, *J. Catal.*, 1994, **150**, 217–220.

14. A. Trovarelli, C. Deleitenburg, G. Dolcetti and J. L. Lorca, *J. Catal.*, 1995, **151**, 111–124.

15. W.-J. Shen, K.-W. Jun, H.-S. Choi and K.-W. Lee, *Korean J. Chem. Eng.*, 2000, **17**, 210–216.

16. A. T. Pasteur, S. J. Dixon-Warren and D. A. King, *J. Chem. Phys.* 1995, **103**, 2251–2260.

17. J. Ludwig, D. G. Vlachos, A. C. T. van Duin and W. A. Goddard, *J. Phys. Chem. B*, 2006, **110**, 4274–4282.

18. W. Dong, V. Ledentu, P. Sautet, A. Eichler and J. Hafner, *Surf. Sci.*, 1998, **411**, 123–136.

19. T. Mitsui, M. K. Rose, E. Fomin, D. F. Ogletree and M. Salmeron, *Nature*, 2003, **422**, 705–707.

20. M. Pozzo and D. Alfè, *Int. J. Hydrogen Energ.*, 2009, **34**, 1922–1930.

21. C.-H. Kim, J. Lee and D. L. Trimm, *Topi. Catal.*, 2003, **22**, 319–324.

22. N. Koizumi, X. Jiang, J. Kugai and C. Song, *Catal. Today*, 2012, **194**, 16–24.

23. X.-L. Liang, X. Dong, G.-D. Lin and H.-B. Zhang, *Appl. Catal. B: Environ.*, 2009, **88**, 315–322.

24. J. Qu, X. Zhou, F. Xu, X.-Q. Gong and S. C. E. Tsang, *J. Phys. Chem. C*, 2014, **118**, 24452–24466.

25. N. Iwasa, H. Suzuki, M. Terashita, M. Arai and N. Takezawa, *Catal. Lett.*, 2004, **96**, 75–78.

26. A. Ciftci, S. Eren, D. A. J. M. Ligthart and E. J. M. Hensen, *Chem. Cat. Chem.*, 2014, **6**, 1260–1269.

27. S. D. Gardner, G. B. Hoflund, M. R. Davidson, H. A. Laitinen, D. R. Schryer and B. T. Upchurch, *Langmuir*, 1991, **7**, 2140–2145.

28. G. L. Haller and D. E. Resasco, in *Advances in Catal.*, eds. H. P. D.D. Eley and B. W. Paul, Academic Press, 1989, vol. 36, pp. 173–235.

29. T. Fujitani, M. Saito, Y. Kanai, T. Watanabe, J. Nakamura and T. Uchijima, *Appl. Catal. A: Gen.*, 1995, **125**, L199–L202.

30. K. Samson, M. Śliwa, R. P. Socha, K. Góra-Marek, D. Mucha, D. Rutkowska-Zbik, J. F. Paul, M. Ruggiero-Mikołajczyk, R. Grabowski and J. Słoczyński, *ACS Catal.*, 2014, **4**, 3730–3741.

31. Z.-s. Hong, Y. Cao, J.-f. Deng and K.-n. Fan, *Catal. Lett.*, 2002, **82**, 37–44.

32. J. Wu, M. Saito, M. Takeuchi and T. Watanabe, *Appl. Catal. A: Gen.*, 2001, **218**, 235–240.

33. T. Takeguchi, K.-i. Yanagisawa, T. Inui and M. Inoue, *Appl. Catal. A: Gen.*, 2000, **192**, 201–209.

34. T. Witoon, T. Permsirivanich, N. Kanjanasoontorn, C. Akkaraphataworn, A. Seubsai, K. Faungnawakij, C. Warakulwit, M. Chareonpanich and J. Limtrakul, *Catal. Sci. Technol.*, 2015, **5**, 2347–2357.

35. G. Du, S. Lim, Y. Yang, C. Wang, L. Pfefferle and G. L. Haller, *J. Catal.*, 2007, **249**, 370–379.

36. K. Sun, W. Lu, M. Wang and X. Xu, *Catal. Commun.*, 2004, **5**, 367–370.

37. Z. Li, Z. Zuo, W. Huang and K. Xie, *Appl. Surf. Sci.*, 2011, **257**, 2180–2183.

38. M. Fujiwara, R. Kieffer, H. Ando, Q. Xu and Y. Souma, *Appl. Catal. a: Gen.*, 1997, **154**, 87–101.

39. S. E. Collins, J. J. Delgado, C. Mira, J. J. Calvino, S. Bernal, D. L. Chiavassa, M. A. Baltanás and A. L. Bonivardi, *J. Catal.*, 2012, **292**, 90–98.

40. J. Vizdal, A. Kroupa, J. Popovic and A. Zemanova, *Advan. Eng. Mater.*, 2006, **8**, 164–176.

41. C.-S. Chen, W.-H. Cheng and S.-S. Lin, *Appl. Catal. A: Gen.*, 2003, **238**, 55–67.

42. R. W. Dorner, D. R. Hardy, F. W. Williams and H. D. Willauer, *Appl. Catal. A: Gen.*, 2010, **373**, 112–121.

43. S. Wohlrab, D. Ehrlich, J. Wambach, H. Kuhlenbeck and H. J. Freund, *Surf. Sci.*, 1989, **220**, 243–252.

44. F. Solymosi, *J. Mol. Catal.*, 1991, **65**, 337–358.

45. K. Kitamura Bando, K. Soga, K. Kunimori and H. Arakawa, *Appl. Catal. A: Gen.*, 1998, **175**, 67–81.

46. M. C. Román-Martínez, D. Cazorla-Amorós, C. Salinas-Martínez de Lecea and A. Linares-Solano, *Langmuir*, 1996, **12**, 379–385.

47. J. Giner, *Electrochim. Acta*, 1963, **8**, 857–865.

48. E. Morallón, J. L. Vázquez, J. M. Pérez and A. Aldaz, *J. Electroanal. Chem.*, 1995, **380**, 47–53.

244   *H. Bahruji et al.*

49. A. Rodes, E. Pastor and T. Iwasita, *J. Electroanal. Chem.*, 1994, **369**, 183–191.
50. M. C. Román-Martínez, D. Cazorla-Amorós, A. Linares-Solano and C. Salinas-Martínez de Lecea, *Appl. Catal. A: Gen.*, 1996, **134**, 159–167.
51. B. Hammer and J. K. Nørskov, in *Advances in Catal.*, eds. H. K. Bruce and C. Gates, Academic Press, Cambridge, 2000, vol. 45, pp. 71–129.
52. A. Ruban, B. Hammer, P. Stoltze, H. L. Skriver and J. K. Nørskov, *J. Mol. Catal. a: Chem.*, 1997, **115**, 421–429.
53. C. A. Menning and J. G. Chen, *J. Chem. Physics*, 2009, **130**, 174709.
54. J. R. Kitchin, J. K. Nørskov, M. A. Barteau and J. G. Chen, *Phys. Revi. Lett.*, 2004, **93**, 156801.
55. M. D. Porosoff and J. G. Chen, *J. Catal.*, 2013, **301**, 30–37.
56. X. Jiang, N. Koizumi, X. Guo and C. Song, *Appl. Catal. B: Environ.*, 2015, **170–171**, 173–185.
57. L. Liu, H. Zhao, J. M. Andino and Y. Li, *ACS Catal.*, 2012, **2**, 1817–1828.
58. B. Vijayan, N. M. Dimitrijevic, T. Rajh and K. Gray, *J. Phys. Chem. C*, 2010, **114**, 12994–13002.
59. T. Ohno, N. Murakami, T. Koyanagi and Y. Yang, *J. CO₂ Utilization*, 2014, **6**, 17–25.
60. B.-J. Liu, T. Torimoto and H. Yoneyama, *J. Photochem. and Photobiol. A: Chem.*, 1998, **113**, 93–97.
61. K. Sayama and H. Arakawa, *J. Phys. Chem.*, 1993, **97**, 531–533.
62. T. Tanaka, Y. Kohno and S. Yoshida, *Res. Chem. Intermed.*, 2000, **26**, 93–101.
63. B. Srinivas, B. Shubhamangala, K. Lalitha, P. Anil Kumar Reddy, V. Durga Kumari, M. Subrahmanyam and B. R. De, *Photochem. Photobiol.*, 2011, **87**, 995–1001.
64. Z. Goren, I. Willner, A. J. Nelson and A. J. Frank, *J. Phys. Chem.*, 1990, **94**, 3784–3790.
65. O. Ishitani, *J. Photochem. Photobiol. A: Chem.*, 1993, **72**, 269–271.
66. Y. Kohno, H. Hayashi, S. Takenaka, T. Tanaka, T. Funabiki and S. Yoshida, *J. Photochem. and Photobiol. A: Chem.*, 1999, **126**, 117–123.
67. M. Anpo, H. Yamashita, Y. Ichihashi, Y. Fujii and M. Honda, *J. Phys. Chem. B*, 1997, **101**, 2632–2636.
68. O. Ozcan, F. Yukruk, E. U. Akkaya and D. Uner, *Top. Catal.*, 2007, **44**, 523–528.
69. C.-C. Yang, Y.-H. Yu, B. van der Linden, J. C. S. Wu and G. Mul, *J. Am. Chem. Soc.*, 2010, **132**, 8398–8406.
70. J. Hong, W. Zhang, J. Ren and R. Xu, *Anal. Methods*, 2013, **5**, 1086–1097.
71. T. Yui, A. Kan, C. Saitoh, K. Koike, T. Ibusuki and O. Ishitani, *ACS Appl. Mater. Inter.*, 2011, **3**, 2594–2600.

72. O. Ola, M. Maroto-Valer, D. Liu, S. Mackintosh, C.-W. Lee and J. C. S. Wu, *Appl. Catal. B: Environ.*, 2012, **126**, 172–179.

73. Q.-H. Zhang, W.-D. Han, Y.-J. Hong and J.-G. Yu, *Catal. Today*, 2009, **148**, 335–340.

74. M. M. Halmann and M. Steinberg, *Greenhouse Gas Carbon Dioxide Mitigation: Sci. and Technol.*, Taylor & Francis, Abingdon, 1998.

75. M. A. Scibioh and B. Viswanathan, in *Photo/Electrochem. & Photobiology in the Environment, Energy and Fuel*, ed. S. Kaneco, Research Signpost, Kerela, 2002, pp. 1–46.

76. G. R. Dey and K. Kishore, in *Photo/Electrochem. & Photobiology in the Environment, Energy and Fuel*, ed. S. Kaneco, Research Signpost, Kerela, India, 2002, pp. 357–388.

77. P. Sabatier and J. B. Senderens, *C. R. Acad. Sci., Paris Ser. C.*, 1899, **128**, 1173.

78. T. Mizuno, A. Naitoh and K. Ohta, *J. Electroanal. Chem.*, 1995, **391**, 199–201.

79. M. Azuma, K. Hashimoto, M. Hiramoto, M. Watanabe and T. Sakata, *J. Electrochem. Soc.*, 1990, **137**, 1772–1778.

80. Y. Hori, K. Kikuchi and S. Suzuki, *Chem. Lett.*, 1985, **14**, 1695–1698.

81. Y. Hori, A. Murata, K. Kikuchi and S. Suzuki, *J. Chem. Soc. Chem. Commun.*, 1987, DOI: 10.1039/C39870000728, 728–729.

82. H. Noda, S. Ikeda, Y. Oda, K. Imai, M. Maeda and K. Ito, *Bull. Chem. Soci. Jpn*, 1990, **63**, 2459–2462.

83. S. Ikeda, T. Takagi and K. Ito, *Bull. Chem. Soci. Jpn.*, 1987, **60**, 2517–2522.

84. D. P. Summers, S. Leach and K. W. Frese Jr, *J. Electroanal. Chem. Inter. Electrochem.*, 1986, **205**, 219–232.

85. K. W. Frese and S. Leach, *J. Electrochem. Soc.*, 1985, **132**, 259–260.

86. M. Azuma, K. Hashimoto, M. Watanabe and T. Sakata, *J. Electroanal. Chem. Inter. Electrochem.*, 1990, **294**, 299–303.

87. A. Bandi, *J. Electrochem. Soc.*, 1990, **137**, 2157–2160.

88. Y. Hori, K. Kikuchi, A. Murata and S. Suzuki, *Chem. Lett.*, 1986, DOI: 10.1246/cl.1986.897, 897–898.

89. R. L. Cook, R. C. MacDuff and A. F. Sammells, *J. Electrochem. Soc.*, 1987, **134**, 2375–2376.

90. D. W. DeWulf, T. Jin and A. J. Bard, *J. Electrochem. Soc.*, 1989, **136**, 1686–1691.

91. J. J. Kim, D. P. Summers and K. W. Frese Jr, *J. Electroanal. Chem. Inter. Electrochem.*, 1988, **245**, 223–244.

92. H. Noda, S. Ikeda, Y. Oda and K. Ito, *Chem. Lett.*, 1989, **18**, 289–292.

93. R. L. Cook, R. C. MacDuff and A. F. Sammells, *J. Electrochem. Soc.*, 1988, **135**, 1320–1326.
94. Y. Hori, A. Murata and R. Takahashi, *J. Chem. Soc., Faraday Trans. 1: Phys. Chem. Condensed Phases*, 1989, **85**, 2309–2326.
95. K. Hara, A. Kudo and T. Sakata, *J. Electroanal. Chem.*, 1995, **391**, 141–147.
96. A. P. Abbott and C. A. Eardley, *J. Phys. Chem. B*, 2000, **104**, 775–779.
97. Y. B. Vassiliev, V. S. Bagotsky, N. V. Osetrova, O. A. Khazova and N. A. Mayorova, *J. Electroanal. Chem. Inter. Electrochem.*, 1985, **189**, 271–294.
98. K. Ito, S. Ikeda and M. Okabe, *Denki Kagaku*, 1980, **48**, 247.
99. K. Ito, S. Ikeda, T. Iida and H. Niwa, *Denki Kagaku*, 1981, **49**, 106.
100. S. Kaneco, K. Iiba, K. Ohta, T. Mizuno and A. Saji, *J. Electroanal. Chem.*, 1998, **441**, 215–220.
101. S. Kaneco, K. Iiba, K. Ohta, T. Mizuno and A. Saji, *Electrochim. Acta*, 1998, **44**, 573–578.
102. S. Kaneco, H. Katsumata, T. Suzuki and K. Ohta, *Energy Fuels*, 2006, **20**, 409–414.
103. C. Amatore and J. M. Saveant, *J. Am. Chem. Soc.*, 1981, **103**, 5021–5023.
104. K. Ito, S. Ikeda, T. Iida and A. Nomura, *Denki Kagaku*, 1982, **50**, 463.
105. K. Ito, S. Ikeda, N. Yamauchi, T. Iida and T. Takagi, *Bull. Chem. Soc. Jpn.*, 1985, **58**, 3027–3028.
106. K. W. Frese, in *Electrochemical. and Electrocatalytic Reactions of Carbon Dioxide*, eds. B. P. Sullivan, K. Krist and H. E. Guard, Elsevier Science, Amsterdam, 1993, p. 191.
107. Y. Hori, H. Wakebe, T. Tsukamoto and O. Koga, *Surf. Sci.*, 1995, **335**, 258–263.
108. Y. Hori, A. Murata, R. Takahashi and S. Suzuki, *J. Am. Chem. Soc.*, 1987, **109**, 5022–5023.
109. A. Brouzgou, F. Tzorbatzoglou and P. Tsiakaras. Direct Alcohol Fuel Cells: Challenges and Future Trends. In *IEEE Proceedings of the 2011 3rd International Youth Conference on Energetics (IYCE)*. 2011.
110. F. Vigier, S. Rousseau, C. Coutanceau, J.-M. Leger and C. Lamy, *Top. Catal.*, 2006, **40**, 111–121.
111. M. Winter and R. J. Brodd, *Chem. Rev.*, 2004, **104**, 4245–4270.
112. S. Song, V. Maragou and P. Tsiakaras, *J. Fuel Cell Sci. Technol.*, 2006, **4**, 203–209.
113. Y. Nie, L. Li and Z. Wei, *Chem. Soc. Rev.*, 2015, **44**, 2168–2201.
114. F. A. de Bruijn, V. A. T. Dam and G. J. M. Janssen, *Fuel Cells*, 2008, **8**, 3–22.
115. C. Lamy, E. M. Belgsir and J. M. Léger, *J. Appl. Electrochem.*, 2001, **31**, 799–809.

116. Z. X. Liang and T. S. Zhao, *Catalysts for Alcohol-fuelled Direct Oxidation Fuel Cells*, Royal Society of Chemistry, London, 2012.
117. M. R. Tarasevich, O. V. Korchagin and A. V. Kuzov, *Russ. Chem. Revs.*, 2013, **82**, 1047–1065.
118. J. H. Kim, S. M. Choi, S. H. Nam, M. H. Seo, S. H. Choi and W. B. Kim, *Appl. Catal. B: Environ.*, 2008, **82**, 89–102.
119. M. J. González, C. T. Hable and M. S. Wrighton, *J. Phys. Chem. B*, 1998, **102**, 9881–9890.
120. I. Kim, O. H. Han, S. A. Chae, Y. Paik, S.-H. Kwon, K.-S. Lee, Y.-E. Sung and H. Kim, *Angewa. Chem. Int. Ed.*, 2011, **50**, 2270–2274.
121. S. García-Rodríguez, T. Herranz and S. Rojas, in *New and Future Developments in Catalysis.*, ed. S. L. Suib, Elsevier, Amsterdam, 2013, pp. 33–67.
122. P. B. Balbuena and V. R. Subramanian, eds., *Theory and Experiment in Electrocatalysis*, Springer, Berlin, 2010.
123. J. L. Cohen, D. J. Volpe and H. D. Abruna, *Phys. Chem. Chem. Phys.*, 2007, **9**, 49–77.
124. C. Countanceau, B. Baranton and C. Lamy, in *Theory and Experiment in Electrocatalysis*, eds. P. B. Balbuena and V. R. Subramanian, Springer, New York, London, 2010, vol. 50.
125. B. Beden, C. Lamy, A. Bewick and K. Kunimatsu, *J. Electroanal. Chem. Inter. Electrochem.*, 1981, **121**, 343–347.
126. M. Haruta, N. Yamada, T. Kobayashi and S. Iijima, *J. Catal.*, 1989, **115**, 301–309.
127. R. J. Davis, *Sci.*, 2003, **301**, 926–927.
128. N. M. Marković and P. N. Ross Jr, *Surf. Sci. Rep.*, 2002, **45**, 117–229.
129. S. Ye, in *PEM Fuel Cell Electrocatalysts and Catalyst Layers*, ed. J. Zhang, Springer London, 2008, ch. 16, pp. 759–834.
130. B. N. Grgur, N. M. Markovi, C. A. Lucas, P. N and R. JR, *J. Serb. Chem. Soc.*, 2001, **66**, 785–797.
131. S. Gilman, *J. Phys. Chem.*, 1964, **68**, 70–80.
132. N. M. Marković, in *Handbook of Fuel Cells*, John Wiley & Sons Ltd., Hoboken, 2010, DOI: 10.1002/9780470974001.f204027m.
133. T. Iwasita, in *Handbook of Fuel Cells*, John Wiley & Sons, Ltd., Hoboken, 2010, DOI: 10.1002/9780470974001.f206047.
134. T. Frelink, W. Visscher and J. A. R. van Veen, *Surf. Sci.*, 1995, **335**, 353–360.
135. H. Baltruschat, S. Ernst and N. Bogolowski, in *Catal. in Electrochemzistry.*, John Wiley & Sons Inc., Hoboken, 2011, ch. 9, pp. 297–337.
136. J. Zhang, ed., *PEM Fuel Cell Electrocatalysts and Catalyst Layers-Fundamentals and Applications*, Springer, London, 2008.

137. H. A. Gasteiger, N. Markovic, P. N. Ross and E. J. Cairns, *J. Phys. Chem.*, 1994, **98**, 617–625.

138. J. S. Spendelow, P. K. Babu and A. Wieckowski, *Curr. Opin. Solid State Mater. Sci.*, 2005, **9**, 37–48.

139. K. Wang, H. A. Gasteiger, N. M. Markovic and P. N. Ross Jr., *Electrochimi. Acta*, 1996, **41**, 2587–2593.

140. H. A. Gasteiger, N. M. Markovic and P. N. Ross, *J. Phys. Chem.*, 1995, **99**, 8945–8949.

141. G. A. Tritsaris and J. Rossmeisl, *J. Phys. Chem. C*, 2012, **116**, 11980–11986.

142. J. A. Kumar, P. Kalyani and R. Saravanan, *Int. J. Electrochem. Sci.*, 2008, **3**, 961–969.

143. F. Vigier, C. Coutanceau, F. Hahn, E. M. Belgsir and C. Lamy, *J. Electroanal. Chem.*, 2004, **563**, 81–89.

144. J. M. Léger, S. Rousseau, C. Coutanceau, F. Hahn and C. Lamy, *Electrochim. Acta*, 2005, **50**, 5118–5125.

145. J. Rossmeisl, P. Ferrin, G. A. Tritsaris, A. U. Nilekar, S. Koh, S. E. Bae, S. R. Brankovic, P. Strasser and M. Mavrikakis, *Energy Environ. Sci.*, 2012, **5**, 8335–8342.

146. M. Watanabe and S. Motoo, *J. Electroanal. Chem. Inter. Electrochem.*, 1975, **60**, 275–283.

147. L. Dubau, F. Hahn, C. Coutanceau, J. M. Léger C. Lamy, *J. Electroanal. Chem.*, 2003, **554–555**, 407–415.

148. D. D. Hibbitts, PhD, Nature and Reactivity of Oxygen and Hydroxide on Metal Surfaces, University of Virginia, 2012.

149. P. Liu, A. Logadottir and J. K. Nørskov, *Electrochim. Acta*, 2003, **48**, 3731–3742.

150. Y. Ando, K. Sasaki and R. Adzic, *Electrochem. Commun.*, 2009, **11**, 1135–1138.

151. M. T. M. Koper, T. E. Shubina and R. A. van Santen, *J. Phys. Chem. B*, 2002, **106**, 686–692.

152. M. Watanabe and S. Motoo, *J. Electroanal. Chem. Inter. Electrochem.*, 1975, **60**, 267–273.

153. M. T. M. Koper, A. P. J. Jansen, R. A. van Santen, J. J. Lukkien and P. A. J. Hilbers, *J. Chem. Phy.*, 1998, **109**, 6051–6062.

154. O. Petrii, *J Solid State Electrochem.*, 2008, **12**, 609–642.

155. J. L. Gómez de la Fuente, M. V. Martínez-Huerta, S. Rojas, P. Hernández-Fernández, P. Terreros, J. L. G. Fierro and M. A. Peña, *Appl. Catal. B: Environ.*, 2009, **88**, 505–514.

156. C. Alegre, L. Calvillo, R. Moliner, J. A. González-Expósito, O. Guillén-Villafuerte, M. V. M. Huerta, E. Pastor and M. J. Lázaro, *J. Power Sources*, 2011, **196**, 4226–4235.
157. G. García, J. Florez-Montaño, A. Hernandez-Creus, E. Pastor and G. A. Planes, *Journal of Power Sources*, 2011, **196**, 2979–2986.
158. T. Maiyalagan, T. O. Alaje and K. Scott, *Journal of Physical Chemistry C*, 2012, **116**, 2630–2638.
159. H. Liu, C. Song, L. Zhang, J. Zhang, H. Wang and D. P. Wilkinson, *Journal of Power Sources*, 2006, **155**, 95–110.
160. L. A. Kibler, A. M. El-Aziz, R. Hoyer and D. M. Kolb, *Angewandte Chemie International Edition*, 2005, **44**, 2080–2084.
161. L. A. Kibler, *Chem. Phys. Chem.*, 2006, **7**, 985–991.
162. M. Li, P. Liu and R. R. Adzic, *J. Phys. Chem. Lett.*, 2012, **3**, 3480–3485.
163. K. Bergamaski, E. R. Gonzalez and F. C. Nart, *Electrochim. Acta*, 2008, **53**, 4396–4406.
164. H. Wang, Z. Jusys and R. J. Behm, *J. Phys. Chem. B*, 2004, **108**, 19413–19424.
165. E. Antolini, *J. Power Sources*, 2007, **170**, 1–12.
166. Y. Wang, S. Zou and W.-B. Cai, *Catalysts*, 2015, **5**, 1507.
167. S. Chen and M. Schell, *Electrochim. Acta*, 2000, **45**, 3069–3080.
168. E. Pastor and T. Iwasita, *Electrochim. Acta*, 1994, **39**, 547–551.
169. G. A. Camara, R. B. de Lima and T. Iwasita, *Electrochem. Commun.*, 2004, **6**, 812–815.
170. N. Fujiwara, K. A. Friedrich and U. Stimming, *J. Electroanal. Chem.*, 1999, **472**, 120–125.
171. J. P. I. Souza, F. J. Botelho Rabelo, I. R. de Moraes and F. C. Nart, *J. Electroanal. Chem.*, 1997, **420**, 17–20.
172. S. Beyhan, C. Coutanceau, J.-M. Léger, T. W. Napporn and F. Kadırgan, *Int. J. Hydrogen Energy*, 2013, **38**, 6830–6841.
173. W. J. Zhou, S. Q. Song, W. Z. Li, Z. H. Zhou, G. Q. Sun, Q. Xin, S. Douvartzides and P. Tsiakaras, *J. Power Sources*, 2005, **140**, 50–58.
174. F. Delime, J. M. Léger and C. Lamy, *J. Appl. Electrochem.*, 1999, **29**, 1249–1254.
175. S. S. Gupta, S. Singh and J. Datta, *Mater. Chem. Phys.*, 2009, **116**, 223–228.
176. W.-P. Zhou, S. Axnanda, M. G. White, R. R. Adzic and J. Hrbek, *J. Phys. Chem. C*, 2011, **115**, 16467–16473.
177. F. L. S. Purgato, P. Olivi, J. M. Léger, A. R. de Andrade, G. Tremiliosi-Filho, E. R. Gonzalez, C. Lamy and K. B. Kokoh, *J. Electroanal. Chem.*, 2009, **628**, 81–89.

178. Z. D. Wei, L. L. Li, Y. H. Luo, C. Yan, C. X. Sun, G. Z. Yin and P. K. Shen, *J. Phys. Chem. B*, 2006, **110**, 26055–26061.
179. S. Tillmann, G. Samjeské, K. A. Friedrich and H. Baltruschat, *Electrochim. Acta*, 2003, **49**, 73–83.
180. S. Mukerjee and J. McBreen, *J. Electrochem. Soc.*, 1999, **146**, 600–606.
181. H. Wang, Z. Jusys and R. J. Behm, *J. Power Sources*, 2006, **154**, 351–359.
182. V. Del Colle, J. Souza-Garcia, G. Tremiliosi-Filho, E. Herrero and J. M. Feliu, *Phys. Chem. Chem. Phys.*, 2011, **13**, 12163–12172.
183. A. B. Anderson, E. Grantscharova and P. Shiller, *J. Electrochem. Soc.*, 1995, **142**, 1880–1884.
184. F. Vigier, C. Coutanceau, A. Perrard, E. M. Belgsir and C. Lamy, *J. Appl. Electrochem.*, 2004, **34**, 439–446.
185. J. Souza-Garcia, E. Herrero and J. M. Feliu, *Chem. Phys. Chem.*, 2010, **11**, 1391–1394.
186. S. Rousseau, C. Coutanceau, C. Lamy and J. M. Léger, *J. Power Sources*, 2006, **158**, 18–24.
187. M. J. Giz, G. A. Camara and G. Maia, *Electrochem. Commun.*, 2009, **11**, 1586–1589.
188. F. H. B. Lima and E. R. Gonzalez, *Electrochim. Acta*, 2008, **53**, 2963–2971.
189. A. Oliveira Neto, M. J. Giz, J. Perez, E. A. Ticianelli and E. R. Gonzalez, *J. Electrochem. Soc.*, 2002, **149**, A272–A279.
190. J. P. I. de Souza, S. L. Queiroz, K. Bergamaski, E. R. Gonzalez and F. C. Nart, *J. Phys. Chem. B*, 2002, **106**, 9825–9830.
191. A. Rabis, P. Rodriguez and T. J. Schmidt, *ACS Catal.*, 2012, **2**, 864–890.
192. Y. H. Chu and Y. G. Shul, *Int. J. Hydrogen Energy*, 2010, **35**, 11261–11270.
193. M. Li, A. Kowal, K. Sasaki, N. Marinkovic, D. Su, E. Korach, P. Liu and R. R. Adzic, *Electrochim. Acta*, 2010, **55**, 4331–4338.
194. A. Kowal, M. Li, M. Shao, K. Sasaki, M. B. Vukmirovic, J. Zhang, N. S. Marinkovic, P. Liu, A. I. Frenkel and R. R. Adzic, *Nat. Mater.*, 2009, **8**, 325–330.
195. E. A. de Souza, M. J. Giz, G. A. Camara, E. Antolini and R. R. Passos, *Electrochim. Acta*, 2014, **147**, 483–489.
196. L. C. Silva-Junior, G. Maia, R. R. Passos, E. A. de Souza, G. A. Camara and M. J. Giz, *Electrochim. Acta*, 2013, **112**, 612–619.
197. M. Zhu, G. Sun, S. Yan, H. Li and Q. Xin, *Energy Fuels*, 2008, **23**, 403–407.
198. C. Carrareto Caliman, L. M. Palma and J. Ribeiro, *J. Electrochem. Soc.*, 2013, **160**, F853-F858.
199. X.-H. Jian, D.-S. Tsai, W.-H. Chung, Y.-S. Huang and F.-J. Liu, *J. Mater. Chem.*, 2009, **19**, 1601–1607.

200. G. García, N. Tsiouvaras, E. Pastor, M. A. Peña, J. L. G. Fierro and M. V. Martínez-Huerta, *Int. J. Hydrogen Energy*, 2012, **37**, 7131–7140.
201. Z.-B. Wang, G.-P. Yin and Y.-G. Lin, *J. Power Sources*, 2007, **170**, 242–250.
202. M. Martínez-Huerta, N. Tsiouvaras, G. García, M. Peña, E. Pastor, J. Rodriguez and J. Fierro, *Catalysts*, 2013, **3**, 811–838.
203. E. Lee, A. Murthy and A. Manthiram, *Electrochim. Acta*, 2011, **56**, 1611–1618.
204. S. Guo, S. Zhang and S. Sun, *Angew. Chem. Int. Ed.*, 2013, **52**, 8526–8544.
205. T. S. Olson, S. Pylypenko, J. E. Fulghum and P. Atanassov, *J. Electrochem. Soc.*, 2010, **157**, B54–B63.
206. *Fuel Cell Technology–Reaching Towards Commercialization*, Springer-Verlag, London, 2006.
207. C. Song and J. Zhang, in *PEM Fuel Cell Electrocatalysts and Catalyst Layers*, ed. J. Zhang, Springer London, 2008, ch. 2, pp. 89–134.
208. N. Markovic, H. Gasteiger and P. N. Ross, *J. Electrochem. Soc.*, 1997, **144**, 1591–1597.
209. J. K. Nørskov, J. Rossmeisl, A. Logadottir, L. Lindqvist, J. R. Kitchin, T. Bligaard and H. Jónsson, *J. Phys. Chem. B*, 2004, **108**, 17886–17892.

Chapter 8

Quantum and Statistical Mechanical Simulations for Porous Catalyst Modelling

Jayesh S. Bhatt*, Arunabhiram Chutia[†,‡], C. Richard A. Catlow[†,‡,§] and Marc-Olivier Coppens*

*Department of Chemical Engineering, University College London, Torrington Place London WC1E 7JE, UK

†Department of Chemistry, University College London, Gower Street, London WC1E 6BT, UK

‡UK Catalysis Hub, RCaH, Rutherford Appleton Laboratory, Harwell Oxford, Oxfordshire OX11 OFA, UK

§School of Chemistry, Cardiff University, Cardiff CF10 3AT, UK

1 Introduction

Fundamentally, catalysis is a molecular-level process. Experimental investigations of catalysis usually aim to either probe right down to the nanoscopic level of the system or measure the ensemble average of molecular processes in a macroscopic system. There are, however, several essential properties that are inaccessible or difficult to access by experiments such as the electronic charge distribution, molecular structures, potential energy surface and the resulting dynamics of the atoms in real time. To fill this gap and help interpret experiments, theoretical methods are required. However, given that catalysis involves an N-body system of constituent particles, there is a limit on the range of analytical solutions that can be obtained. Thus, one has to rely on various computational techniques.

Designing heterogeneous, porous catalysts for practical applications requires not only the ability to identify active sites within the porous medium and optimise intrinsic activity and selectivity, but also the reduction of diffusion limitations and deactivation within the catalyst particle.[1,2] The processes of atomic transport and catalyst evolutions span timescales that are several orders of magnitude larger than the typical timescales of molecular vibrations and also span length scales that are much larger than the size of typical molecules.[3,4] No single modelling technique can access all these scales within reasonable computational times. For instance, quantum chemical techniques, coupled with spectroscopy, could simulate potential active sites and their immediate vicinity, but cannot simulate much larger domains, even *via* high-end computational resources. Monte Carlo (MC) techniques have the ability to simulate systems up to the macro-scale, but they lack an *ab initio* basis and must be parameterised in an adequate way. Molecular dynamics (MDs) simulations, due to their ability to simulate both the micro- and meso-scales, have proven to be an excellent bridge between these scales. Calculations based on transition state theory (TST) and network models further complement these techniques. A recent review by Van Speybroeck *et al.* presents an overview of computational techniques and their application in catalysis.[5] They presented the various techniques as a function of system size, reproduced in Figure 1. It is therefore clear that an ideal strategy would involve a multiscale approach, whereby one technique is used to obtain results for one subdomain of the macroscopic system and the output is employed to parameterise another model suitable for the next level of domain. Bottom-up as well as top-down approaches have been used depending on the problem being solved. Such a strategy can not only help decipher experimental data, but can also give indicative results that may guide the future course of experiments.

In this chapter, we review three techniques of molecular modelling that are relevant to designing the next generation of catalysts: density functional theory (DFT), MD and MC simulations, with focus on kinetic MC. We refer to Keil *et al.* and Smit and Maesen[6] for comprehensive reviews on

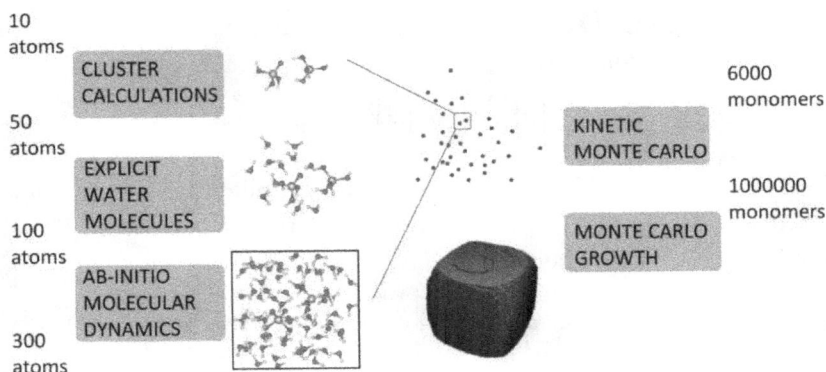

Figure 1: Computational methods available in function of the system size. Reproduced from Ref. [5] with permission from Royal Society of Chemistry.

the application of molecular simulation techniques to diffusion in zeolites. More recent reviews on computational methods in heterogeneous catalysis[7] and for complex reaction mechanisms[8,9] have summarised the fundamentals of this field. Our aim is to provide a tutorial review of some representative work undertaken towards heterogeneous catalyst analysis and design, after briefly outlining the theoretical basis of each technique discussed.

2 Fundamentals of DFT

2.1 The Hohenberg–Kohn theorems

In 1964, Pierre Hohenberg and Walter Kohn proposed and proved that for a non-degenerate ground state, all the ground state electronic properties are determined by the ground state electron probability density (ρ_0), which could be mathematically represented as[10,11]:

$$\rho_0(\mathbf{r}) \rightarrow v_{\text{ext}} \rightarrow \psi_0 \rightarrow \text{all properties.} \tag{8.1}$$

Here, v_{ext} is the potential energy of interaction between electrons and the nuclei, and ψ_0 is a unique ground state wave function.[12] The electron density is dependent on three spatial variables ($\mathbf{r}$) only whereas, the wave function depends on $3N$ variables for a system of N electron.

Thus, the ground state energy could be written as a functional of the ground state electron density,

$$E_0 = E_{v_{ext}}[\rho_0] = \bar{T}[\rho_0] + \bar{V}_{ne}[\rho_0] + \bar{V}_{ee}[\rho_0]. \qquad (8.2)$$

Here, $\bar{T}, \bar{V}_{ne}$ and $\bar{V}_{ee}$ are the kinetic energy, potential energy due to nucleus–electron and electron–electron interactions, respectively. The $\bar{V}_{ne}$ term in Eq. (8.2) could be represented as[15]

$$\bar{V}_{ne} = \left\langle \psi_0 \left| \sum_{i=1}^{N} v(\mathbf{r}_i) \right| \psi_0 \right\rangle = \int \rho_0(\mathbf{r}) v(\mathbf{r}) d\mathbf{r}, \qquad (8.3)$$

where $v(r_i)$ is the nuclear–electron potential energy for the ith electron. Combining Eqs. (8.2) and (8.3) yields

$$E_0 = E_{v_{ext}}[\rho_0] = \bar{T}[\rho_0] + \int \rho_0(\mathbf{r}) v(\mathbf{r}) d\mathbf{r} + \bar{V}_{ee}[\rho_0]. \qquad (8.4)$$

In their variational theorem, Hohenberg and Kohn have furthermore proved that, the true ground state electron density minimises the energy functional $E_v[\rho_{tr}]$, where ρ_{tr} is the trial density function that satisfies $\int \rho_{tr}(\mathbf{r}) d\mathbf{r} = N$. For simplicity, from now we will omit 0 from ρ for all other equations.

The term $\bar{T}[\rho_0] + \bar{V}_{ee}[\rho_0]$ in Eq. (8.4) is unknown. While the Hohenberg–Kohn theorem tells us that if we know the ground-state electron density $\rho(\mathbf{r})$, we can determine the ground state wave function, it does not tell how to calculate E_0 from ρ_0. It also does not tell how to obtain ρ_0 without finding the wavefunction.[10,13] In 1965, Kohn and Sham revisited this problem, which is briefly outlined below.

2.2 The Kohn–Sham method

Kohn and Sham considered a fictitious system of N non-interacting electrons, which experiences an external potential energy $v_s(\mathbf{r}_i)$.[13] It was further assumed that the electron density of this reference system $\rho_s(\mathbf{r})$ is equal to the exact ground-state electron density $\rho(\mathbf{r})$ of the system of N interacting electron system under consideration. By mapping the real system on the fictitious system, Kohn and Sham have been able to introduce one electron orbitals (wavefunctions) in a method analogous to that of Hartree–Fock theory. Moreover, their approach allows to reduce errors that would arise from approximations to the unknown energy term $\bar{T}[\rho_0]+\bar{V}_{ee}[\rho]$ by separating out the largest well-defined terms $\bar{T}[\rho_s]+\bar{V}_{ee}[\rho_s]$.

Indeed, the kinetic energy of the real system can be expressed as:

$$\bar{T}[\rho_0]\equiv\Delta\bar{T}[\rho]+\bar{T}_s[\rho], \qquad (8.5)$$

using the $\Delta\bar{T}[\rho]$, the difference of the kinetic energies between real and fictitious systems (index s is reserved here for the fictitious systems).

The electron–electron interaction energy can be represented as a sum of two terms, $\Delta\bar{V}_{ee}[\rho]$, which is unknown, and classical electrostatic potential energy of the charge distribution $\rho(\mathbf{r})$.

$$\bar{V}_{ee}[\rho]\equiv\Delta\bar{V}_{ee}[\rho]+\frac{1}{2}\iint\frac{\rho(\mathbf{r}_1)\rho(\mathbf{r}_2)}{r_{12}}d\mathbf{r}_1 d\mathbf{r}_2. \qquad (8.6)$$

Here the term ½ is added so that the repulsion energy between the charges is not counted twice.

On substituting Eqs. (8.5) and (8.6) in Eq. (8.4) we have:

$$E_{v_{ext}}[\rho_0]=\Delta\bar{T}[\rho]+\bar{T}_s[\rho]+\int\rho(\mathbf{r})v(\mathbf{r})d\mathbf{r}+\Delta\bar{V}_{ee}[\rho]$$
$$+\frac{1}{2}\iint\frac{\rho(\mathbf{r}_1)\rho(\mathbf{r}_2)}{r_{12}}d\mathbf{r}_1 d\mathbf{r}_2, \qquad (8.7)$$

which can be rewritten as

$$E_{v_{ext}}[\rho_0]=\bar{T}_s[\rho]+\int\rho(\mathbf{r})v(\mathbf{r})d\mathbf{r}+\frac{1}{2}\iint\frac{\rho(\mathbf{r}_1)\rho(\mathbf{r}_2)}{r_{12}}d\mathbf{r}_1 d\mathbf{r}_2+E_{xc}[\rho], \qquad (8.8)$$

where,

$$E_{xc}[\rho]=\Delta\bar{T}[\rho]+\Delta\bar{V}_{ee}[\rho]. \qquad (8.9)$$

The term E_{xc} is an unknown term, referred to as the *exchange–correlation* (*xc*) *energy functional*, and it must be approximated.[13] Equation (8.9) can be further written as

$$E_{xc}=E_x+E_c, \qquad (8.10)$$

where E_x is the exchange energy and E_c is the correlation energy.

The fictitious reference system *s* can be described using an antisymmetrised product (Slater determinant) of the lowest-energy Kohn–Sham spin-orbitals whose spatial part ϕ_i is an eigenfunction of the one-electron operator h_i.

$$h_i\phi_i = \varepsilon_i\phi_i. \qquad (8.11)$$

The one-electron Hamiltonian is obtained by varying the energy (Eq. (8.8)) with respect to Kohn–Sham orbitals under the constrained $\int \rho\, d\mathbf{r} = N$:

$$h_i = -\frac{1}{2}\nabla_1^2 - \sum_a \frac{Z_a}{r_{1a}} + \int \frac{\rho(\mathbf{r}_2)}{r_{12}} d\mathbf{r}_2 + v_{xc}(r_1,\sigma) \qquad (8.12)$$

where 1 and 2 are the spatial and spin (σ) variables of the first and second electrons, electron-nuclear potential is summed over all nuclei with charge Z_a and exchange-correlation potential v_{xc} is the variational derivative of Eq. (8.9). Equation (8.11) with Hamiltonian (8.12) constitutes the *Kohn–Sham equations*. The equations so formed are mathematically equivalent to the Hartree–Fock equations with a modified potential form; therefore, all methods developed in quantum chemistry to solve the Hartree–Fock equations numerically are applicable to the Kohn–Sham equations.

2.3 Exchange and correlation functionals

In this section, we shall briefly discuss commonly used exchange–correlation functionals. This discussion is especially vital as, (1) the exact form of xc is still unknown and (2) the appropriate choice of xc is very important for an accurate description of the geometrical and electronic properties of materials. Even though the xc energy is a small part of the total energy of a typical system, it plays the vital role in binding atoms together. Perdew[14] coined it as "nature's glue". This quantum mechanical phenomenon arises as electrons move in such a way to avoid one another, which in turn, lowers the expectation value of the electron–electron Coulomb interaction. The exchange energy in the *xc* functional (Eq. (8.10)) is a consequence of the system obeying the Pauli principle and in particular (but not only) corrects the spurious self-interaction of an electron, and the correlation energy ($E_c = E_{xc} - E_x$) accounts for the remaining effects of spatial and spin

Chemical accuracy

Unoccupied $\psi_\alpha(\acute{r})$	Exact exchange and exact partial correlation
Occupied $\psi_\alpha(\acute{r})$	Exact exchange and compatible correlation
$\tau(r), \nabla^2$	Meta-Generalized Gradient Approximation
$\nabla n(r)$	Generalized Gradient Approximation
$n(r)$	Local Density Approximation

Hartree world

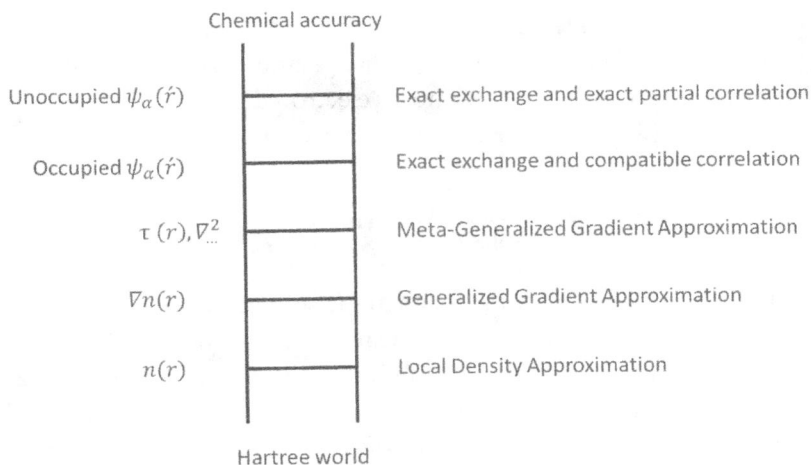

Figure 2: Jacob's ladder of density functional approximations.[15]

correlation in the many electron system. The evolution of various xc functionals from the simplest to the most sophisticated ones could be represented by "Jacob's ladder", as described by Perdew and Schmidt (see Figure 2).[15] On this ladder, various forms of exchange and correlation functionals are found; as we climb higher rungs the accuracy and sophistication of the xc functionals increases. The main objective of this is to formulate the most suitable functional to attain the desired chemical accuracy. Furthermore, the accuracy and sophistication of the higher rungs are complemented by the simplicity and transparency of the lower ones.

2.3.1 The Local density approximation (LDA)

The LDA is sometimes referred to as the "mother of all approximations" in DFT.[15] The basic idea of the method is to relate the exchange and correlation potential at a given point in space to the charge density only at the same point. The Hohenberg–Kohn theorem in fact, tells us that the potential depends on the charge density everywhere. The mathematical form of the LDA is given by:

$$E_{xc}^{LDA}\left[\rho\right] = \int \rho\left(\mathbf{r}\right)\varepsilon_{xc}\left(\rho\right)d\mathbf{r}, \tag{8.13}$$

where the integral is taken over all space, *dr* stands for (*dx, dy, dz*). Further, ε_{xc} is postulated to be equal to the exchange and correlation energy per particle of an electron gas with uniform electron density ρ. Since E_x dominates E_{xc} in Eq. (8.10) here we describe the exchange part only:

$$E_x^{LDA} = -\frac{3}{2}\left(\frac{3}{4\pi}\right)^{\frac{1}{3}} \int \left(\rho_\alpha^{\frac{4}{3}}(\mathbf{r}) + \rho_\beta^{\frac{4}{3}}(\mathbf{r}) \right) d\mathbf{r}. \tag{8.14}$$

where, α and β in Eq. (8.14) refers to the spin of the electrons. In contrast to Eq. (8.14) there is no simple expression for the correlation part, which is however known numerically from the quantum Monte Carlo simulations of Ceperley and Alder[16] that have been parameterized by Vosko, Wilk and Nusair, (VWN)[17] and Perdew and Wang (PWC).[18] This functional is good for atoms and molecules and especially adequate for a slowly varying electron density $\rho(\mathbf{r})$.

2.3.2 Generalised-gradient approximation (GGA)

The LDA exchange and correlation functional based on uniform electron–gas models works well for predicting molecular geometries, vibrational frequencies and dipole moments. However, the prediction of accurate dissociation energies requires functionals to go beyond LDA and to take into account a slowly varying $\rho(\mathbf{r})$. This requirement led to the development of the GGA functionals, in which a dimensionless reduced gradient $x_\sigma \frac{|\nabla\rho^\sigma|}{(\rho^\sigma)^{\frac{4}{3}}}$ is considered.[19] The general form of the GGA exchange is:

$$E_x^{GGA}[\rho,x] = \int \rho^{\frac{4}{3}} F(x) d\mathbf{r}, \tag{8.15}$$

where $F(x)$ is chosen to obey the gradient expansion approximation first introduced by Hohenberg and Kohn:

$$E_x^{GEA} = -\int \rho^{\frac{4}{3}} \left[\frac{3}{4}\left(\frac{3}{\pi}\right)^{\frac{1}{3}} + \frac{7}{432\pi\left(3\pi^2\right)^{\frac{1}{3}}} x^2 + \cdots \right] d\mathbf{r}. \tag{8.16}$$

Thus, the xc potential at a given point now depends both on the density at this point and its derivatives, which approximates the density in close locality of this point hence referred to as semi-local potentials. However, coefficients in Eq. (8.14) are modified to satisfy known limiting situations or fit empirical data. One of the most commonly used forms of exchange functionals is B88,[20] which is given by:

$$E_x^{88} = E_x^{LDA} - b \sum_{\sigma=\alpha,\beta} \int \rho_\sigma^{\frac{4}{3}} \frac{\chi_\sigma^2}{1+6bx_\sigma \sinh^{-1}x_\sigma} d\mathbf{r}. \qquad (8.17)$$

Here, b = 0.0042 is the single parameter used to fit the exact atomic energies.

Other popular GGA exchange functionals have been proposed over the years for example by Perdew and co-workers including PW91, PBE, RPBE (also known as RevPBE) and PBESol.[21–24] In recent years, some theoretical studies have shown that conventional LDA and GGA cannot describe accurately the strong on-site Coulomb interaction of localised electrons for example, in f-electrons in UO_2. In order to account for the underestimation of the intraband Coulomb interactions, Liechtenstein *et al.* and Dudarev *et al.*, proposed the use of Hubbard U parameter and J (exchange energy) parameters.[24(b-c)] In the former approach, both U and J values are considered as independent corrections while in the later only a single effective $U_{eff} = U - J$ parameter is considered. Mathematically, DFT+U total energy $(E_{DFT}+U))$ can be written as:

$$E_{DFT+U} = E_{DFT} + \frac{U-J}{2} \sum \left[Tr\left(\rho^\sigma - \rho^\sigma \rho^\sigma \right) \right]$$

here, ρ^σ is the density matrix.

2.3.3 Hybrid functionals

While pure local and semilocal exchange and correlation functionals in DFT proved to be very successful, already in 1993 Becke observed that GGA, which is an improvement over LDA, still has over-binding tendency. To overcome this problem, he proposed a fully non-local B3PW91 exchange and correlation functional, which has the form:

$$E_{xc}^{B3PW91} = E_{xc}^{LDA} + a\left(E_x^{exact} - E_x^{LDA}\right) + b\Delta E_x^{88} + c\Delta E_c^{PW91}, \qquad (8.18)$$

where a = 0.20, b = 0.72 and c = 0.81; and the Hartree–Fock like exact exchange term E_x^{exact}, removes some of the self-interaction energy error inherent in GGA functionals.

This functional preserves the uniform electron gas (UEG) limit and reduces the PW91 gradient correlation ΔE_c^{PW91} and Becke88 or B88 gradient correction ΔE_x^{88}.[25] As an alternative to this, functional Lee–Yang–Parr correlation has been also proposed instead of PW91 and referred to as B3LYP,[26] which is one of the most popular functionals in quantum chemistry. The major difference between B3PW91 and B3LYP is related to the slowly varying density limit. The reason for the high accuracy of B3LYP is due to its eight empirical parameters. While this works very well with smaller molecules, it fails to reproduce the correct exchange and correlation energy for the homogenous electron gas and, as a result, accuracy of the predictions decreases with increasing size of molecules. According to recent reports, it fails for metallic systems.[27–29]

The hybrid functionals discussed in this sections are sometimes referred to as global hybrids. In global hybrids, a percentage of local exchange is replaced by Hartree–Fock exchange.[30] The hybrid functionals (rung 4 of Jacob's ladder) take into account occupied Kohn–Sham orbitals everywhere in space but differ from each other by the mathematical form in which they combine, a distinct family of such functionals of particular flavour that are popular in solid state studies are overviewed in the section below.

To complete our climb up the Jacob's ladder, we only mention the highest state of the art level of double hybrids, which include into account unoccupied orbitals. In these approaches, GGA (or meta-GGA) description of electron correlation is replaced by post Hartree–Fock like terms that account for the probability of a dynamical occupation of higher energy (unoccupied) orbitals.[31] Some examples of double hybrids include B2PLYP,[32] XYG3.[33] In these functionals, the Kohn–Sham orbitals can be used in a self-consistent or an *a posteriori* manners in the calculations of the correlation terms. One of the problems with these approaches is a reliance on the MP2 energy expression, which contains a singular term dependent inversely on the one-electron energy gap between occupied and unoccupied states, which complicates the treatment of zero-gap systems.

2.3.4 Range separated hybrids

Now we turn to the methods that are especially popular in modelling electronic properties of semi-conducting and insulating materials. These techniques offer an alternative approach to include non-locality into exchange part as proposed by Savin.[34,35] In this approach, the two-electron operator $1/r_{ij}$ in the molecular Hamiltonian is separated into the short-range and long-range parts, that is,

$$\frac{1}{r_{ij}} = \frac{1-g(r_{ij})}{r_{ij}} + \frac{g(r_{ij})}{r_{ij}}. \tag{8.19}$$

Here, $g(r_{ij})$ could be for example the standard error function $\mathrm{erf}(\mu r_{ij})$ and μ is a parameter that determines the ratio between short- and long-range contributions. The first and the second terms in the right-hand side of the equation are, respectively, short- and the long-range.

There are two approaches via which the short- and long-range parts in Eq. (8.17) are included. In the first approach, 100% HF exchange is used for the long-range orbital–orbital and DFT exchange interaction is included in the short-range limit. These functionals are referred to as long-range-corrected (LC) functionals.[36] A number of such functionals have been implemented in several popular quantum chemical packages. For example, LC-ωPBE is the long range corrected version of ωPBE.[37–39] Another popular and closely related example is CAM-B3LYP.[36] In the second approach, a small amount of HF exchange is used for the short-range limit along with the DFT short-range, and the HF long-range is replaced by the DFT long-range. Based on this scheme, Heyd and Scuseria developed the HSE03 range-separated hybrid functional[40]

$$E_{xc}^{HSE} = aE_x^{HF,SR}(\omega) + (1-a)E_x^{\omega PBE,SR}(\omega) + E_x^{\omega PBE,LR}(\omega) + E_c^{PBE}. \tag{8.20}$$

Here, $E_x^{HF,SR}$ is the short-range HF exchange, $E_x^{\omega PBE,SR}$ and $E_x^{\omega PBE,LR}$ are the short- and long-range components of PBE exchange functional, E_c^{PBE} is the PBE correlation functional, $\omega = 0.15\ a_0^{-1}$ is the splitting parameter and $a = 1/4$ is the HF mixing constant. This functional is based on the hybrid PBE0 exchange–correlation functional developed independently by Ernzerhof and Adamo.[41,42]

While DFT has been a very powerful tool to explore and predict interesting properties of materials to design newer catalyst, other theoretical methods are often required to model larger length scales. We shall briefly summarise them in the following sections.

3 Classical Molecular Dynamics: A Force Field Approach

MD simulations involve obtaining trajectories of atoms and molecules in real time and space by solving the appropriate time-dependent equations of motion. MD is a powerful technique to investigate dynamic phenomena such as diffusion, surface formation, faceting, surface adsorption and phase transformation. In classical molecular dynamics, Newton's laws of motion are applied simultaneously to a system of N interacting particles (atoms). This method started being used in a noticeable manner during the 1970s, when computers accessible to scientists had become just powerful enough to handle the computational demands of this technique within a reasonable time-frame. *Ab initio* MD simulations are also possible, in which principles of quantum mechanics are incorporated to obtain the time evolution of the system. Classical MD is naturally much faster than *ab initio* MD and the former can cover much larger length- and timescales than the latter. Hence, we shall restrict the discussion in the following text to classical MD simulations.

3.1 Force fields

A key ingredient at the root of a classical MD simulation is the set of force fields used to describe interatomic interactions. These need to be obtained, in principle, for every unique pair of atomic species in the system as a function of the spatial coordinates. Although *ab initio* calculations can provide the detailed potential surface, its functional form may be too complicated for use in a classical MD simulation. One highly desirable feature of a force field is that its form should be simple enough to be evaluated quickly at every step of a simulation. Hence, force fields for molecular simulations are usually divided into two parts: bonded and non-bonded interactions. That is to say, the total potential energy, U_{Total}, is expressed as

$$U_{\text{Total}} = U_{\text{Bonded}} + U_{\text{Non-bonded}}. \tag{8.21}$$

Bonded interactions represent the chemical bonds between various atoms within an entity (e.g. a single molecule or an entire crystal) and are expressed as

$$U_{\text{Bonded}} = U_{\text{Stretching}} + U_{\text{Bending}} + U_{\text{Torsion}} + U_{\text{Improper}}, \tag{8.22}$$

where $U_{\text{Stretching}}$ represents bond stretching, U_{Bending} describes valence angle potentials for the bending of the bonds among three atoms, U_{Torsion} is the dihedral angle potential incorporating the twisting movement among atoms and U_{Improper} represents dihedral potentials used to restrict the geometry of molecules (e.g. to maintain the planar geometry of a molecule). Each of these four terms has the form of a harmonic potential, although when rigid bonds are assumed for simplicity, some of these terms may be replaced by constraints.

Non-bonded interactions are usually expressed as a sum of the electrostatic and van der Waals interactions:

$$U_{\text{Non-bonded}} = U_{\text{Electrostatic}} + U_{\text{Van der Waals}}. \tag{8.23}$$

Here, $U_{\text{Electrostatic}}$ is the energy due to the usual Coulombic interaction, while $U_{\text{Van-der-Waals}}$ can take any of a several forms available. More sophisticated bonded potentials for bonded interaction includes the Morse,[45] or Tersoff potentials,[49] while for metals the embedded atom model[46,47] and the Finnis–Sinclair potential[48] are used.

3.2 MD trajectories

Once the full form of the force field for the entire system is known, the force, $\mathbf{f}_j$, acting on atom j due to atom i is evaluated as

$$\mathbf{f}_j = -\frac{1}{r_{ij}}\left[\frac{\partial U(r_{ij})}{\partial r_{ij}}\right]\mathbf{r}_{ij}, \tag{8.24}$$

where $\mathbf{r}_i$ and $\mathbf{r}_j$ are the Cartesian coordinates of the two atoms with respect to an arbitrarily defined origin in the simulation cell, while

$$\mathbf{r}_{ij} = \mathbf{r}_j - \mathbf{r}_i \quad \text{and} \quad r_{ij} = |\mathbf{r}_{ij}| = \mathbf{r}_j - \mathbf{r}_i. \tag{8.25}$$

From these forces, the instantaneous acceleration of each atom, $\mathbf{a}_j$, is obtained via Newton's second law of motion:

$$\mathbf{a}_j = \frac{\mathbf{f}_j}{m}. \tag{8.26}$$

Obtaining the atomic trajectories from these values of acceleration is not quite as trivial as a simple numerical integration of Eq. (8.24). An explicit integration based on Euler's method turns out to be numerically too unstable.[50] Instead, several sophisticated methods have been developed to propagate the trajectories. In most of these schemes, the starting point is the Taylor expansion of the position **r** at time t:

$$\mathbf{r}(t+\Delta t) = \mathbf{r}(t) + \mathbf{v}(t) + \frac{1}{2!}\mathbf{a}(t)(\Delta t)^2 + \frac{1}{3!}\frac{d^3\mathbf{r}(\tau)}{dt^3}\bigg|_{\tau=t}(\Delta t)^3 + \cdots \quad (8.27)$$

Here, $\mathbf{v}(t)$ is the first time derivative of the position and hence is the velocity. A popular integration scheme, devised by Verlet,[51] involves obtaining the Taylor expansion in not just the forward time step Δt, but also in reverse time step $-\Delta t$ and then adding the two resultant expressions. This cancels out all the odd-order derivatives due to opposite signs. If one then truncates the series at the second order, a compact form for trajectory propagation emerges, which does not involve the velocity:

$$\mathbf{r}(t+\Delta t) = 2\mathbf{r}(t) - \mathbf{r}(t-\Delta t) + \mathbf{a}(t)(\Delta t)^2. \quad (8.28)$$

However, if one wanted to control the simulation temperature, it would be desirable to involve the velocity and couple it with the evolution of the position. There are more sophisticated schemes available, such as the leapfrog algorithm, in order to achieve this.

The time step in a MDs simulation is an important parameter that must be chosen judiciously. A large time step will lead to numerical instability during the solution of the propagation equations. On the other hand, an unnecessarily small time step will evolve the system too slowly, slowing down the simulation speed. As a general rule, the time step should be comfortably smaller than the typical timescale of the fastest movement present in the system, e.g. the vibrations of atoms in a molecule or in a crystal. A step of 1 femtosecond works well for many MD systems, although in cases where chemical bonds are held rigid and only diffusional motion is sought, a larger time step may be acceptable.

Another important consideration in MD is the size of the simulation cell. In a large majority of cases, MD simulations are performed with periodic boundary conditions, since a finite size of the system enables one to define parameters such as temperature and pressure. While periodic boundaries allow one, in principle, to mimic an infinite system, reasonably accurate results are only obtained if the simulation box size is kept two or three times larger than the typical correlation length scales involved. The Ewald summation technique was devised nearly a century ago, as a theoretical tool, to incorporate electrostatic interactions beyond the simulation box.[52,53] It is widely used in modern simulation packages to treat interactions of the inverse square nature, although more refined versions of this technique, such as the smoothed particle mesh Ewald (SPME),[54] are also available due to its ability to treat not just the inverse square law interactions, but also the r^{-6} term that appears in the Lennard–Jones potential. MD simulation involves an initial equilibration process in which the system attains thermal equilibrium followed by a production run.

4 MC Methods

Central to a MC simulation applicable to chemistry and materials science is the concept of phase space, which is defined by the two vectors containing the 3D components of the positions (**q**) and momenta (**p**) of all the particles in a given system:

$$\mathbf{q} = (x_1, y_1, z_1, x_2, y_2, z_2, \ldots.), \tag{8.29}$$

$$\mathbf{p} = (p_{x,1}, p_{y,1}, p_{z,1}, p_{x,2}, p_{y,2}, p_{z,2}, \ldots.). \tag{8.30}$$

Unlike MD, in a MC simulation there are no equations of motion to be integrated. Instead, the entire energy surface of the system is probed by randomly choosing points in this phase space. The phase space contains information about all the possible states of the system and hence the expectation value of some property B of the system is given by

$$\langle B \rangle = \iint B(\mathbf{q},\mathbf{p}) P(\mathbf{q},\mathbf{p}) d\mathbf{q} d\mathbf{p}. \tag{8.31}$$

Here, P is the probability of being at a phase point $(\mathbf{q}, \mathbf{p})$. This probability is connected to the energy associated with this phase point through

$$P(\mathbf{q},\mathbf{p}) = Z^{-1} \exp\left[\frac{-E(\mathbf{q},\mathbf{p})}{k_B T}\right], \tag{8.32}$$

where P is the system partition function, k_B is Boltzmann's constant and T is the system temperature.

A majority of the phase space is usually of no use for a simulation. That is to say, the energy barrier at most points in space would be insurmountably high. It would be a waste of time to try and probe such points and that is the key point exploited by an MC simulation in order to render fast and efficient results. The Metropolis sampling method probes only the most probable states in the system by choosing configurations with probabilities $\exp(-E/k_B T)$ and weighing them evenly. A typical MC simulation therefore follows the following recipe:

1. Set up the initial atom coordinates, usually by idealised approximations.
2. Pick an atom i randomly and perturb its position randomly to obtain a new position.
3. Calculate the change of potential energy, ΔU, associated with this displacement.
4. If $\Delta U < 0$, accept the new coordinates and go to step 2.
5. If $\Delta U \geq 0$, select a random number R in the range $[0,1]$ and:

   (a) If $e^{-\Delta U/k_B T} < R$, accept the new coordinates and go to step 2.
   (b) If $e^{-\Delta U/k_B T} \geq R$, reject the move, keep the old coordinates and go to step 2.

If a property B is dependent on the positions of the particles, the Metropolis algorithm renders the average of this property according to

$$\langle B \rangle = \frac{1}{M} \sum_{i=1}^{M} B(\mathbf{q}_i), \tag{8.33}$$

where M is the total number of positions accepted and sampled according to the above recipe. There is no intrinsic time coordinate in this method and the system merely jumps from state to state in order to optimise the speed of the calculations.

For problems that require explicit time dependence without involving the level of details achieved in DFT or MD, a popular choice of method is *kinetic Monte Carlo* (kMC) simulation. While the ordinary MC requires energies as inputs, in kMC, rates of different possible "events" associated with a given process are required as inputs. Quite often, these are obtained through the TST. While in the Metropolis scheme the decision of whether or not to accept a move depends on the energy difference between the states, in kMC one calculates rates that depend on the energy barrier between the states. One needs to know all the possible events that could occur during the process and the rates associated with these events, so that relative probabilities can be evaluated.

Various schemes to implement kMC philosophy exist.[55–59] A very simple scheme based on rejection kMC proceeds as follows:

1. At $t = 0$, set up the system according to the initial conditions.
2. Start the time loop.
3. Calculate the rates, r_i ($i = 1, \cdots, N$), corresponding to all the N possible events.
4. Calculate the sum of the rates $= \sum_{i=1}^{N} r_i$. Each segment $(r_{i-1}, r_i]$ on the $(0, R]$ range now corresponds uniquely to the event i.
5. Obtain a uniform random number, $u \in (0, 1]$.
6. Map the random number range $(0, 1]$ onto the rates range $(0, R]$. Pick a possible event at random by seeing which value of i satisfies the relation $r_{i-1} < uR \le r_i$.
7. Carry out the event i.
8. Recalculate all the r_i based on the new configuration.
9. Obtain a new random number, $u' \in (0, 1]$.
10. Update the time according to $t_{new} = t_{old} - \dfrac{\log u'}{R}$.
11. Go back to step 3.

One potential problem that needs to be guarded against while performing kMC is that during the course of the simulation, the system may visit unforeseen states that were not planned at the beginning of the simulation and hence the corresponding transition rates were not incorporated in the calculations. Hence, careful planning is required before attempting the simulation.

5 Applications to Catalysis

5.1 Applications of DFT

Now that we have discussed the basic concepts of DFT, in the following sections, we will briefly summarise the application of DFT in catalysis with a focus on zeolites as catalyst.

Zeolites are microporous crystalline aluminosilicates, which play an important role in industrial catalysis, for example in catalytic cracking, alkylation, methanol to gasoline (MTG) conversion, decomposition of NO, partial oxidation of hydrocarbons and alcohols, epoxidation of olefins (related titanosilicates) and the hydroxylation of aromatics or ammoxidation of ketones.[60-62] The reason for their overwhelming success in catalysis is related to their high surface area, adsorption capacity, and pore size of molecular dimensions leading to size and shape selectivity. Their active sites can be engineered to facilitate activation of reactants inside the channels, for example, due to strong electric fields in combination with electronic confinement of guest molecules.[63] Much computational work has focused on understanding zeolite catalysis in view of their industrial applications or to explore their fundamental electronic properties. Reviewing this field is beyond the scope of this article. However, the interested readers may refer to a recent review by van Speybroeck *et al.* on the most currently available computational tools and their applications to zeolites.[5] Here, we summarise some of the major applications of DFT in understanding the chemistry of zeolites, taking MTG as an example.

Significant theoretical work was undertaken in the late 1980s and early 1990s to clarify the mechanism of MTG conversion reaction in ZSM-5 zeolites using Hartree–Fock theory.[64-67] However, DFT being computationally economical it provided better scope to consider more realistic clusters. In 1993, Gale *et al.* employed non-local DFT to study the binding of methanol to cluster models for aluminosilicates.[62] This study was undertaken to clarify the mechanism of methanol activation. They used a finite-cluster model approach to show that for the smaller cluster models the methoxonium ion corresponded to the transition state for the hydrogen exchange between two framework oxygen atoms. The geometrical parameters and calculated zero-point energy were in good agreement with experimental findings. Several experiments in the early 1990s suggested that methanol (MeOH) is first dehydrated to dimethyl ether (DME) and an equilibrium mixture of MeOH, and DME is then converted to olefins, aliphatics and aromatics up to C_{10}. Some reports also suggested the role of CO in the conversion of MTG on HZSM-5. However, the mechanism involved in these reactions was not fully understood.[68-71] Theoretical studies during this period of time showed formation of methoxy groups from dissociative adsorption of a single methanol molecule.[72,73] In their studies, Blaszkowski and van Santen employed DFT to determine transition states, adsorption and dissociative complexes of Brønsted-acid-activated methanol. They used two differently sized tritetrahedral clusters, $H_3SiOHAl$ $(OH)_2OSiH_3$ and $H_3SiOHAlH_2OSiH_3$, to represent the acid sites in zeolites. These clusters were first used to optimise the structures using the LDA as parametrised by Vosko *et al.*[17] For the final optimised structures, non-local correlation and exchange corrections due to Perdew[74] and Becke[20] were included in the final total LDA energy. The reason for this was that LDA without non-local correction proved to be inadequate for predicting accurate binding energies for hydrogen transfer reactions.[75] They also concluded that the previously reported chemisorbed methoxonium ion was in fact a transition state rather than a chemisorbed species, which was in agreement with previous theoretical studies as well.[62] Sinclair and Catlow employed DFT to study the formation and deprotonation of surface methyl oxonium ions, $-OCH_3$, at the aluminosilicate Brønsted acid sites.[76] They showed that acid catalysed deprotonation of methyl groups during the MTG conversion can be used to produce surface-stabilised carbenes. The DFT derived activation barriers were in reasonable agreement with the experimental findings, suggesting methyl group deprotonation as the rate-determining step during methanol conversion. Blaszkowski *et al.* went on to study the routes of the conversion of MTG with acidic zeolites by determining the transition states, adsorption complexes of reactants, intermediates and products, and the reaction barriers using DFT.[77] They used small $HOHAl(OH)_2OH$ and $HO(CH_3)Al(OH)_2OH$ clusters to represent the acidic zeolite and methoxy surface respectively. In this study, it was put forward that ethanol and ethyl methyl ether are formed from surface methoxy groups, methanol and/or dimethyl ether, and they were the first intermediates containing C–C bonds. This step was found to be the rate determining step. Ethylene was then easily formed from ethanol and ethyl methyl ether involving a much lower activation energy. In the consecutive steps, ethylene reacts with other alkenes to give higher olefins, aromatics and alkanes. In 1998, Frash *et al.* used B3LYP functionals to fully optimise $H(OH)AlH_2(OH)$ and $H_3Si(OH)AlH_2(OSiH_3)$ clusters and tested the sensitivity of activation energies with HF and MP2 single point energies to investigate the cracking of hydrocarbons on zeolite catalysts.[78] They mapped out a very complex potential energy surface for this reaction and identified three different reaction pathways. For all of them, both the initial and the final states were identified as alkoxy species where the carbon atoms were directly covalently bonded to an oxygen atom of the zeolite. Another interesting approach taken around this time was to use the quantum chemical cluster models, which were embedded in a classical potential that described the interactions with the zeolite framework. This method is referred to as the QM/MM (quantum mechanical/molecular mechanical) embedded cluster method.[79] In this method, the region of interest is described by quantum chemical techniques and the part that is deemed less important is described by a classical potential. This approach was employed by Sinclair *et al.*, for example, in hydrocarbon chemisorption in chabazite.[79] They showed that the chemisorption of ethene, propene and isobutene are sensitive to the ability of these hydrocarbons in displaying a carbenium-ion-like energy ordering. While most of these studies were carried out with gas phase cluster models, there were also interesting developments for pursuing quantum chemical calculations using solvation models, such as the conductor-like screening model (COSMO) and polarisable continuum model (PCM).[80–82] Mora–Fonz *et al.*, for example, employed the COSMO approach in addition to explicitly considering water molecules in order to calculate free energies as well as energies of reaction, which were found to be in good agreement with experimental data on free energies of de-protonation and condensation of the $Si(OH)_4$ clusters.[83–85]

During this period, there were also developments in methods involving plane wave based DFT codes for extended systems and their application in understanding zeolite chemistry. In 1996, Segall *et al.*, for example, used plane wave pseudopotential calculations to demonstrate that these techniques could be used for large systems such as zeolites to evaluate local atomic properties such as proton transfer from the Brønsted acid site of zeolite to the methanol.[86] Other interesting works, such as the nature of Brønsted acid site and adsorption of methanol in zeolite catalyst using plane wave codes were also undertaken.[60,87,88] In the work related to methanol adsorption, a number of possible geometries for adsorption of methanol were considered. It was put forward that that the nature of adsorbed species can depend on the particular zeolite structure, i.e. in chabazite the stable structure of methanol was found to be proto-nated while in a sodalite structure, methanol was simply found to be physisorbed.

In 2004, Tuma and Sauer reported a unique QM/QM embedding scheme to introduce local corrections at post-Hartree–Fock level to DFT calculations.[89] In this scheme, they used DFT in a plane wave approach as "low"-level method and added local corrections at post-Hartree–Fock level using second-order Møller-Plesset perturbation theory (MP2). They used this scheme to understand the proton jump reaction in the zeolite HSSZ-13 (see Figure 3). They concluded that the energy barriers and rate constants are significantly changed by MP2/DFT corrections, i.e. the electronic energy barriers increase from 68 kJ/mol to 81 kJ/mol for dry zeolite and from 22 kJ/mol to 30 kJ/mol for hydrated zeolite. Since the inception of this method has been employed in recent years to investigate various other sys-tems such as protonation of isobutene in zeolite ferrierite, zeolite catalysed methylation reactions and benzene ethylation over H-ZSM-5,[90–92] this approach was shown to provide more accurate results with van der Waals contributions for small molecules binding to surfaces and for biomolecules.

Recently, O'Malley *et al.* combined DFT with inelastic neutron scatter-ing (INS) and quasielastic neutron scattering (QENS) to report the com-plete conversion of methanol to framework methoxy species in a commercial H-ZSM-5 sample at room temperature.[93] DFT calculations were used to calculate the vibrational spectra of the empty protonated zeolite (Figure 4(A)) and the framework bound methoxy species (Figure 4(C)). The calculated spectra were then weighted for the neutron scattering cross-section to generate theoretical INS spectra. Figure 4(D) shows the experimental and theoretical INS spectra. There is very close agreement for both systems with calculated absorption bands of the OH stretch and deformation in the empty HY system calculated at 3624 cm^{-1} and 1117 cm^{-1}, respectively. The calculated methoxylated HY spectrum also gives close agreement with experiment, with the CH$_3$ stretch at 3055 cm^{-1} and the CH bend and CH$_3$ rocking mode observed at 1,428 and 1,126 cm^{-1}, respectively. This agreement, coupled with the immobilisation in the QENS measurements, confirms the room temperature conversion to methoxy in the H-ZSM-5 framework in contrast to the inactivity observed in zeolite HY.

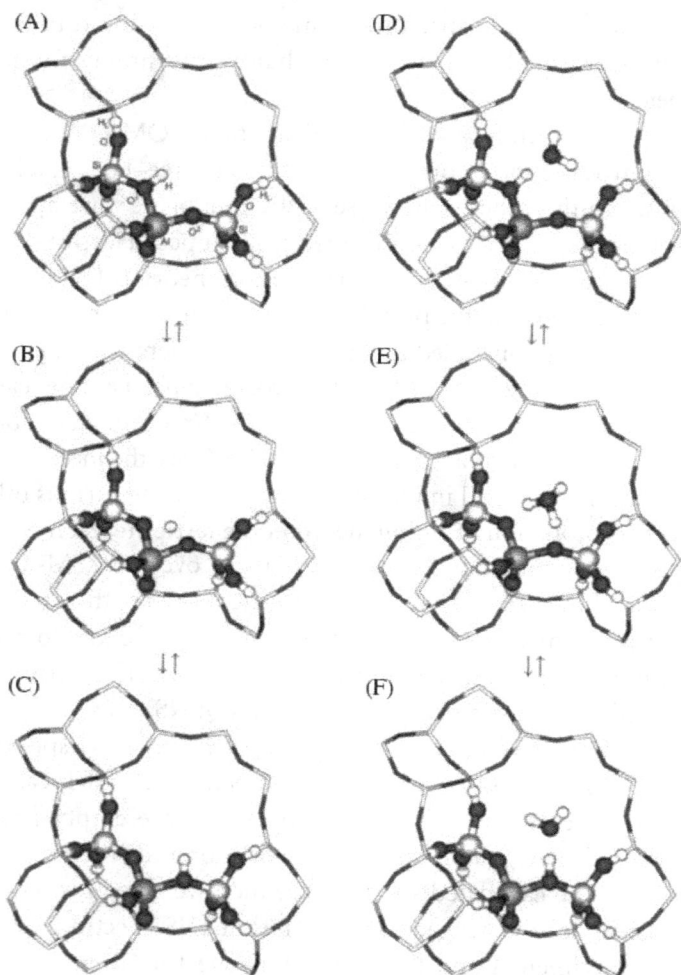

Figure 3: (A–F) Proton jump reaction between oxygen sites O1 and O2 in the zeolite HSSZ-13 without (A–C) and in the presence of a water molecule (D–F). Structures for intial and final states (A, C, D, F) and transition state structures (B, E) are shown, the ball-and-stick parts represent the 3T cluster model for the MP2/DFT mechanical embedding calculations. Reproduced from Ref. [89] with permission from Elsevier.

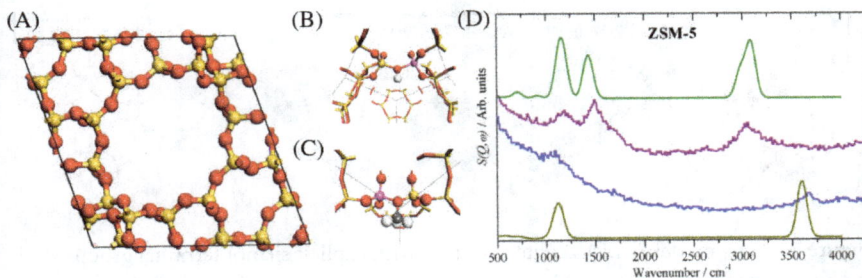

Figure 4: Active site configurations of the DFT calculation used of the protonated zeolite (A) and the methoxylated framework (B) in the primitive unit cell of HY (C) and (D) experimental spectra of the methoxylated ZSM-5 system (--), and the empty zeolite system (--) combined with the calculated spectra of the dehydrated zeolite (--) and the surface methoxylated zeolite (--). The very close agreement between the calculated experimental spectra support the case for framework methoxylation. Reproduced from Ref. [93] with permission from Royal Society of Chemistry.

5.2 Applications of MD

One of the reasons why MD is a logical complement to *ab initio* simulations is that MD simulations can be used to investigate the diffusion behaviour of reacting species through the catalyst. MD simulations of surface adsorption of methane in a model Si-LTA zeolite structure with explicit silanol terminal groups at 300 K (Figure 5) have shown the existence of preferential zones of interaction around the partially exposed oxygen atoms of the surface silanol groups (Figure 6).[94] Newsome and Coppens recently performed MD simulations to study the effect of the Si/Al ratio and Na^+ cations that balance the negative charge on the framework Al.[95] These cations interact with quadrupolar guest molecules, such as CO_2 and N_2. Again, preferential zones were observed, where the guest molecules spend most of their time (Figure 7). It was therefore hypothesised that such preferential zones would enable one to perform coarse-grained simulations, which would allow one to simulate larger systems and longer time scales. It was also found that the Na^+ cations have a larger effect on CO_2 diffusivity than on N_2 diffusivity. Both self-diffusion (relevant to catalysis) and corrected diffusion (relevant to transport) were studied.

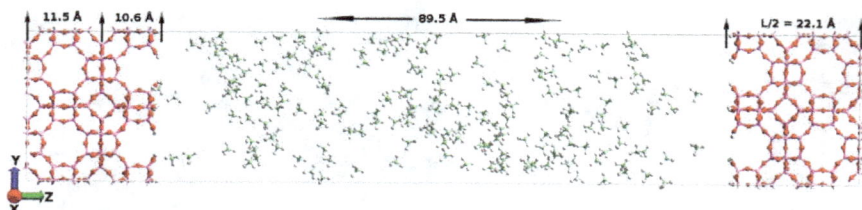

Figure 5: The model Si-LTA zeolite structure with explicit silanol terminal groups studied by Combariza and Sastre.[94] Methane gas is shown as green and gray balls in the middle part. Reprinted with permission from Ref. [94]. Copyright (2015) American Chemical Society.

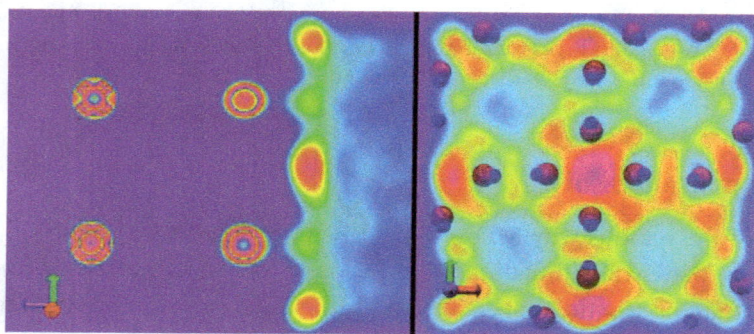

Figure 6: Density map of methane in the system studied by Combariza and Sastre.[94] The left hand panel shows the density variation perpendicular to the zeolite surface, while the right hand panel looks top-down onto the surface plane. Methane concentration regions are distinguishable as red and pink regions. The silanol concentrations are shown as red (O) and blue (H) spheres in the right hand panel. Reprinted with permission from Ref. [94]. Copyright (2015) American Chemical Society.

MD simulations of the diffusion of hydrocarbons such as CH_4, C_2H_6, C_3H_8 and n-C_4H_{10} through MFI-type silicalite crystals have been performed for more than two decades.[96–100] In these simulations, the host membrane is usually held rigid by holding the crystal atoms fixed in space, while the guest molecules are allowed to move without restriction. Comparisons of simulation results with membrane permeation experiments showed that the permeabilities through a defect-less perfect membrane obtained using

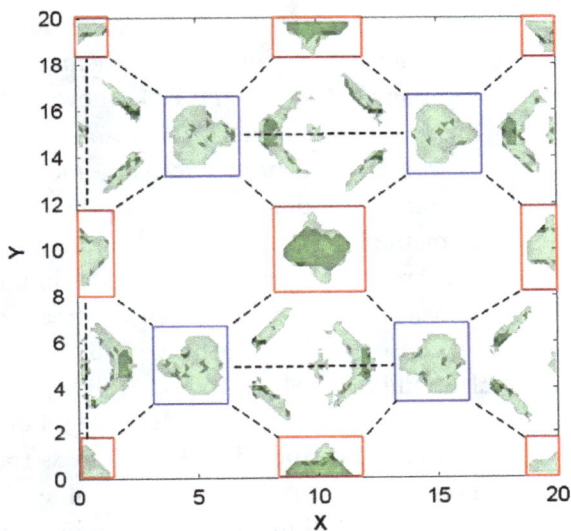

Figure 7: The green zones show the stable basins identified from the MD simulation of 4 CO_2/Unit Cell of silicalite-1 at 200 K, using threshold density 0.01 $Å^{-3}$, as observed by Newsome and Coppens.[95] Coordinates are in Å. Red boxes indicate straight channel basins; blue boxes indicate sinosoidal channel basins. Reprinted from Ref. [95]. Copyright (2015), with permission from Elsevier.

MD simulations were an order of magnitude larger than those reported in experiments.[101] Several possible reasons for this large inconsistency have been suggested, including different methods used in crystal synthesis and sample generation during experiments, non-applicability of microscopic simulations to explain macroscopic experimental results and possible grain boundary effects in real catalysts.[102–107]

It has long been argued that pore entrances to zeolite crystallites exhibit surface resistance that leads to discontinuity in the pore network of the catalyst, so that for crystals smaller than 1 μm, this can exceed intracrystalline resistance by several orders of magnitude.[108,109] The effect of surface resistance on permeation through a single crystal membrane of silicalite was investigated by Ahunbay *et al.* using the Dual Control Volume-Grand Canonical Molecular Dynamics (DCV-GCMD) method.[110] In essence, this method involves setting up three regions within the simulation cell: two control volumes positioned outside and in the middle of the membrane, plus a transport region between them. Results for three spherical gas molecules, namely CH_4, Ar and CH_4, showed that for larger molecules, not only the magnitude of surface resistance is larger, but also the resistance extends to a larger distance from the membrane surface. Again, a rigid crystal frame was assumed in these simulations.

The DCV-GCMD method was subsequently used by Newsome and Sholl to show that earlier results obtained using this technique may have been influenced by non-isothermal effects, such as streaming velocities and collisional energy transfer.[111] Subsequent work by the same authors incorporated a Local Equilibrium Flux Method (LEFM), that aimed to correct for the shortcomings of the DCV-GCMD method even while the crystal lattice was held rigid, and studied interfacial mass transfer resistance due to internal grain boundaries in twinned silicalite.[112] They calculated that the resistance to CH_4 diffusion due to grain boundaries was significantly larger than the intracrystalline resistance.

As alluded to earlier, another promising use of zeolites is in separation technology. Separating hydrogen from other gases is a particularly attractive prospect in energy related applications. For such technology, surface resistance may not necessarily be a bad thing and may be used cleverly to design a separation matrix. An ideal zeolite structure is not well suited for hydrogen separation. It has been shown experimentally that highly selective separation of hydrogen may be possible if the zeolite membrane's pore mouths are chemically modified.[113–115] At the same time, such a modification must not adversely impact the hydrogen flux. This problem was investigated by Jee *et al.* by performing MD simulations of CH_4 and H_2 diffusion through zeolite membranes whose pore mouths had been modified in such a way as to mimic vapour deposition of Si and O atoms near the surface of the crystal.[116] Again, the DCV-GCMD method was used along with the LEFM correction, while keeping the zeolite crystal rigid. It was shown that under careful modification of the surface, it is possible to create large resistance to CH_4 diffusion while having a minimal impact on H_2 flux, hence separating the two gases.

Metal-organic frameworks (MOFs) are a promising set of materials for separation processes, catalysis and gas storage.[117–119] A subset of MOFs are the zeolitic imidazolate frameworks (ZIFs), in which transition metal (TM) ions such as Fe, Zn, Co and Cu are connected by organic imidazole linkers.[120,121] Wongsinlatam and Remsungnen recently performed MD simulations of CO_2 diffusion in ZIFs by using force field parameters based on *ab initio* binding energies of $CO_2-[C_7H_5N_2]$ complexes.[122] Both rigid-lattice and flexible-lattice simulations were performed for comparison. CO_2 molecules were found to be located mostly around the hydrogen atom of imidazolate rings.

In recent years, the role of the host framework in simulations, namely rigid lattice vs. flexible lattice, has been scrutinised. Zimmermann *et al.* performed MD simulations of methane diffusion in 1D pores of AFI-, LTL- and MTW-type zeolites using both these types of frameworks.[123] The authors argued that although a flexible lattice would mimic the physical reality more accurately, that choice is not guaranteed to give more accurate diffusivities in MD simulations; in fact, it may give misleading results if the force field is not chosen carefully. However, more recently O'Malley and Catlow have studied diffusion of the longer *n*-alkanes (n-C_8 to n-C_{20}) in silicalite and compared the results of rigid and flexible frameworks (Figure 8).[124,125] They have shown that the "pore breathing" intrinsic to a flexible lattice assists diffusion and that the diffusivities in zeolites with a flexible lattice are significantly closer to experimental values than with a rigid lattice.

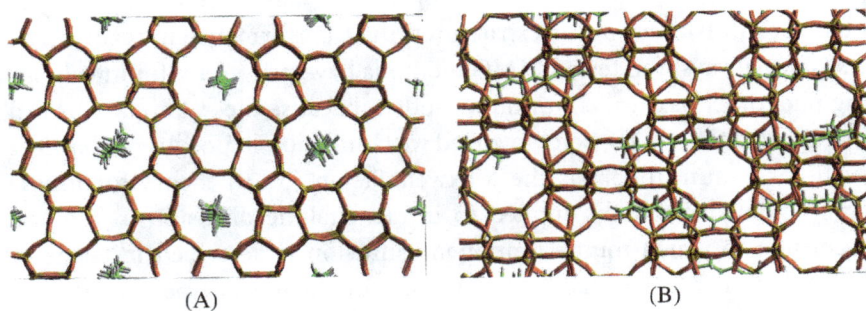

(A)   (B)

Figure 8: A view of the silicalite system studied by O'Malley and Catlow viewed down the (A) [010] direction and (B) [001] direction.[125] The chain-like octane molecules are shown in green and gray. Reproduced from Ref. [125] with permission of the PCCP Owner Societies.

5.3 Applications of kMC simulations

While MD simulations are typically performed on a single crystallite and possibly incorporate the surroundings in their immediate vicinity, kinetic or dynamic MC simulations can span much larger times and spatial dimensions due to their ability to skip through times when no events of interest take place. This makes it possible, for instance, to simulate an entire pore network for up to microseconds and even milliseconds with the current generation of computational resources, thus covering macroscopic distances of relevance to real applications. The construction of the pore network may involve either resorting to some phenomenological model or a more detailed model, depending on the information available from textural analysis.

Scanning electron micrographs of porous amorphous SiO_2, Al_2O_3 and TiO_2 used as catalyst supports frequently show pellets of agglomerates.[126] Zalc *et al.* constructed a model of such structures by packing spheres with different distributions of microsphere radii.[126] Unimodal, Gaussian and bimodal distributions were realised, while variation in pellet porosity was achieved by varying the microsphere radii and by random removal of spheres from highly densified packings. Gas and surface diffusion within such structures was studied by placing tracer molecules within the voids and allowing them to move randomly within the pore space while avoiding the regions occupied by the spheres.[127]

Although several theoretical approaches have been suggested to quantify diffusion in porous structures, their performance in real materials is not always well tested. KMC methods have proved useful in validating phenomenological continuum models, because these simulations can incorporate the stochasticity involved with diffusion in porous media. One such continuum theory is the Maxwell–Stefan (M–S) theory for multi-component diffusion.[129] applied to intracrystalline diffusion in zeolites, according to which multi-component diffusion is described in terms of two types of interactions: the interaction of each guest species with the host matrix and the intermolecular interaction within the gases. It has been used with a fair degree of success to predict multi-component diffusion in silicalite-1.[130,131] However, KMC simulations of binary self-diffusion on heterogeneous cubic lattices and in ZSM-5 (MFI containing Al) by Coppens *et al.* found that the M–S approach is inaccurate when correlation effects induced by adsorption of species on strong adsorption sites are large.[132–135] By remaining for relatively long time on a strong adsorption site, a molecule temporarily blocks access to other molecules, thus inducing negative correlation effects in the motion of these molecules. Instead, a theoretical method, based on the effective medium approximation (EMA) and percolation theory was proposed, which accounts for effects due to strong adsorption and lattice heterogeneities. Predictions of this model are in better agreement with kMC simulations, because they explicitly account for the pore network structure, which the Stefan–Maxwell approach does not.

KMC simulations have also been used to illuminate the role of surface barriers. Heinke and Kärger constructed a discrete MC system to represent Zn(tbip), which is a member of the metal organic framework (MOF) family, as shown in Figure 9.[128] The model MOF consists of 1D channels, in which a guest molecule diffuses by performing random jumps between adjacent segments in the x direction. Guest molecules could enter the crystal through only a small number of open mouths; the rest of the pores is blocked at the external surface. Once inside the crystal, a guest molecule diffuses by attempting jumps randomly in all six directions. Not all

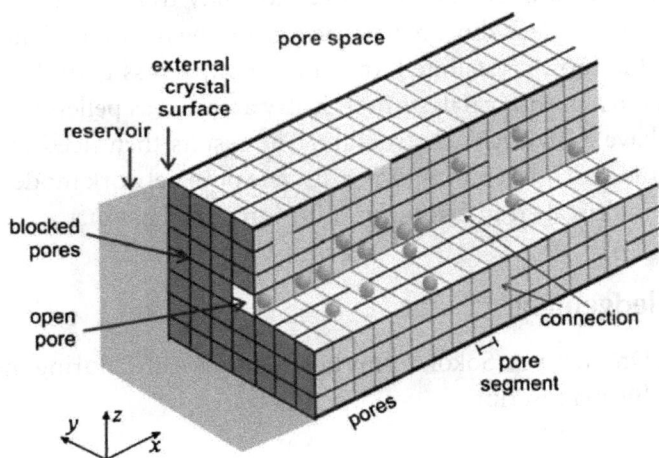

Figure 9: The discretised model MOF system simulated by Heinke and Kärger.[128] Reprinted figure with permission from Ref. [128]. Copyright (2015) by the American Physical Society.

segments inside the crystal were connected and a jump was considered successful only if it was attempted in a direction where there was an adjacent connecting segment. It was found that mass transfer of the guest molecules in this manner was in agreement with experimental observations for short-chain alkanes in Zn(tbip) and could be well described by an effective medium approach.[136]

6 Conclusions

Fast computing platforms are allowing for increasingly detailed, quantitative insight in catalysis, from atomistic scales, thanks to quantum chemical methods, to the mesoscale of porous catalyst grains, thanks to statistical mechanical approaches, such as MD and kMC. The approaches were illustrated for zeolites in particular, where molecular confinement effects are strong and continuum approaches developed for bulk fluids, such as the Stefan–Maxwell equations for multi-component diffusion, may cease to be applicable when the medium is heterogeneous and pore network connectivity is low. Statistical methods are particularly useful to probe heterogeneity, such as intra-crystalline defects, a distribution of different types of adsorption sites and the effects of partially blocking species, such as the counter-ions to Al in zeolites, as well as to study transport across zeolite grain boundaries. This short review focused on the micro- and meso-scale, mostly at the intra-crystalline level and, briefly, across grain boundaries. However, most porous catalysts in industry are used as pellets or composites, and have a hierarchical structure. The results then need to be integrated in macroscopic continuum or discrete pore network models, as part of a truly multiscale framework for diffusion and reaction.

Acknowledgement

We thank Dr. Alexey A. Sokol for his helpful discussions during the preparation of this manuscript.

References

1. C. N. Satterfield, *Mass Transfer in Heterogeneous Catalysis*, MIT Press, Cambridge, MA, 1970.

2. N. Wakao and J. M. Smith, *Chem. Eng. Sci.*, 1962, **17**, 825–834.
3. H. H. Lee, *Heterogeneous Reactor Design*, Butterworth, Boston, 1985.
4. S. Raimondeau and D. G. Vlachos, *Chem. Eng. J.*, 2002, **90**, 3–23.
5. V. Van Speybroeck, K. Hemelsoet, L. Joos, M. Waroquier, R. G. Bell and C. R. A. Catlow, *Chem. Soc. Rev.*, 2015, **44**, 7044–7111.
6. B. Smit and T. L. M. Maesen, *Chem. Rev.*, 2008, **108**, 4125–4184.
7. F. J. Keil, *Top. Curr. Chem.*, 2012, **307**, 69–107.
8. M. Salciccioli, M. Stamatakis, S. Caratzoulas and D. G. Vlachos, *Chem. Eng. Sci.*, 2011, **66**, 4319–4355.
9. M. Stamatakis and D. G. Vlachos, *ACS Catal.*, 2012, **2**, 2648-2663.
10. W. K. P. Hohenberg, *Phys. Rev. B*, 1964, **136**, B 864–B 871.
11. W. Kohn and L. J. Sham, *Phys. Rev.*, 1965, **140**, 1133–1138.
12. A. D. Becke, *J. Chem. Phys.*, 2014, **140**, 18A301-1–18A301-18.
13. I. N. Levine, *Quantum Chemistry*, Prentice Hall, Upper Saddle River, New Jersey, 5th edn., 2000.
14. S. Kurth and J. Perdew, *Int. J. Quantum Chem.*, 2000, **77**, 814–818.
15. J. P. Perdew and K. Schmidt, *AIP Conf. Proc.*, 2001, **577**, 1–20.
16. D. M. Ceperley and B. J. Alder, *Phys. Rev. Lett.*, 1980, 45, 566–569.
17. S. H. Vosko, L. Wilk and M. Nusair, *Can. J. Phys.*, 1980, **58**, 1200–1211.
18. J. P. Perdew and Y. Wang, *Phys. Rev. B*, 1992, **45**, 13244–13249.
19. A. J. Cohen, P. Mori-Sanchez and W. Yang, 2012, **112**, 289–320.
20. A. D. Becke, *Phys. Rev. A*, 1988, **38**, 3098–3100.
21. J. Perdew, J. Chevary, S. Vosko, K. Jackson, M. Pederson, D. Singh and C. Fiolhais, *Phys. Rev. B*, 1993, **48**, 4978–4978.
22. J. P. Perdew, K. Burke, M. Ernzerhof, D. of Physics and N. O. L. 70118 J. Quantum Theory Group Tulane University, *Phys. Rev. Lett.*, 1996, **77**, 3865–3868.
23. B. Hammer, L. Hansen and J. Nørskov, *Phys. Rev. B*, 1999, **59**, 7413–7421.
24. (a) J. P. Perdew, A. Ruzsinszky, G. I. Csonka, O. a. Vydrov, G. E. Scuseria, L. a. Constantin, X. Zhou and K. Burke, *Phys. Rev. Lett.*, 2008, **100**, 136406–1–136406–4. (b) A. I. Liechtenstein, V. I. Anisimov, J, Zaanen, Phys. Rev. B, 1995, 52, 5467-5471, (c) S. L. Dudarev, G. A. Botton, S. Y. Savrasov, C. J. Humphreys and A. P. Sutton, *Phys. Rev. B* 1998, **52**, 1505–1509 and references there in.
25. C. Lee, W. Yang and R. G. Parr, *Phys. Rev. B*, 1988, **37**, 785–789.
26. A. Becke, *J. Chem. Phys.*, 1993, **98**, 5648–5652.
27. J. Paier, M. Marsman and G. Kresse, *J. Chem. Phys.*, 2007, **127**, 024103-1-024103-10.
28. J. Tao, J. P. Perdew, V. N. Staroverov and G. E. Scuseria, *Phys. Rev. Lett.*, 2003, **91**, 146401.

29. J. P. Perdew, S. Kurth and P. Blaha, 1999, **82**, 2544–2547.
30. A. D. Becke, *J. Chem. Phys.*, 1993, **98**, 1372–1377.
31. Y. Zhao, B. J. Lynch and D. G. Truhlar, *J. Phys. Chem. A*, 2004, **108**, 4786–4791.
32. S. Grimme, *J. Chem. Phys.*, 2006, **124**, 034108-1-034108-16.
33. Y. Zhang, X. Xu and W. a Goddard, *Proc. Natl. Acad. Sci. U. S. A.*, 2009, **106**, 4963–4968.
34. T. Leininger, H. Stoll, H.-J. Werner and A. Savin, *Chem. Phys. Lett.*, 1997, **275**, 151–160.
35. J. Toulouse, F. Colonna and A. Savin, *Phys Rev. A.* 2004, **70**, 062505-1–062505-18 (something amiss?).
36. T. Yanai, D. P. Tew and N. C. Handy, *Chem. Phys. Lett.*, 2004, **393**, 51–57.
37. O. A. Vydrov, J. Heyd, A. V. Krukau and G. E. Scuseria, *J. Chem. Phys.*, 2006, **125**, 074106.
38. O. A. Vydrov, G. E. Scuseria and J. P. Perdew, *J. Chem. Phys.*, 2007, **126**, 154109.
39. O. A. Vydrov and G. E. Scuseria, *J. Chem. Phys.*, 2006, **125**, 234109.
40. J. Heyd and G. E. Scuseria, *J. Chem. Phys.*, 2004, **121**, 1187–1192.
41. M. Ernzerhof and G. E. Scuseria, *J. Chem. Phys.*, 1999, **110**, 5029–5036.
42. C. Adamo and V. Barone, *J. Chem. Phys.*, 1999, **110**, 6158–6170.
43. J. E. Lennard-Jones, *Proc. R. Soc. Lond. A*, 1924, **106**, 463–477.
44. R. A. Buckingham, *Proc. R. Soc. Lond. A. Math. Phys. Sci.*, 1938, **168**, 264–283.
45. P. M. Morse, *Phys. Rev.*, 1929, **34**, 57–64.
46. M. S. Daw and M. I. Baskes, *Phys. Rev. B*, 1984, **29**, 6443–6453.
47. S. M. Foiles, M. I. Baskes and M. S. Daw, *Phys. Rev. B*, 1986, **33**, 7983–7991.
48. M. W. Finnis and J. E. Sinclair, *Philos. Mag. A*, 1984, **50**, 45–55.
49. J. Tersoff, *Phys. Rev. B*, 1989, **39**, 5566–5568.
50. C. J. Cramer, *Essentials of Computational Chemistry*, John Wiley & Sons, 2nd edn., 2004.
51. L. Verlet, *Phys. Rev.*, 1967, **159**, 98–103.
52. P. Ewald, *Ann. Phys.*, 1921, **369**, 253–287.
53. M. P. Allen and D. J. Tildesley, *Computer Simulation of Liquids*, Clarendon Press, Oxford, 1989.
54. U. Essmann, L. Perera, M. L. Berkowitz, T. Darden, H. Lee and L. G. Pedersen, *J. Chem. Phys.*, 1995, **103**, 8577–8593.
55. D. T. Gillespie, *J. Comput. Phys.*, 1976, **22**, 403–434.
56. D. T. Gillespie, *J. Phys. Chem.*, 1977, **81**, 2340–2361.
57. A. F. Voter, *Phys. Rev. B*, 1986, **34**, 6819–6829.

58. P.-L. Cao, *Phys. Rev. Lett.*, 1994, **73**, 2595–2598.
59. E. J. Dawnkaski, D. Srivastava and B. J. Garrison, *J. Chem. Phys.*, 1995, **102**, 9401–9411.
60. R. Shah, J. D. Gale and M. C. Payne, *J. Phys. Chem.*, 1996, **100**, 11688–11697.
61. F. Tielens, M. Calatayud, S. Dzwigaj and M. Che, *Microporous Mesoporous Mater.*, 2009, **119**, 137–143.
62. J. D. Gale, C. R. a. Catlow and J. R. Carruthers, *Chem. Phys. Lett.*, 1993, **216**, 155–161.
63. A. Corma, *Chem. Rev.*, 1997, **97**, 2373–2419.
64. R. Vetrivel, C. R. A. Catlow and E. A. Colbourn, *J. Phys. Chem.*, 1989, **93**, 4594–4598.
65. J. Sauer, C. M. Kölmel, J.-R. Hill and R. Ahlrichs, *Chem. Phys. Lett.*, 1989, **164**, 193–198.
66. J. Sauer, H. Horn, M. Häser and R. Ahlrichs, *Chem. Phys. Lett.*, 1990, **173**, 26–32.
67. G. J. Kramer, R. A. van Santen, C. A. Emeis and A. K. Nowak, *Nature*, 1993, **363**, 529–531.
68. M. Anderson and J. Klinowski, *J. Am. Chem. Soc.*, 1990, **112**, 10–16.
69. G. J. Hutchings, D. F. Lee and M. Lynch, *Appl. Catal. A Gen.*, 1993, **106**, 115–123.
70. B. Sulikowski and J. Klinowski, *Appl. Catal. A Gen.*, 1992, **89**, 69–75.
71. E. J. Munson, A. A. Kheir, N. D. Lazo, M. E. Moellenhoff and F. J. Haw, *J. Am. Chem. Soc.*, 1991, **113**, 2783–2784.
72. F. Haase and J. Sauer, *J. Am. Chem. Soc.*, 1995, **117**, 3780–3789.
73. S. R. Blaszkowski and R. A. van Santen, *J. Phys. Chem.*, 1995, **99**, 11728–11738.
74. J. P. Perdew, Phys. Rev B, 1986, **33**, 8822–8824.
75. L. Fan and T. Ziegler, *J. Am. Chem. Soc.*, 1992, **114**, 10890–10897.
76. P. E. Sinclair and C. R. A. Catlow, *J. Phys. Chem. B*, 1997, **101**, 295–298.
77. S. R. Blaszkowski and R. A. van Santen, *J. Am. Chem. Soc.*, 1997, **119**, 5020–5027.
78. M. V Frash, V. B. Kazansky, A. M. Rigby and R. A. van Santen, *J. Phys. Chem. B*, 1998, **102**, 2232–2238.
79. P. E. Sinclair, A. de Vries, P. Sherwood, C. R. A. Catlow and R. A. van Santen, *J. Chem. Soc., Faraday Trans.*, 1998, **94**, 3401–3408.
80. A. Klamt, *J. Phys. Chem.*, 1995, **99**, 2224–2235.
81. M. Cossi, V. Barone, R. Cammi and J. Tomasi, *Chem. Phys. Lett.*, 1996, **255**, 327–335.
82. J. Tomasi, B. Mennucci and R. Cammi, *Chem. Rev.*, 2005, **105**, 2999–3093.

83. M. J. Mora-Fonz, C. R. A. Catlow and D. W. Lewis, *J. Phys. Chem. C*, 2007, **111**, 18155–18158.

84. M. J. Mora-Fonz, C. R. A. Catlow and D. W. Lewis, *Angew. Chemie - Int. Ed.*, 2005, **44**, 3082–3086.

85. M. J. Mora-Fonz, C. R. A. Catlow and D. W. Lewis, *Phys. Chem. Chem. Phys.*, 2008, **10**, 6571–6578.

86. M. D. Segall, *Mol. Phys.*, 1996, **89**, 571–577.

87. R. Shah, M. C. Payne, M.-H. Lee and J. D. Gale, *Science*, 1996, **271**, 1395–1397.

88. R. Shah, J. Gale and M. Payne, *Phase Transitions*, 1997, **61**, 67-81.

89. C. Tuma and J. Sauer, *Chem. Phys. Lett.*, 2004, **387**, 388–394.

90. C. Tuma and J. Sauer, *Phys. Chem. Chem. Phys.*, 2006, **8**, 3955–3965.

91. N. Hansen, T. Kerber, J. Sauer, A. T. Bell and F. J. Keil, *J. Am. Chem. Soc.*, 2010, **132**, 11525–11538.

92. J. Sauer, *Chem. Rev.*, 1989, **89**, 199–255.

93. A. J. O'Malley, S. Parker, A. Chutia, M. R. Farrow, I. P. Silverwood, V. Garcia Sakai and C. R. A. Catlow, *Chem. Commun.*, 2016, **18**, 17294–17302.

94. A. F. Combariza and G. Sastre, *J. Phys. Chem. C*, 2011, **115**, 13751–13758.

95. D. Newsome and M.-O. Coppens, *Chem. Eng. Sci.*, 2015, **121**, 300–312.

96. R. June, A. Bell and D. Theodorou, *J. Phys. Chem.*, 1990, **94**, 1508–1516.

97. R. L. June, A. T. Bell and D. N. Theodorou, *J. Phys. Chem.*, 1990, **94**, 8232–8240.

98. R. L. June, A. T. Bell and D. N. Theodorou, *J. Phys. Chem.*, 1992, **96**, 1051–1060.

99. E. J. Maginn, A. T. Bell and D. N. Theodorou, *J. Phys. Chem.*, 1993, **97**, 4173–4181.

100. R. Nagumo, H. Takaba and S. Nakao, *J. Phys. Chem. B*, 2003, **107**, 14422–14428.

101. D. Newsome and M.-O. Coppens, *To be Publ.*

102. J. Karger and D. M. Ruthven, *Zeolites*, 1989, **9**, 267–281.

103. D. M. Ruthven, *Stud. Surf. Sci. Catal.*, 1995, **97**, 223–234.

104. D. Newsome and D. S. Sholl, *J. Phys. Chem. B*, 2006, **110**, 22681–22689.

105. J. Kärger, D. M. Ruthven and D. N. Theodorou, *Diffusion in Nanoporous Materials*, Wiley-VCH, Weinheim, 2012.

106. N. E. R. Zimmermann, S. P. Balaji and F. J. Keil, *J. Phys. Chem. C*, 2012, **116**, 3677–3683.

107. A. R. Teixeira, C.-C. Chang, T. Coogan, R. Kendall, W. Fan and P. J. Dauenhauer, *J. Phys. Chem. C*, 2013, **117**, 25545–25555.

108. M. Kocirik, P. Struve, K. Fiedler and M. Bulow, *J. Chem. Soc., Faraday Trans. 1*, 1988, **88**, 3001–3013.

109. D. M. Ford and E. D. Glandt, *J. Membr. Sci.*, 1995, **107**, 47–57.

110. M. G. Ahunbay, J. R. Elliott Jr and O. Talu, *J. Phys. Chem. B*, 2004, **108**, 7801–7808.

111. D. A. Newsome and D. S. Sholl, *J. Phys. Chem. B*, 2005, **109**, 7237–7244.

112. D. A. Newsome and D. S. Sholl, *J. Phys. Chem. B*, 2006, **110**, 22681–22689.

113. M. Hong, J. L. Falconer and R. D. Noble, *Ind. Eng. Chem. Res.*, 2005, **44**, 4035–4041.

114. T. Masuda, N. Fukumoto, M. Kitamura and S. R. Mukai, *Microporous Mesoporous Mater.*, 2001, **48**, 239–245.

115. S. Gopalakrishnan, Y. Yoshino, M. Nomura, B. N. Nair and S. Nakao, *J. Membr. Sci.*, 2007, **297**, 5–9.

116. S. E. Jee, A. J. H. McGaughey and D. S. Sholl, *Mol. Simul.*, 2009, **35**, 70–78.

117. J. R. Li, Y. Ma, M. C. McCarthy, J. Sculleya, J. Yub, H.-K. Jeongb, P. B. Balbuenab and H.-C. Zhou, *Coord. Chem. Rev.*, 2011, **255**, 1791–1823.

118. D. Saha, Z. Bao, F. Jia and S. Deng, *Environ. Sci. Technol.*, 2010, **44**, 1820–1826.

119. A. G. Wong-Foy, A. J. Matzger and O. M. Yaghi, *J. Amer. Chem. Soc.*, 2006, **128**, 3494–3495.

120. K. S. Park, Z. Ni, A. P. Cote, J. Y. Choi, R. Huang, F. J. Uribe-Romo, H. K. Chae, M. O'Keeffe and O. M. Yaghi, *Proc. Natl. Acad. Sci.*, 2006, **103**, 10186–10191.

121. J. C. Tan, T. D. Bennett and A. K. Cheetham, *Proc. Natl. Acad. Sci.*, 2010, **107**, 9938–9943.

122. W. Wongsinlatam and T. Remsungnen, *J. Chem.*, 2013, **2013**, 415027.

123. N. E. R. Zimmermann, S. Jakobtorweihen, E. Beerdsen, B. Smit and F. J. Keil, *J. Phys. Chem. C*, 2007, **111**, 17370–17381.

124. A. J. O'Malley and C. R. A. Catlow, *Phys. Chem. Chem. Phys.*, 2013, **15**, 19024–19030.

125. A. J. O'Malley and C. R. A. Catlow, *Phys. Chem. Chem. Phys.*, 2015, **17**, 1943–1948.

126. J. M. Zalc, S. C. Reyes and E. Iglesia, *Chem. Eng. Sci.*, 2003, **58**, 4605–4617.

127. J. M. Zalc, S. C. Reyes and E. Iglesia, *Chem. Eng. Sci.*, 2004, **59**, 2947–2960.

128. L. Heinke and J. Kärger, *Phys. Rev. Lett.*, 2011, **106**, 74501.

129. R. B. Bird, W. E. Stewart and E. N. Lightfoot, *Transport Phenomena*, Wiley, Singapore, 1960.

130. R. Krishna and D. Paschek, *Phys. Chem. Chem. Phys.*, 2002, **4**, 1891–1898.

131. R. Krishna and R. Baur, *Sep. Purif. Technol.*, 2003, **33**, 213–254.
132. M.-O. Coppens, A. T. Bell and A. K. Chakraborty, *Chem. Eng. Sci.*, 1999, **54**, 3455–3463.
133. V. Iyengar and M.-O. Coppens, *Chem. Eng. Sci.*, 2004, **59**, 4747–4753.
134. M.-O. Coppens and V. Iyengar, *Nanotechnology*, 2005, **16**, S442–S448.
135. X. Liu, D. Newsome and M.-O. Coppens, *Microporous Mesoporous Mater.*, 2009, **125**, 149–159.
136. C. Tuck, *Effective Medium Theory*, Oxford University Press, Oxford, 1999.

Chapter 9

Life Cycle Assessment of Emerging Catalyst Technologies: The Case of Polymer Electrolyte Membrane Fuel Cells

Sara Evangelisti, Carla Tagliaferri, Dan Brett and Paola Lettieri

Department of Chemical Engineering University College London, Torrington Place, London

Climate change and depletion of fossil fuel resources urge the development of new technologies to be able to produce energy from alternative sources and with higher efficiency. Life cycle assessment (LCA) is one of the most developed and widely used environmental assessment tools for comparing alternative technologies on the basis of their resource consumptions and potentially harmful emissions. In this work, we tackle the methodological challenges of applying an LCA approach to the production of a new catalyst for a polymer electrolyte membrane (PEM) fuel cell as a case study.

Catalyst manufacture plays a key role in the development of new fuel cell technologies; however, there is the need to give evidence of the improved environmental impacts of these new technologies both at laboratory and industrial scale.

1 Introduction

The urgency of tackling climate change is pushing policy makers and industrial sectors to investigate new technologies for the reduction of emissions and fuel consumption, especially in the energy sector. In the UK, for example, the energy sector alone was responsible of more than 30% of the total greenhouse gases (GHGs) emitted to the atmosphere in 2013.[1]

European and national directives are driving this change, setting targets for reduction of carbon emissions in the coming years. At European level, the "Europe 2020" strategy sets the target to reduce GHG emissions to 20% below 1990 levels. By 2050, the EU commitment is to reduce the emissions to 80–95% below 1990 levels.[2] Due to the high impact of the energy sector, the emissions of this industry need to be cut by 40% and 87% by 2030 and 2050, respectively.[3] In the UK, the Climate Change Programming aims to cut carbon emissions by 80% from 1990 levels by 2050 through the 2008 Climate Change Act.[4]

Amongst the industrial sectors, the chemical industry plays an important role in the UK economy, being responsible of approximately 16% share of UK manufacturer sales, with a contribution to the UK's GDP of 1.4%.[5–6] As reported by Gilbert *et al.*[7] in their report on *The Chemical Industry in UK*, more than one-third of the chemical industry output is used as inputs in other manufacturing industries, bringing the upstream chemical industry and downstream chemical-using sectors to an overall GDP contribution of 21%.[8]

Figure 1 shows the overall downstream sectors which use chemical products. Amongst all, the production of improved and new chemical catalysts can play a central role in tackling emissions reduction. For example, it has been estimated that catalysts and related process improvements could reduce energy intensity for some chemical products by 20–40% as a whole by 2050.[9] The 2013 IEA roadmap for "energy and greenhouse gas emission reduction in the chemical industry via catalytic process" examined the potential of reducing the impact implementing catalytic cracking (as opposed to steam cracking) for large-volume chemical production, as an example of emerging catalyst-related technology in the chemical industry. By 2050, the energy saving potential of catalytic cracking would be 2.3 EJ and the corresponding GHG savings would be 143 $MtCO_2$-eq.[9]

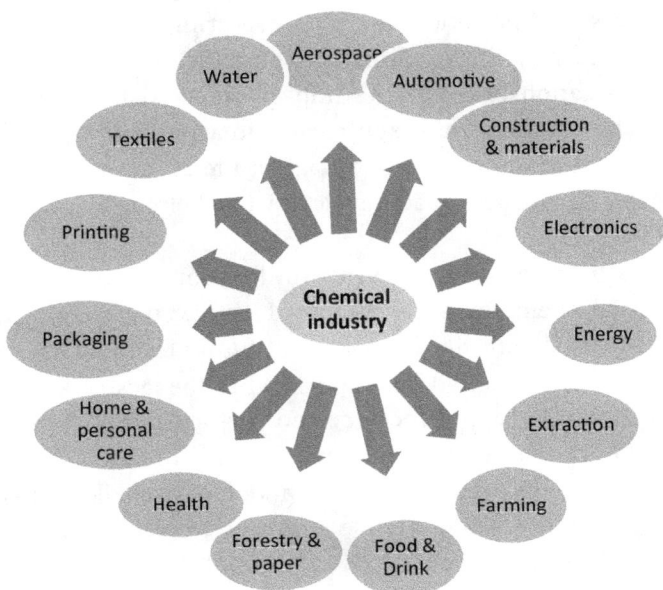

Figure 1: Direct relationships between chemical industry and downstream sectors (adapted from Ref. [7]).

Currently, around 90% of chemical processes involve the use of catalysts and related processes to enhance production efficiency and reduce energy use, thereby curtailing GHG emission levels. Freedonia Research[10] estimated that the global market for catalysts is expected to reach $20.6 billion in 2018, with a growth of 4.8% per year: about 75% of those are used in the chemical processing and 25% in the petroleum refining industry.

For effective policies on emerging technologies, the expected reductions on the environmental burdens need to be assessed from a wide system perspective.[11] This includes an investigation of the complete life cycle of the emerging technology, ensuring that all the environmental aspects are considered in the assessment.

Life Cycle Thinking is an approach for the evaluation of environmental burdens of products and services that aims at identifying single steps as well as the whole picture of an entire product or a service system. In regards to a product, it starts with raw material extraction and processing, then considers transportation and manufacturing, distribution, use and ends with re-use, recycling or ultimate disposal. The overall idea of making a holistic evaluation of a system's impact can be defined as Life Cycle Thinking. The key aim of this approach is to avoid burdens shifting that means "minimizing impacts at one stage of life cycle, or in a geographic region, or in a particular impact category, while helping to avoid increases elsewhere".[12]

LCA has been recognised as a useful tool for studying the environmental impacts of emerging technologies. However, methodological problems are associated with the assessment of future technologies which are not yet fully commercialised, mainly related to the lack of knowledge on the product system (i.e. its efficiency and performances), time and scale aspects, and uncertainties on the manufacturing process.[13–15]

This chapter aims to explore the methodological challenges associated with the environmental impact assessment of emerging technologies, in particular in the catalyst sector. A description of the LCA methodology is presented in Sec. 2, together with the problems associated with its application to future technologies. Section 3 presents different approaches which can be applied to solve the gap of knowledge and forecasting assessment. Finally, Sec. 4 shows an example of LCA of emerging technology, applied to the catalyst production for fuel cells.

2 Assessing Future Catalyst-related Technologies Through LCA

The LCA approach seeks to quantify all the physical exchanges with the environment, such as inputs of natural resources and energy, and outputs in the form of emissions to air, water and soil. This information is organised in a balance sheet, or life cycle inventory, for the system under study. After the inventory is completed, emissions and resources are related to various environmental problems, with different ways of classification and characterisation, based on indicators associated to different environmental burdens; for example, resources depletion, climate change, acidification, etc. Finally, the different environmental impacts related to the life cycle may be put on the same scale through weighting.[16]

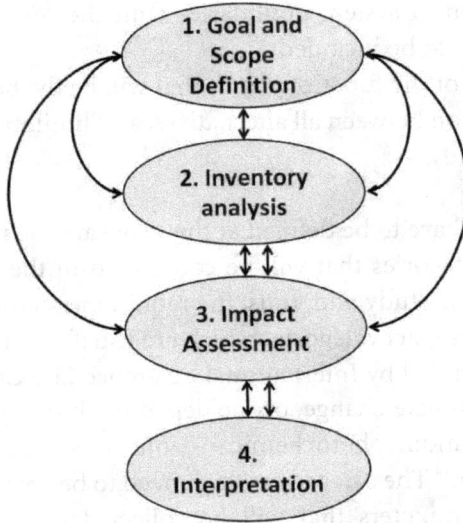

Figure 2: Phases of a LCA study (adapted from Ref. [17]).

2.1 LCA

LCA is an international standardised method, and the International Standard Organisation (ISO) provides a rigorous approach for improving decision support in environmental management in the norm ISO 14040.[17] The LCA methodology consists of four stages, as shown in Figure 2:

- Goal and scope definition;
- Inventory analysis;
- Impact assessment;
- Interpretation.

2.1.1 Goal and scope definition

The first stage presents the purpose of the study. This includes:

- Why the LCA is to be carried out and what decision makers are to be informed by the results;

- The description of system boundaries with the processes and operations which are to be included;
- The definition of the functional unit that will be the basis for comparison and common between all alternatives, and limitations and assumptions of the study.

Other things that have to be defined at this stage are the types of environmental impact categories that will be considered in the analysis and the level of detail in the study and, thus, the requirements on the data. There is a list of default impact categories which are usually considered in a LCA study as recommended by International Reference Life Cycle Data System (ILCD), such as climate change, ozone depletion, human toxicity, acidification, eutrophication, photochemical ozone creation, ecotoxicity and resource depletion.[18] The categories which need to be analysed in the study determine the parameters that will be collected during the inventory phase. The standard stresses that the goal and scope of a LCA study must be clearly defined and consistent with the intended application.

2.1.2 Inventory analysis

In this phase, a system model is built according to the requirements of the goal and scope definition. The system model is a flow model of a technical system, sometimes called techno-sphere, where the system boundaries are shown, and it represents the incomplete mass and energy balance of the overall system. It is "incomplete" because only the relevant environmental flows are considered. The system model is usually represented by a flow chart, where the processes/activities included in the analysis are detailed, as well as the system boundaries. After that, the data are collected for each single process. Finally, the amounts of resources and emissions of the system are calculated based on the functional unit. This allows completing the inventory of all relevant environmental interventions occurring along the process, thus, "every human intervention, physical, chemical or biological", as defined by Guineé *et al.*[19]

This phase is complicated when the process under analysis is a multi-functional process. In LCA, a multifunctional process is defined as an activity that fulfils more than one function, such as a combined heat and power plant generating electricity and heat at the same time.[20] It is then necessary to find a rational basis for allocating the environmental burdens between the processes. The problem of allocation in LCA has been the topic of much debate.[21-22] The ISO standards[17] recommend that the environmental benefits of recovered resources should be accounted for by broadening the system boundaries to include the avoided burdens of conventional production.[23] The same approach is recommended for product labelling provided that it can be proved that the recovered material or energy is actually put to the use claimed.[24] This approach, usually called system expansion, is also applied in this work.

Following the methodological approach of Clift *et al.*[21] for Integrated Waste Management (IWM), a pragmatic distinction is usually made between Foreground and Background, considering the former as "the set of processes whose selection or mode of operation is affected directly by decisions based on the study" and the latter as "all other processes which interact with the Foreground, usually by supplying or receiving material or energy". The burdens evaluated here are considered with three categories[21]: *direct burdens*, associated with the use phase of the process/service; *indirect burdens*, due to upstream and downstream processes (e.g. energy provision for electricity or diesel for transportation); and *avoided burdens* associated with products or services supplied by the process (e.g. energy or secondary material produced by the system).

2.1.3 Impact assessment

In this phase, the environmental impacts related to the emissions and resources collected in the inventory phase are described and evaluated in aggregated indicators. The first step of the Life Cycle Impact Assessment (LCIA) phase is classification, where the environmental interventions listed during the inventory phase are qualitatively classified according to the type of environmental impact category they contribute to. Then the environmental interventions are quantified based on a common unit specific for each category, allowing aggregation to a single score (the category indicator). This step is called characterisation. Such calculations are based on scientific models of cause — effect chains in the natural systems using characterisation factors defined while modelling the cause-effect chain.

The definition of characterisation factors is based on the physical–chemical mechanism of how a different substance can contribute to different impact categories, based on a specific characterisation model. The category indicator, the characterisation model and the characterisation factors derived from the model constitute the characterisation method.[19]

An important part of an impact assessment phase is the choice of the characterisation method applied to describe and quantify the environmental burdens. These methods are sometimes called impact assessment methods. They are based on scientific methods coming from chemistry, toxicology, ecology, etc. The general categories of environmental impacts under consideration in a LCA study are resource use, human health and ecological consequences.[16] They may be defined close to the intervention (the midpoint or problem-oriented approach) or, alternatively, they may be defined at the level of category endpoints (the endpoint or damage approach). For most impact categories in LCIA, the category indicator describes events in the early stage of the cause–effect chain (mid-point characterisation method). This means that the potential rather than the actual effects of the pollutant are described. In this way, the environmental problems of a specific geographical location are avoided and, as a matter of fact, global impacts are better dealt with than local impacts in LCAs.

The results of the characterisation phase can be normalised based on selected reference values, in a way that the impact indicators can be compared amongst different categories. Sometime the results need a further aggregation and interpretation. This can be done in the weighting step, with formalised quantitative weighting procedures. While classification and characterisation are compulsory steps in a LCA study,[17] normalisation and weighting are optional.

2.1.4 Interpretation

Interpretation phase is defined by the standard as the "phase of life cycle assessment in which the findings of either the inventory analysis or the impact assessment, or both, are combined consistent with the defined goal and scope in order to reach conclusions and recommendations".[17] In this phase, the results obtained in the previous phases are assessed and conclusions are drawn, based on the initial goal and scope of the analysis. An important step of this phase is to test the robustness of the model through sensitivity analysis, uncertainty analysis, and data quality assessments.

2.1.5 Consequential vs. attributional LCA

The LCA community agrees on recognising two main types of LCA: attributional (or accounting) and consequential (or change-oriented). Attributional LCA focuses on the environmentally relevant physical flows to and from a life cycle system or its subsystems. This type of LCA answers questions such as "which environmental impacts can be associated with this product/service" Consequential LCA, on the other hand, focuses on how environmentally relevant flows will change in response to possible decisions.[25] It answers such questions as "What would happen if...?" Some authors argue that consequential LCA should be applied to decision making, when uncertainties in decision outweigh the insights gained within the attributional LCA analysis.[26] Other authors prefer an attributional study because of its relevance in understanding the product chain and the weak points in a wider picture.[27] Attributional and consequential LCA can both be applied for modelling of the future, past or current systems. Some authors have shown the difference in results when applying attributional and consequential analysis on the same product.[28] The choice of attributional or consequential LCA is reflected in different methodological choices of the analysis.[16] In regards to data type, average data represents the average environmental burdens for producing a unit of product/service in the system. Marginal data represents the effects of a small change in the output of goods and/or services from a system on the environmental burdens of the same system.

The decision to use marginal data can be significant when electricity production technologies are being modelled. This is because the technologies used in this field are under development due to the awareness on climate change and fossil fuel scarcity. This is related with temporal effects as well. Short term effects are changes in the utilisation of the exiting production capacity in present production plants. Long-term effects involve changes in the production capacity and/or technologies.[27]

As shown by Earles and Halog,[29] consequential LCA is recommended as a tool to use for the assessment of future technologies, which results can affect policies and strategic environmental planning.

2.1.6 Challenges in LCA of emerging technologies

When assessing emerging technologies, practitioners are often dealing with processes which are not fully developed for mass production, with limited availability of data. This means that the assessments are mainly based on little information, such as research publications, technology press releases, prototypes and patent records.[30] Therefore, it is essential to identify the issues related with the assessment of future technologies and try to tackle them in order to make the results obtained valid for a higher level of technology maturity, or for large scale production.

Pehnt,[13] identifies four main methodological issues associated with LCA of future technologies:

- *System expansion* related problem. In order to account for the environmental benefits of the recovered resources (i.e. electricity production in a fuel cell system), the environmental impact of the substituted systems need to be evaluated. This can be increasingly difficult in an emerging technology assessment, due to the long-term perspective of the study. For example, to account for the avoided burdens associated with the use of future fuel cell vehicles (FCV) compared with internal combustion engine vehicles (ICEV), it is necessary to evaluate the gain of diesel engine efficiencies in the future, and the increase of biogenic share in the diesel.
- Increasing difficulties associated with the *relationships between the background and the foreground systems*. In the assessment of future technologies (foreground system) which are not yet presently developed, new fuel supply chains (background system), which can affect different sectors, need to be considered.
- Need to consider the *infrastructures*. When assessing the environmental impact of systems which are used in everyday life (such as energy systems), the environmental burden associated with the manufacturing phase is usually negligible. This cannot be ignored in emerging

technology studies. In particular, the development of renewable energy systems will require new fuel supply systems and new infrastructures for the production of the new technology. For example, the use of FCV in the future will require the development of the hydrogen supply chain at industrial scale; moreover, the production of higher efficiency fuel cell systems will ask for production of higher amount of precious metals, which need to be accounted for.

- *Data gaps.* Pehnt[13] identified different causes for data gaps. Those referred to emerging technologies are summarised in Table 1. If data gaps are caused by confidentiality or complexity of the system, possible solutions can be taken in the first steps of the LCA in order to overcome

Table 1: Types of data gaps in the assessment of emerging technologies.

Type of data gap	Description	Application	Possible solution
Confidentiality of data	Data is known by the manufacturers, but not fully available due to confidentiality	Innovative or competitive products, e.g. fuel cell system	Agreements of confidentiality with the developer, aggregation of data in the publications
Complexity of data	Complex processes for which data acquisition is too intricate and time consuming	Integrated industrial processes e.g. platinum supply chain	Simplified assumptions and forecasting methods
Products at an early stage of market development	Processes or technologies not yet fully developed at industrial scale, or only tested in lab	New products e.g. new catalyst for fuel cell	See Sec. 3
Context of product unknown	Background system difficult to forecast	Evolving markets e.g. biodiesel supply chain; smart grids	See Sec. 3
Lack of knowledge	Due to lack of knowledge processes, components etc. cannot be considered	e.g. Nafion production for membranes	Systematic data collection; use of proxy

Source: Adapted from Ref. [13].

the gaps. However, if the data gap is mainly due to lack of knowledge, only limited action can be taken during the data collection step (in the inventory phase) and not all the gaps can be fulfilled at the end of the analysis.

The choice of the approach needs to be decided in the goal and scope definition phase, depending on the final aim of the LCA study. However, assessments of emerging technologies are usually focused on the hot spot analysis, i.e. identifying the most "hot" phase of the whole process in terms of environmental impact, understanding the potential criticalities of the process in the future. For this kind of analysis, the use of proxy datasets is suggested for initial screening.[31]

While the work of Pehnt[13] was mainly focused on LCA modelling of new energy systems, Miller and Keoleian[32] analysed the methodological challenges connected with the LCA of "transformative technologies" which are, amongst the emerging technologies, the ones that change the system or systems with which they are associated, i.e. catalysts for fuel cell. Miller and Keoleian[32] identified 10 key factors that can have a considerable impact on the LCA results of a transformative technology. These factors are divided in *intrinsic, indirect* or *external* factors. *Intrinsic* factors are the ones directly related with the technology under development. *Indirect* factors are associated with the background systems which are interacting with the new technology. Finally, *external* factors are those which influence the transformative technology but are independent from it. Table 2 reconciles the factors selected by Miller and Keoleian[32] with the different methodological issues identified by Pehnt[13] in LCA of emerging technologies.

The factors identified by Miller and Keoleian[32] are now described.

Efficiency and functionality change factor is referred to the expected improvement of performances in an emerging technology. This is related even to the efficiency of the manufacturing process of the new product, which becomes important to assess in an LCA perspective. In fact, the production of a new technology is expected to have a lower efficiency compared with a well-established technology (i.e. in terms of electricity and chemical consumed during the manufacturing). The effects of "manufacturing readiness" of a technology on LCA results has been proven by Gavankar *et al.*[30] and it is later discussed in Sec. 3. Moreover, the expected

Table 2: Methodological issues and key factors influencing the results of LCA of transformative technologies.

Methodological issues associated with emerging technologies (Pehnt[13])	Factor influencing LCA results of transformative technologies (based on Miller and Keoleian[32])		
	Intrinsic factor	Indirect factor	External factor
System expansion related problem	Efficiency and functionality change; Geographical effects	Technology substitution	
The relationships between the background and the foreground systems	Resource criticality	Supply chain effects; Rebound effects	Policy and regulatory effects
Infrastructures	Infrastructure change		Policy and regulatory effects
Data gap	Efficiency and functionality change	Supply chain effects; Rebound effects; Behaviour change	Exogenous system effects

Source: Adapted from Refs. [32] and [13].

improvements in terms of efficiency and life time of a new technology are going to affect the systems assumed in the system expansion and the overall LCA recommendations.

Geographical effects on LCA results have been extensively demonstrated.[33-35] When assessing emerging technologies, the effect of the geographic location on the potential future market of the technology needs to be considered. The same technology can be used in different extension when considering more than one reference country. Moreover, spatial location can affect the background systems as well: for example, the burdens associated with the supply of 1 kWh of electricity in the UK have a carbon footprint which is about 6 times the one associated with the production of 1 kWh$_{el}$ in France.[36]

The changes related to the *infrastructure* required by emerging technologies have been already described. Inclusion of the environmental impacts associated with the infrastructures is expected to influence the short-term LCA results of emerging technologies, and this needs to be assessed.

Resource criticality is a particularly important factor for emerging technologies and is associated with the introduction of a "new or additional stress on a resource".[32] Although the extraction and then the availability of a specific resource is referred to the background system, the quantity of the resource used in the process is directly related to the emerging technology and for this reason it is considered an intrinsic factor. An example is the availability of platinum group metals (PGM) in the future, which are the fundamental products used in catalysts for fuel cells.

The *substituted technology* replaced by the emerging product represents the baseline of the LCA study. The correct definition of this is critical for the LCA results and need to be assessed based on some other modelling technique, i.e. economic models. This factor is strictly associated with the system expansion approach suggested by the authors, and it is more often discussed in consequential LCA studies. However, it becomes important also for comparative attributional LCAs.

Behaviour change and *rebound effects* are related indirect factors associated with the sociological changes induced by the new technology. While the first is directly related to changes in behaviour connected with the interaction with the technology, the rebound effect is related to different consumption patterns of the technology due to change in the efficiency. Both of them are quite difficult to evaluate. Although rebound effect can be assessed based on the expected performance improvements of the technology (i.e. higher efficiency FCVs means less fuel and therefore cost saving for the users), behaviour change effects are highly variable and uncertain.

Indirect effects related to change in the *supply chain* caused by the new technology can result in data gaps in the inventory. For example, the increased efficiency of FCVs may lead to an increase of hydrogen demand due to an expansion of the FCV market. This can have an effect on the hydrogen supply chain, alter the request of fossil fuel (i.e. natural gas) for fulfilling the demand or determine a shift in the processes used to produce hydrogen. This gap can be overcome using simplified assumptions, or forecasting methods to identify the most likely scenario.

Finally, external factors (such as *exogenous system effects* and *policy and regulatory effects*) are the ones that are still affecting the LCA of an emerging technology, although they are completely independent from it. Other techniques can be used to predict them, like scenario analysis or dynamic LCA.[37] A good understanding of the policy framework in which the technology will be collocated is required to tackle these factors, which are highly connected with the geographical effects due to the changing of policies within different countries. As an example, the use of FCVs in the future may be incentivised by specific regulation on vehicle emission limits, as well as deregulation in the hydrogen supply market.[38]

3 From Lab Scale to Industrial Scale

This section investigates the possible approaches to address the uncertainties and data gaps associated with emerging technologies which are not yet fully developed for an industrial scale. Considering the life cycle of a technology, from the manufacturing to the disposal phase, very often the uncertainties are related to the first and the last phase, while the use phase is usually derived from target data from manufacturers, policies and regulations, and system modelling.

Pehnt[13] developed a procedure which can be followed to assess the environmental impact of an emerging technology when data gaps are found during the inventory phase. This is shown in Figure 3. The forecasting methods proposed so far in the literature have been adapted from cost analysis and market forecasting. Weidema[39] listed the forecasting methods which can be adapted to LCA based on the application areas of LCA and the time of the forecast. Here we propose a list of forecasting methods based on the works of Pehnt[13] and Weidema[39]:

- *Extrapolation of trends and historical data* to extend them into the future. This method is based on the assumption that what happened in the past is expected to happen also in the future. The main effort here is the collection of past data. Weidema recommends this method for short-to-medium term forecast, when typical developments of the assessed process/product are expected. A common technique belonging to this group is the *regression analysis*, which determines the relationships between different variables.
- *Extrapolation from different data sources* (or *adaption method*), consists of adopting comparable projects or processes in which the data to be

```
┌─────────────────────────┐
│  Preliminary Inventory of the │
│       status quo system  │
└─────────────────────────┘
            ↓
┌─────────────────────────┐
│   Identify the data gaps │
└─────────────────────────┘
            ↓
┌─────────────────────────┐
│    Relevance analysis    │
└─────────────────────────┘
            ↓
┌─────────────────────────┐
│  Selection of data gaps with │
│  environmental relevance on │
│    LCA results and time  │
│        dependency        │
└─────────────────────────┘
            ↓
┌─────────────────────────┐
│  Forecast of LCI data with │
│  independent forecasting │
│         methods          │
└─────────────────────────┘
```

Results from different methods are consistent → Data from forecasting can be used

Results from different methods are not consistent → Sensitivity analysis using the different data from forecasting

Results are not influenced by the forecasting method

Results are influenced by the forecasting method → Data can be used only referring to the forecasting method used

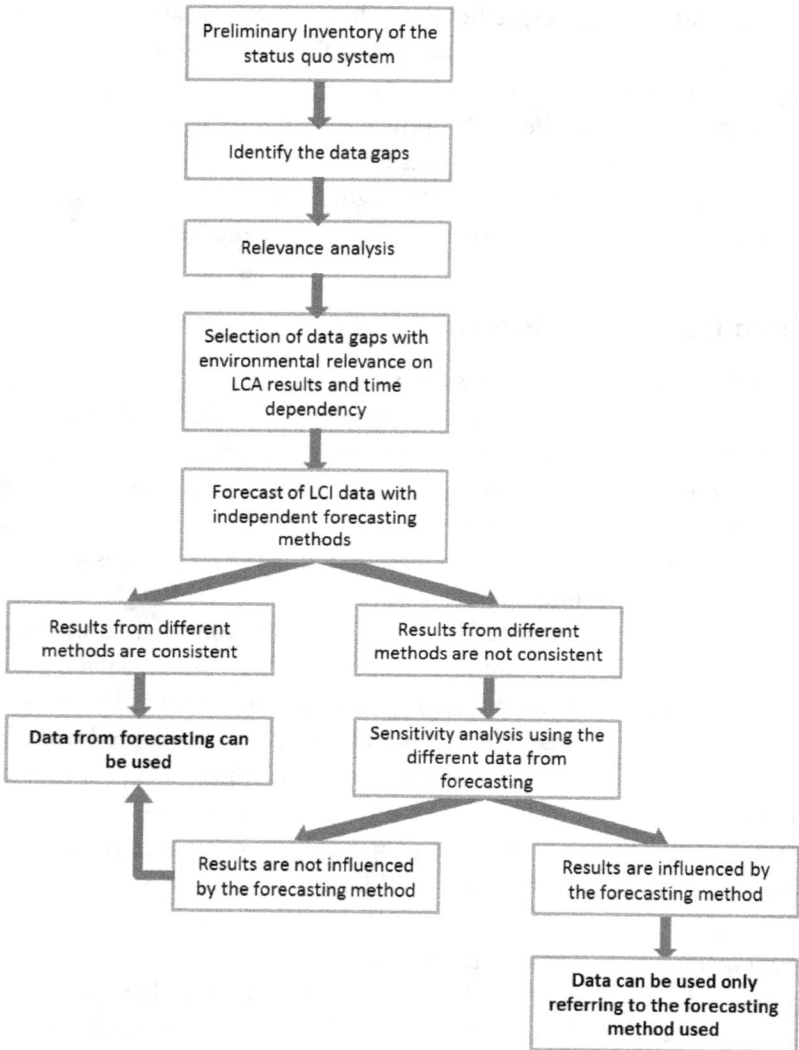

Figure 3: Procedure to perform a LCA of emerging technologies with inventory data gaps (adapted from Ref. [33]).

forecasted is available and transferred it to the analysed process or products. Mila-i-Canals *et al.*[31] stated that finding surrogate data (which are data sets that are sufficiently similar to the "missing" process/material/product) may often be a good compromise between creating a new dataset (time and resource consuming) and leaving the data gap. They identified two approaches to define the surrogate data: use of *proxy* data sets and *data extrapolation*. The first approach consists of using existing data sets of alternative products which have a similar environmental impact to the original products. The selection of the alternative products is usually based on the physical and chemical compositions or the manufacturing processes of the original products. This choice is strongly related to the knowledge and experience of the LCA practitioner and there are limited possibilities to validate such choice. The second approach, i.e. data extrapolation, generates a new sets of data from data sets "outside the range of their original validity".[31] In this case, a deep understanding of the extrapolated systems is necessary. This approach can be resource and time consuming, although characterised by a lower level of uncertainties compared with the proxy approach.

- *Modelling* can be applied when no empirical data are available. It helps identifying the mechanisms of combined effects which will determine the future process/product and which are assumed to be based on similar mechanisms of past events. It can be applied for long-term analysis when expected trends of the system are known; when less predictable processes and complex systems are analysed, modelling can be valid only for short/medium term.

- *Participatory methods* (or subjective assessment methods), which are based on experts and stakeholders opinions, through surveys, brainstorming, or focus groups. This method is quite time and cost consuming and it is generally adapted in large LCA projects. The results obtained are also influenced by the way the interviews are carried out and the choice of experts and stakeholders involved.

- *Scenario analysis* is a more complex method which combines aspects of the other methods with the aim of creating different scenarios which describe potential futures. It is a time and resource consuming method which needs to be applied when complex and less predictable systems are analysed.

- *Exploratory and normative methods* are identified by Weidema[39] for processes upon which the decision makers have a large degree of influence. These are more qualitative methods rather than analytical methods, which identify possible futures based on imaginative techniques (exploratory methods) or try to set future trends and determine how to obtain them (normative methods). Both methods require a deep knowledge of the framework of applicability of the system under analysis, in terms of policy and market targets to meet.

The choice of the best forecasting method depends on different factors. Mainly the availability of data determines the suitability of the technique to use; the time horizon also influences the choice of the method. Moreover, the goal of the LCA and the identification of the potential stakeholder are important parameters to consider, as well as the financial availability which can make some of the methods unfeasible. Often, a combination of different methods is the best way to proceed.

4 Example: LCA of Catalysts in PEM Fuel Cells

In this section, the manufacturing of catalysts used in PEM fuel cell electrodes is used as an example, to show the approach to take when conducting a LCA of emerging technologies. The electro-catalysts used in low-temperature fuel cells play a critical role in determining the performance (efficiency) and durability of the device. Owing to the wide-spread use of PGMs, the electrodes are also a major cost. There is intense worldwide research effort to replace and/or reduce the amount of PGM in PEMFCs, and the materials cost associated with electrode fabrication is well-known; however, the environmental cost and life-cycle impact of electro-catalyst manufacture needs to be properly understood.

A range of catalyst fabrication routes exist. Figure 4 shows a common manufacturing process related to the carbon pre-treatment and catalyst manufacturing phase, to obtain 1 kg of catalyst on carbon support.

First, the carbon support is prepared. Potassium hydroxide is added to hot water to produce 1 mol of KOH. To this, carbon is added and the mixture is stirred at 70–100°C for about 5 h, before heat is removed and the mixture is stirred overnight. The carbon is filtered out and washed extensively, until the pH of the filtrate returns to neutral. Filtering is usually performed using a vacuum filtration assembly. Finally, the washed carbon is dried at 80–105°C under air, in an oven. In the catalyst manufacturing phase, the treated carbon is dispersed in hot water by stirring. After that, the required amount of platinum salt (typically H_2PtCl_6), to give the desired

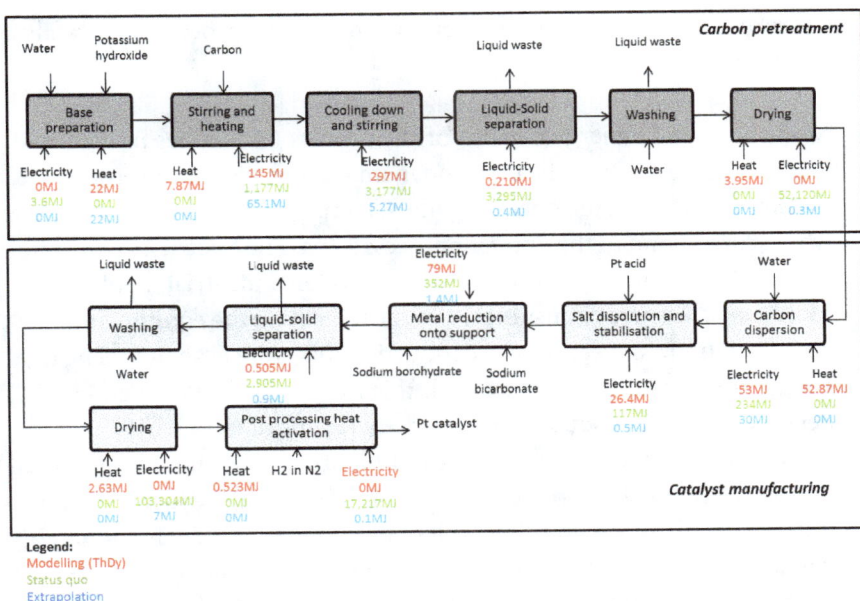

Figure 4: Production process for 1 kg of Pt catalyst on carbon support for PEM fuel cells.

Note: 1. FU = 1 kg catalyst as output.

weight of platinum on carbon (i.e. 40%) is added to the hot water/carbon mixture and stirred at ~70–100°C for 30 min. At this point, the pH of the reaction mixture, still at ~70–100°C and stirring, is adjusted using a sodium bicarbonate solution; pH 7 is then maintained for 30 min. The reaction mixture is reduced by adding a concentrated solution of sodium borohydride; this is left stirring at ~70–100°C for 1 h before allowing cooling. Once cooled down, the reaction mixture is filtered and thoroughly washed. Filtering is performed using a vacuum filtration assembly. The washed catalyst is then dried at 80–105°C under air, in an oven. Finally, the catalyst is treated in a tube furnace at 150–200°C under 10% hydrogen in nitrogen.

Three different data sets are considered to estimate the energy demand of the catalyst production process, in terms of electricity and heat demand. The status quo represents the energy consumption of the process as it is now due to the laboratory nature of the equipment used, which is equal to 183,974 MJ of electricity per kg of catalyst produced. This figure includes process electricity and heat requirements (i.e. the heat is provided by electric oven). Then two different forecasting methods are applied to estimate the future energy consumption of an industrial process for catalyst production:

- Modelling of the energy balance of the process (red values in Figure 4). This takes into account the thermodynamic properties of the process and system, considering the operating mass, temperature and pressure values which correspond to the different process steps. The electricity consumption is based on typical machines used for industrial applications. Moreover, the energy consumptions due to heating the equipment materials, radiation, convection and conduction losses are evaluated. The total energy consumption estimates with this method are 304 MJ and 90 MJ per kg of catalyst produced for electricity and heat, respectively.
- Extrapolation method (blue values in Figure 4). The figures are based on the trend for the production volume of catalyst for fuel cells in the future (based on 50,000 fuel cell stack per year).[40] Then the energy demand associated with this production volume is calculated based on the present lab process. The total energy consumption forecasted with this method is 111 MJ for electricity and 22 MJ for heat per kg of catalyst produced.

Figure 5 shows the total primary energy consumption forecasted with the two methods compared with the status quo. When forecasting methods

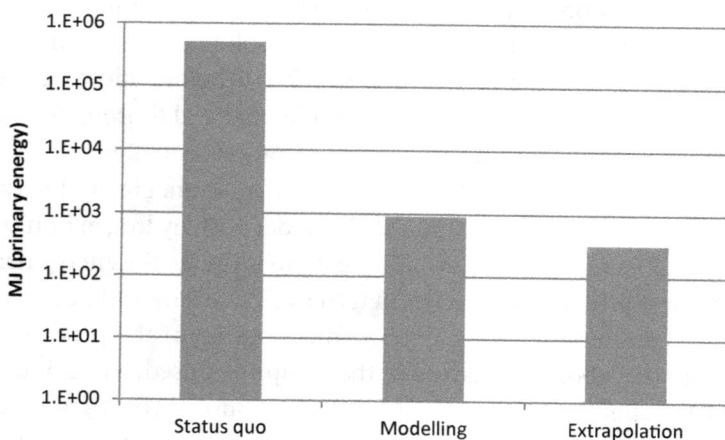

Figure 5:   Primary energy consumption for 1 kg Pt catalyst production for fuel cell applications when different forecasting methods are applied.

are applied for future production of catalyst for FCVs application, a reduction of the energy requirements of about three orders of magnitude is achieved for both modelling and extrapolation. Although the results from the forecasting methods applied are not the exactly the same, they are comparable. Values reported in literature studies show primary energy requirements of 120,000 MJ (corresponding to a production of 1,000 stack per year)[41] and 68,000 MJ (corresponding to a price target of 30 $/kWel from the stack)[42] per kg of Pt catalyst produced. However, they refer to different production volumes and manufacturing processes. Following the procedure shown in Figure 2, the results can then be used to fill the gaps in the LCI of catalyst manufacturing for fuel cell application.

5 Conclusions

Emerging catalyst-related technologies may reduce the global environmental impact of specific sectors in the future. In particular, the energy sector is expected to undertake a significant change due to the development of new technologies. The environmental impact associated with these new technologies needs to be properly assessed to identify potential improvements of the future industrial scale production processes. LCA has been recognised as the best tool for environmental system analysis in order to predict the benefits/burdens associated with the new technologies.

A proper standardised approach for applying LCA to emerging technologies is presently missing. Practitioners are often dealing with processes which are not fully developed for mass production, with limited availability of data. Therefore, it is essential to identify the issues related with the assessment of future technologies and try to tackle them in order to make the results obtained valid for technologies at a higher level of maturity or for large scale production.

An example of LCA of emerging technology is presented in this chapter, based on the catalyst production for polymer exchange membrane fuel cells. The first step is to identify the potential issues associated with the analysed technology; then forecasting methods in order to fill the data gap in the life cycle inventory need to be applied. Results from the exemplary emerging technology show that different forecasting methods give different results. However, if the results converged, they can be used for the analysis. Otherwise, a sensitivity analysis is recommended in order to ensure the robustness of the findings of the LCA.

References

1. DECC (Department of Energy & Climate Change), 2013, UK Greenhouse Gas Emissions, final figures, 2015. Available at: https://www.gov.uk/government/uploads/system/uploads/attachment_data/file/407432/20150203_2013_Final_Emissions_statistics.pdf. Accessed date 30-5-2015.
2. EC (European Commission), 2010. EUROPE 2020 — A strategy for smart, sustainable and inclusive growth, COM(2010) 2020 final. Available at: http://eur-lex.europa.eu/LexUriServ/LexUriServ.do?uri=COM:2010:2020:FIN:EN:PDF. Accessed date 30-5-2015.
3. EC (European Commission), 2015. Climate Action. Accessed 28.05.2015. Available at: http://ec.europa.eu/clima/policies/roadmap/perspective/index_en.htm. Accessed date 30-5-2015.
4. UK Government, *Climate Change Act*, The Stationery Office Limited, Belfast, 2008.
5. BIS (Business, Innovation and Skills), 2012. Industrial strategy: UK sector analysis. Available at: http://www.bis.gov.uk/assets/biscore/economics-and-statistics/i/12-1140-industrial-strategy-uk-sector-analysis. Accessed date 10-5-2015.
6. ONS (Office for National Statistics), 2012. UK Manufacturers' Sales by product (PRODCOM) for 2011. Statistics, O.f.N. Available at http://www.ons.gov.uk/ons/dcp171778_293762.pdf. Accessed date 10-5-2015.
7. P. Gilbert, M. Roeder and P. Thornley, 2013. The chemical industry in the UK — market and climate change challenges. Tyndall Manchester report.
8. Oxford Economics, 2010. The economic benefits of the chemistry research in the UK.
9. IEA, 2013. Technology Roadmap Energy and GHG Reductions in the Chemical Industry via Catalytic Processes.
10. Freedonia Research, 2014. World Catalysts Industry study with forecasts for 2018 & 2023. Study #3217.
11. I.C. Chen, Y. Fukushima, Y. Kikuchi and M. Hirao, *Int. J. Life Cycle Assess.*, 2012, **17**, 119–125.
12. JRC — IES (Joint Research Centre — Institute for Environment and Sustainability), 2011. Supporting environmentally sound decisions for bio-waste management: A practical guide to Life Cycle Thinking (LCT) and Life Cycle Assessment (LCA), eds., S. Manfredi and R. Pant [Online]. Available at: doi:10.2788/53942. Accessed date 10-5-2015.

13. M. Pehnt, *Int. J. LCA,* 2003, **8**(5): 283–289.
14. K. Jonasson, and B. Sandén, 2004. Time and scale aspects in life cycle assessment of emerging technologies: case study on alternative transport fuels, Chalmers University of Technology, CPM-report 2004:6, ISSN 1403–2694.
15. B. P. Weidema, G. Rebitzer and T. Ekvall, eds. 2004. Scenarios in life-cycle assessment, Society of Environmental Toxicology and Chemistry, Pensacola, 2004, p. 88.
16. H. Baumann, and A. M. Tillman, (2004) The Hitch Hiker's Guide to LCA. An orientation in life cycle assessment methodology and application. Lund, Sweden, Studentlitteratur.
17. ISO (2006) ISO 14040:2006 — Environmental management — Life cycle assessment — Principles and framework. Available at: http://www.iso.org/iso/catalogue_detail.htm?csnumber=37456. Accessed date 15-05-2013.
18. ILCD, 2010. European Commission — Joint Research Centre — Institute for Environment and Sustainability: International Reference Life Cycle Data System (ILCD) Handbook — General guide for Life Cycle Assessment — Detailed guidance. First edition March 2010. EUR 24708 EN. Luxembourg. Publications Office of the European Union.
19. J.B. Guinée, M. Gorrée, R. Heijungs, G. Huppes, R. Kleijn, A. de Koning, L. van Oers, A. Wegener Sleeswijk, S. Suh, H. A. U. de Haes, H. de Bruijn, M. A. J. Hujbregts, 2002. Life cycle assessment An operational guide to the ISO standards. Final report. Centre of Environmental Science, Leiden Univeristy.
20. T. Ekvall and G. Finnveden, *J. Clean. Prod.,* 9197–9208.
21. R. Clift, A. Doig, and G. Finnveden, *Trans. IChemE,* 2000, **78**(4), 279–287.
22. R. Heijungs and J. B. Guinée, Waste Management, 2007, **27**(8), 997–1005.
23. O. Eriksson, G. Finnveden, T. Ekvall and A. Bjorklund. *Energy Policy,* 2007, **35**(2), 1346–1362.
24. BSI (British Standard Institute), 2011. PAS 2050:2011. Specification for the Assessment of the Life Cycle Greenhouse GasEmissions of Goods and Services. Available at: http://www.tuinbouw.nl/sites/default/files/page/PAS2050-1_0.pdf. Accessed date 10-5-2015.
25. G. Finnveden, M. Z. Hauschild, T. Ekvall, J. Guinée, R. Heijungs, S. Hellweg, A. Koehler, D. Pennington, and S. Suh. *J. Environ. Manage.,* 2009, **91**(1), 1–21.
26. S. Lundie, *J. Clean. Prod.,* 2005, **13**(3), 275–286.
27. B. P. Weidema, N. Frees and A. Nielsen, *Energy,* 1999, **4**(1), 48–56.
28. M. A. Thomassen, R. Dalgaard, R. Heijungs and I. Boer, *Int. J. Life Cycle Assess.,* 2008, **13**(4), 339–349.
29. J. M. Earles and A. Halog, *Int. J. Life Cycle Assess.,* 2011, **16**(5), 445–453.
30. S. Gavankar, S. Suh and A. A. Keller, *J. Indus. Ecol.* **19**(1), 51–60.

31. L. Mila i, Canals, A. Azapagic, G. Doka, D. Jefferies, H. King, C. Mutel, T. Nemecek, A. Roches, S. Sim, H. Stichnothe, G. Thoma and A. Williams, *J. Indus. Ecol.*, 2011, **15**(5), 707–725.
32. S. Miller, and G. A. Keoleian, *Environ. Sci. Technol.*, 2015, **29**, 3067–3075.
33. M. A. J. Huijbregts, W. Schöpp, E. Verkuijlen, R. Heijungs, and L. Reijnders, *J. Ind. Ecol.*, 2000, **4**(3), 75–92.
34. M. Hauschild, *Int. J. Life Cycle Assess.*, 2006, **11**(1), 11–13.
35. N. Jungbluth, C. Bauer, R. Dones and R. Frischknecht, *Int. J. Life Cycle Assess.*, 2006, **10**(1), 24–34.
36. PE International, 2015. GaBi sustainability software. Available at: http://www.gabi-software.com/uk-ireland/index/. Accessed date 10-5-2015.
37. M. Pehnt, *Renew. Energy,* 2006, **31**(2006) 55–71.
38. M. J. Kang and H. Park, 2011, *Energy Policy,* **39**(6), 3465–3475. Available at: http://dx.doi.org/10.1016/j.enpol.2011.03.045. Accessed date 10-5-2015.
39. B. Weidema, 2003. Market information in life cycle assessment. Environmental Project No. 863. Danish environmental Protection Agency.
40. US Drive, 2013. Fuel Cell Technical Team roadmap. Available at: http://energy.gov/sites/prod/files/2014/02/f8/fctt_roadmap_june2013.pdf. Accessed date 10-5-2015.
41. Battelle, 2013. Manufacturing cost analysis of 10 kW and 25 kW direct hydrogen polymer electrolyte membrane (PEM) fuel cell for material handling applications, prepared for U.S. Department of Energy, DOE Contract No. DE-EE0005250.
42. A. Simons and C. Bauer, 2015. A life-cycle perspective on automotive fuel cells, *Applied Energy* (online). Available at: http://dx.doi.org/10.1016/j.apenergy.2015.02.049. Accessed date 10-5-2015.

Conclusions and Perspective

The chapters in this book have illustrated the challenges, the range and the excitement of contemporary catalytic science, and, we hope, the directions in which the field is developing.

Catalytic systems now embrace the three main areas of heterogeneous, homogeneous and biocatalysis, between which there is increasing interdependence and connectivity. Within the former, we have emphasised the role of supported metal nanoparticulate catalysts, which, with their growing range of functionalities, occupy a central position within the contemporary field, while posing fundamental problems relating to nanoparticle structures and the nanoparticle support interactions. Microporous catalysts — of enduring importance in the field — also remain an active area of contemporary heterogeneous catalytic science. Our accounts of homogeneous catalysis have focused on immobilised systems and on ring-opening polymerisation catalysis. These are two areas where, in fact the application of homogeneous catalysis expertise is vital but also where all areas of catalysis come together to deliver technologies that are of both fundamental and practical importance. Our discussion of biocatalysis has described how artificial metallo-enzymes can be designed and then used to catalyse reactions that extend the repertoire of existing natural biocatalysts. This approach leads to the creation of a broader range of biocatalysts for applications in new types of chemical synthesis.

Catalytic science is increasingly dependent on advanced characterisation techniques, within which, we have exemplified the growing importance of techniques based on the exploitation of synchrotron radiation and neutron scattering and their ability to probe catalytic systems *in situ.* Their importance will continue to grow and will be reinforced by the access to new radiation sources, especially those based on X-Ray Free Electron Laser (XFEL) technology.

We have also seen how computer modelling —now an integral component of catalytic science — can provide detailed information on structure and mechanism at the molecular level which assists and complements experimental studies. Modelling is indeed a vital tool which is increasingly predictive both qualitatively and quantitatively in current catalytic science; and with the continuing growth both in computer power and in methods and algorithms its role will expand.

Catalytic science plays a key role in several fields of applied science including energy technologies and environmental protection and remediation. Catalytic technology will, however, only be usable when it is economically and environmentally viable, for which a rigorous life cycle analysis is necessary. The importance of the field in industry continues to grow with catalysis underpinning a large component of manufacturing industry; and we hope that we have shown how the challenges posed by industrial applications require the input from the exciting developments in fundamental science highlighted in this book.

C. Richard A. Catlow
Paul Collier
Matthew G. Davidson
Christopher Hardacre
Graham J. Hutchings
Nicholas J. Turner

Index

electrochemical co-deposition, 236
electronic effect, 229–230
encapsulation, 127–128, 140–141,
 143, 149–150, 152
energy, 227
energy density, 227
enzyme, 139
ethanol, 225, 228
Euler's method, 266
exchange-correlation, 257
exchange-correlation functionals,
 258

F

Faradaic efficiency, 221–222, 224
Fe_2O_3, 215
force fields, 264
freeze quench, 61
FTIR, 67, 73, 228, 235
 Attenuated Total Reflection
  Infrared (ATR-IR), 62
 DRIFTS, 67–68, 70
 Synchrotron radiation infrared
  (SR-IR), 65
fuel cell, 62, 225–226
 Anode, 225
 Cathode, 225
 efficiency, 225
 proton exchange membrane fuel
  cells (PEMFC), 225, 227
functional, 257

G

Ga_2O_3, 215
galvanic cells, 225
geometric effect, 241
Generalised-gradient approximation
 (GGA), 260
Gibbs free energy, 229

gold, 181–187, 190, 193–195,
 198–199, 204–206
gradient expansion approximation,
 260
grafting, 127–129, 134, 136–138, 151

H

H_2, 225, 227
H_2/O_2 fuel cell, 226
H_2O_2, 239
H-acceptor, 135
Hamiltonian, 258
Hartree–Fock theory, 257, 270
hemin, 140
heterogeneous, 215
HF long-range, 263
HKUST-1, 129, 146–149
Hohenberg–Kohn theorem, 259
hydride H-spillover, 213
hydrogen, 1–9, 11–12, 16–24,
 26–31
hydrogen oxidation reaction (HOR),
 239
hydrolysis of ethyl acetate, 147
hydroxylation, 92, 96–98, 110, 115

I

Immortal polymerization, 162, 174
in situ, 41–42, 51–53, 59, 62, 64–65,
 74, 76, 78–79
in situ FTIR, 230
infrared reflection-absorption
 spectroscopy, 238
initiation, 170–171, 174
intermediates, 228
inventory, 292–296, 300, 302, 304,
 309
IR, 229, 240
isoreticular, 137

www.ingramcontent.com/pod-product-compliance
Lightning Source LLC
Chambersburg PA
CBHW061622220326
41598CB00026BA/3850